T0298320

South American camelids research – Volume I

Georg-August-Universität Göttingen, Germany

Università degli Studi di Camerino, Italy

South American camelids research Volume I

Proceedings of the 4th European Symposium on South American Camelids and DECAMA European Seminar

Göttingen, Germany
7-9 October 2004

edited by:
Martina Gerken
Carlo Renieri

Wageningen Academic
P u b l i s h e r s

ISBN-10: 90-76998-98-1
ISBN-13: 978-90-76998-98-5

Cover photo: Alexander Riek,
University Göttingen,
Germany

First published, 2006

Wageningen Academic Publishers
TheNetherlands, 2006

Preface

South American Camelids are receiving increased interest not only in South America but also on a world-wide scale. They possess some unique features such as their fine fibre and their high adaptivity to many climatic regions across the world. Apart from the important productive aspects, their physical attractiveness also makes them popular as pet animals.

Following the tradition of previous events, the 4th European Symposium covered a broad range of topics related to breeding and husbandry of South American Camelids. The event coincided with the Seminar of the DECAMA[1] EU-Project. This opportunity brought together the combined experience of both South American and European experts. The present collection of papers considers current trends in reproduction, nutrition, health, fibre morphology and genetics. Attention is also given to the new topic of the potential for meat production and commercialisation in South America. The particular advantages of South American Camelids for the sustainable use of fragile ecosystems with native pastures are outlined. Roundtables discussions focus on the interaction between wild and domestic species, the management of alpaca populations outside of South America and health aspects under European conditions.

The Symposium was organised by Prof. Dr. Martina Gerken, University of Göttingen, (Germany), Prof. Dr. Carlo Renieri, University of Camerino (Italy), and Prof. Dr. Dr. M. Gauly, University of Göttingen, in co-operation with the members of the EU-Project DECAMA.

The Symposium was hosted by the Institute of Animal Breeding and Genetics (Institut für Tierzucht und Haustiergenetik) of the Agricultural Faculty of Göttingen University. We thank the administration of the University for providing lecture halls and the Ancient Aula. The Symposium would not have been possible without the excellent support of the students and the members of the Institute and the organisers gratefully acknowledge their support: in particular, we wish to thank Dr. Sabine Bramsmann, Jürgen Dörl, Martina Krumm, Dr. Eva Moors, and Dipl.Ing. Alexander Riek.

We are indebted to the "Der Verein der Züchter, Halter und Freunde von Neuweltkameliden" (Association of Breeders, Producers and Friends of New World Camelids, Germany) which funded the attendance of the main speaker and generously supported the organisation of the Symposium. We also thank all participants who readily agreed to chair sessions or to participate in the Round Tables.

Funding for the participation of the young scientists - Dr. Celso Ayala (Bolivia), Genaro Condori (Bolivia) and Dr. Carlos Pacheco (Peru) was generously provided by the H. Wilhelm Schaumann Foundation, Germany. This support is very gratefully acknowledged.

The present Proceedings were printed from the manuscripts supplied by the authors and we are grateful to the support given by the editors of Wageningen Academic Publishers. The assistance of Dr. Hugh Galbraith, University of Aberdeen (Scotland), and Dipl.Ing. Alexander Riek, Göttingen, in the editorial process is gratefully acknowledged and appreciated. All papers labelled DECAMA - project report results obtained during this EU-funded research.

---

[1] EU-Project within INCO-DEV (ICA4-CT-2002-10014): Sustainable Development of Camelid products and services marketed oriented in Andean Region.

There are still many gaps in the scientific literature with regard to South American Camelids. The organisers hope that this 4th Symposium will stimulate further research and support the sustainable use of these unique species.

Martina Gerken and Carlo Renieri

Table of contents

Roundtables

List of participants   303

DECAMA-project: Sustainable development of camelid products and services marketed oriented in the Andean Region

C. Renieri

Dipartimento di Scienze Veterinarie, Via Circonvallazione 93/95, I-62024 Matelica (MC), Italy; carlo.renieri@unicam.it

Context and objectives

The aim of the project is to improve the quality of life of the rural population through the development of meat production of Domesticated South American Camelids (Lama and Alpaca). The objectives are: the improvement of camelid meat production, the suggestion of new production technologies, the proposal of technical, commercial and socio-economic interventions and services in order to support the development of the meat market and the exploitation of meat by-products, to obtain more added value, the development of scientific interchange and the technological transfer.

Activities

The activities are divided into 4 Work Packages (WP). Work Package 1 (Meat Products); section Fresh meat (1.1.): technological and nutritional characteristics of fresh meat, classification of carcasses, tenderisation of carcasses by electric stimulation; session Processed meat (1.2.), sub section Traditional products (1.2.1.): technological and nutritional characteristics of charqui; sub-section new products and new technologies (1.2.2.): new processed products, new processing methods for charqui, organic products; section Added value (1.3.): by products of meat; section Relationship between meat and milk production (1.4.): human Andean population and milk, chemical composition of milk, lactation curve and relationship with growth, technology of milking.

WP 2 (Market and Services); section Market analysis (2.1.): demand and offer at the production and processing levels, different steps of the chain, final commercial value and productivity of chain, governmental and macro-economical influences on the sector; section Certification systems (2.2.): innovation of products and processing, labels, certification systems, traceability; section Sustainability (2.3.): sustainability level of production, processing and distribution, safeguard natural resources, enhance income production and social responsibility of producers. WP 3 (Exchange of information/experiences): exchange between all the partners, transfer to the industrial sector, to the producers, to different policy decision points. WP 4 (Coordination and evaluation): project management.

Expected results and outcomes

Expected results and outcomes of the project are: certification system for nutritional content, classification carcasses system, standardisation of traditional and new processed products, development of new products and technologies, standardisation of meat conservation and processing, network of NGO's and institutions involved in the Andean region development, analysis of meat market, guidelines for the development of the meat chain, certification of the meat production and processing, decalogue of criteria for sustainability of meat chain, Web site of the project.

References

Homepage, DECAMA, http://www.decama.net/

Artificial breeding in alpacas

Jane Vaughan
Cria Genesis, PO Box 406, Ocean Grove, Vic 3226, Australia; vaughan@ava.com.au

Abstract

Male alpacas produce low volume, low density, high viscosity semen during an ejaculatory period lasting 15 to 20 minutes. There is extreme variation in semen quality both within and between males. It is essential to collect the best quality ejaculate in the first instance to ensure maximum numbers of live sperm are available to inseminate after extension and preservation. The high viscosity of semen makes it difficult to extract semen from the artificial vagina and to mix with extenders. The low density of semen means that few doses per ejaculate are available for artificial insemination. Semen is deposited transcervically at the utero-tubal junction ipsilateral to the dominant follicle 24 hours after induction of ovulation in the female. While pregnancies have been easy to obtain from artificial insemination of fresh semen, it has been more difficult using chilled or frozen semen. The understanding of ovarian function in alpacas has been instrumental in the success of developing non-surgical, transcervical single and multiple ovulation embryo transfer. Females may be flushed every 10-12 days using single-embryo flushing techniques and approximately every month following multiple ovulation. Fifty live births (26 males, 24 females) have occurred over the last 3 years in Australia, following single-embryo flushing. Multiple ovulation and embryo transfer produces an average of 3 embryos per flush (up to 14 embryos per individual) on most farms. Attention to detail in preparation of females is essential for success. Significant factors include use of normally fertile animals, ensuring body condition score averages 2.5-3 (out of 5), animals have access to green grass and adequate selenium intake.

Keywords: alpaca, artificial insemination, embryo transfer

Introduction

The development of artificial breeding technologies in alpacas will increase the use and allow more economic movement of genetically superior animals nationally and internationally. Generation intervals are relatively long in alpacas because males are slow to sexually mature and females exhibit an extended gestation (11.5 months), so conventional breeding results in slow genetic gain. Assisted breeding technologies are being used to improve wool quality more rapidly than would otherwise be possible by natural mating in industries such as Merino sheep and Angora goats. However, the reproductive physiology of alpacas differs to that of other domestic livestock and remains poorly understood, therefore hindering the direct transfer of artificial insemination (AI) and embryo transfer (ET) technologies from ruminants to alpacas.

Artificial insemination

Artificial insemination (AI) in alpacas and llamas is the subject of research in several countries. First attempts at AI occurred in the 1960s in South American camelids (SAC) in Peru (Calderon *et al.,* 1968; Fernandez-Baca *et al.,* 1966). Since then, studies in semen collection and AI have been performed in SAC in Australia (Vaughan *et al.* 2003), North and South America (Ratto *et al.,* 1999; Bravo *et al.,* 1997b), Britain (McEvoy *et al.,* 1992) and Germany (Burgel *et al.,* 1999). Concentrated efforts at commercialising AI in camels have occurred in the last 20 years in an attempt to improve quality and production of milk and meat, in response to higher stakes in camel racing and increased interest in breed conservation (Purohit, 1999) and will directly assist developing similar technologies in alpacas.

Development of AI is aimed at increasing the use of superior alpaca males as multiple doses of semen will theoretically be derived from each ejaculate. Artificial insemination will reduce the need for 'mobile matings' and therefore reduce the risk of injury and transport-associated diseases in elite males and significantly reduce the risk transmission of Johne's disease, lice and other infectious agents among alpaca farms. Artificial insemination will also allow more humane and economic international movement of superior genotypes and increase the efficiency of dissemination of superior alpaca genotypes within countries. It will also allow storage of genetics prior to disease or death of elite males.

There are no established AI services for alpacas and llamas world-wide. The reasons include the difficulty in semen collection, the mucoid characteristic of alpaca semen which precludes the easy adaptation of AI technology from other species, the characteristic of induced ovulation in female alpacas, the lack of techniques to define reproductive excellence in male alpacas, the lack of proven techniques to store alpaca semen in chilled or frozen form and the difficulty of transcervical pipette passage of some females. Research is currently aimed at developing the collection, processing and preservation of alpaca semen and improving the technology for insemination of semen.

Collection of semen

It is difficult to collect semen from camelids because of the mating posture, duration of copulation, intrauterine deposition of semen and the viscous nature of the ejaculate (Bravo *et al.*, 2000a; Bravo & Johnson, 1994). The use of an artificial vagina (AV) in combination with either a receptive female (Huanca & Gauly, 2001; Gauly & Leidinger, 1996), a dummy (Bravo *et al.*, 1997a; Bravo *et al.*, 1997b; Garnica *et al.*, 1993) or half-dummy mount (Lichtenwalner *et al.*, 1996) in South American camelids has yielded the most representative copulation times and mating behaviour in camelids. There is no urine contamination of semen and assessment of sexual behaviour is possible (Lichtenwalner *et al.*, 1996). Therefore, the AV method of semen collection presumably provides semen samples most representative of natural mating.

The configuration of the artificial vagina (AV) used to collect semen must be adequate for each male to assume a natural mating posture to ensure a representative ejaculate is produced. The liner must have a suitable texture and be non-toxic to sperm and the temperature must be maintained at 38-42°C. The AV may be constructed of PVC pipe or reinforced rubber tube (25-30 cm long, 5-7 cm diameter) and lined with a latex liner (Bravo *et al.*, 1997a, b; Lichtenwalner *et al.*, 1996). Various workers have noted the importance of placing a stricture around the latex liner of the AV to simulate the presence of a cervix (Fowler, 1998; Bravo *et al.*, 1997b; Bravo, 1995; Garnica *et al.*, 1993). Warm water is placed between the inner latex liner and the outer tube. The AV is wrapped in a heating pad or warm water is continuously pumped through it (Lichtenwalner *et al.*, 1996) to maintain temperature.

Males are trained to mount a dummy by first exposing them to receptive females and mating pairs of alpacas as a stimulus (Vaughan *et al.*, 2003).

Characteristics of alpaca semen

Ejaculates are not fractionated in alpacas and llamas so semen is uniform in quality from start to end of copulation (Bravo, 2002; Sumar, 1983). There is considerable variation in semen characteristics among different alpaca and llama males, which may pose a problem in obtaining acceptable quality ejaculates from some males (Vaughan *et al.*, 2003; Raymundo *et al.*, 2000; Von Baer & Hellemann, 1999; Gauly & Leidinger, 1996).

The volume of each ejaculate in alpacas averages 1-2 ml with a range of < 1 - 7 ml. Semen is usually described as milky- or creamy-white in colour depending on sperm concentration. Camelid semen is very viscous, making it difficult to handle (*e.g.* removal from AV), difficult to determine parameters such as sperm concentration and sperm motility, and difficult to dilute. The degree of viscosity varies between males and decreases with increasing number of ejaculates on any day (Tibary & Memon, 1999; Fowler, 1998; Bravo *et al.,* 1997a, b; Garnica *et al.,* 1993, 1995).

Sperm concentration reaches up to 300 million per ml in alpacas (Gauly & Leidinger, 1996; Bravo, 1995; Garnica *et al.,* 1995). Wide variations in concentration are attributed to differences in males, semen collection methods and ejaculate number (Tibary & Memon, 1999).

The movement of camelid sperm in undiluted semen is best described as oscillatory as most sperm move back and forth on the spot and only 5 to 10% of individual sperm actively progress forwards (Bravo *et al.,* 1997b). Sperm increase their progressive motility as the ejaculate becomes more liquid (Tibary & Memon, 1999).

Male alpacas produce low volume, low density, high viscosity semen during an ejaculatory period lasting 15 to 20 minutes. There is extreme variation in semen quality both within and between males. It is essential to collect the best quality ejaculate in the first instance to ensure maximum numbers of live sperm are available to inseminate after extension and preservation.

Preservation of alpaca semen

Various methods have been used to liquefy, dilute, extend and freeze camelid semen, with limited success. The difficulties in handling alpaca semen, due to high viscosity, include removing it from the AV, liquefying the ejaculate and mixing with extenders. Semen liquefies naturally 23 hours (range 8-30 hours) after collection (Bravo & Johnson, 1994). Various methods of liquefaction of semen, such as the use of enzymes (Bravo *et al.,* 2000a, b; Callo *et al.,* 1999) and mechanical stirring (A Niasari-Naslaji, personal communication), have been trialled with some success to simplify handling and allow dilution, extension and freezing of semen.

Two major systems of sperm conservation have been developed in domestic species to maintain sperm viability but reduce/arrest metabolism in sperm: cooled or chilled, liquid semen and frozen semen. The main advantage of using liquid semen is that fertility is maintained in most species with low numbers of sperm per AI dose, however, fertility of sperm is only maintained for 3 to 5 days when cooled to 10-21°C or chilled to 4-5°C (Simson, 2001; Yoshida, 2000). Freezing sperm in liquid nitrogen, on the other hand, allows indefinite storage but requires many more sperm per inseminate *e.g.* 1 million bull sperm per dose using liquid semen, compared with 15 million bull sperm per dose using frozen semen (Yoshida, 2000). Rates of cooling and length of storage of sperm affect sperm longevity.

Solutions that are added to semen to assist with sperm preservation are known as 'extenders'. They contain compounds that protect sperm outside the reproductive tract during chilling and freezing, as seminal plasma alone provides limited protection for sperm (Salamon & Maxwell, 1995). Lipoproteins, such as those found in egg yolk and milk, protect sperm against cold shock by stabilising cellular membranes during chilling. Glycerol, the most widely used protective agent for freezing, enters sperm cells and impedes formation of intracellular ice crystals and regulates osmotic balance. Energy is provided to sperm by addition of metabolisable substrates such as glucose, fructose or lactose. Antibiotics may be added to semen extenders to control bacterial growth (Simson, 2001; Salamon & Maxwell, 1995).

Numerous workers have attempted chilling and then freezing camelid semen using a 2-step dilution with glycerol, and slow freezing in liquid nitrogen vapour (Vaughan *et al.,* 2003; Bravo *et al.,* 2000b, 1996). Tris buffer combined with egg yolk and glycerol has been found by some workers to be the most promising combination to maintain sperm viability during dilution, refrigeration, freezing and thawing of camelid semen (Vaughan *et al.,* 2003; Bravo *et al.,* 2002). While pregnancies have been easy to obtain from AI of fresh semen, it has been more difficult using chilled or frozen semen, despite post-thaw activity of up to 40% using Triladyl® (Minitub) and Camel Buffer® (IMV) extenders (Vaughan *et al.,* 2003). Unfortunately, efficient preservation of sperm with good fertilising ability following freezing and thawing has yet to be achieved in alpacas. Freezing may damage sperm by altering membrane structure or inducing osmotic shock, dehydration, salt toxicity, intracellular ice formation, fluctuation in cellular volume/surface area or metabolic imbalance (Simson, 2001). Any alteration to progressive motility, cellular metabolism, cell membranes, acrosomal enzymes or nucleoproteins during freezing will render sperm incapable of fertilising ova.

Method of insemination

Synchronous ovulation and insemination is required to achieve high conception rates. In some species, high conception rates occur when sperm are in the oviduct just before ovulation (Hafez, 1993). Semen deposited transcervically at the utero-tubal junction ipsilateral to the dominant follicle 24 hours after induction of ovulation in the female has resulted in acceptable conception rates in alpacas and llamas (Apaza *et al.,* 1999; Pacheco, 1996; Quispe, 1996).

The sperm concentration required for successful fertilisation and pregnancy is not known in camelids (Brown, 2000). The low density of semen means that few doses per ejaculate are available for AI. Ewes require 120-125 million fresh sperm using transcervical AI for acceptable conception rates, but only 20 million frozen/thawed sperm when intrauterine AI is used (Brown, 2000). Pregnancies have resulted in alpacas and llamas by inseminating 8-26 million sperm transcervically (Aller *et al.,* 1999; Quispe, 1996). Intracornual semen deposition of semen during copulation in camelids may be an adaptation to overcome the relatively low sperm concentrations (Brown, 2000).

Bravo *et al.,* (1997a) compared the techniques of transcervical and laparoscopic AI in alpacas using fresh, undiluted semen and achieved similar conception rates. Transcervical AI would appear to be a simpler, less-invasive procedure, however has limitations of rectal/pelvic capacity to allow transrectal manual stabilisation of the cervix when passing the AI pipette through the cervix, and the difficulty of locating the external os of the cervix during this procedure. Laparoscopic AI requires restraint of the female in a crate, sedation with or without a general anaesthetic, and expensive laparoscopic equipment.

Embryo transfer

Protocols evaluated for embryo transfer in camelids have been adapted from protocols originally developed for cattle, sheep, pigs and horses. Embryo transfer can rapidly increase numbers of crias born to superior females. For example, it is possible to transfer the genes from the top 10% of an alpaca herd (donors) into the bottom 90% of females (recipients). Embryo transfer also allows breeders to determine optimal male/female combinations as multiple sires may be used over the same female in one year. Embryo transfer will give smaller breeders access to elite genes through purchase of embryos and will allow for inter-farm/state/national movement of superior genetics.

The understanding of ovarian function in alpacas has been instrumental in the success of developing non-surgical, transcervical single and multiple ovulation ET. Females exhibit waves of ovarian follicular growth, with new waves emerging every 12 to 22 days (Vaughan et al., 2004). Females are induced ovulators, and ovulate 30 hours after copulation when they have a dominant follicle of at least 6 mm on either ovary (Adams & Ratto, 2001; Bravo et al., 1991). A corpus luteum develops on the ovary at the site of ovulation 3-4 days after mating and secretes progesterone. If conception does not occur, prostaglandin is released from the uterus and induces regression of the corpus luteum 10-12 days after mating (Adams & Ginther, 1989). The embryonic signal for maternal recognition of pregnancy must be transmitted as early as Day 10 after mating in order to 'rescue' the corpus luteum of pregnancy as the corpus luteum is the major source of progesterone throughout pregnancy.

Single ovulation versus multiple ovulation

Single-ovulation embryo transfer of alpacas does not require any hormonal treatment of donor females (Taylor et al., 2000). Donor females are mated once and flushed a week later. Follicle growth in the first 10 days after new wave emergence is consistent regardless of subsequent interwave interval (Vaughan et al., 2004), an observation integral to the success of single-embryo flushing of donor females every 10-12 days. Fifty live births (26 males, 24 females) have occurred over the last 3 years in Australia, following single-embryo flushing performed by the author and Dr David Hopkins in numerous commercial alpaca herds. Another 100 recipient females are carrying embryos which have been flushed and transferred from elite females (Table 1). Donor females have since given birth to crias from matings performed soon after embryo flushing, indicating donor fertility was not interfered with during embryo collection.

Methods of multiple ovulation and embryo transfer (MOET or 'superovulation') are also being examined in alpacas in Australia. Both equine chorionic gonadotrophin and follicle stimulating hormone are currently being tested as agents to stimulate multiple ovulation. Techniques are producing an average of 3 embryos per flush (up to 14 embryos per individual) on most farms. Results have been less reliable on a some farms, presumably due to variations in alpaca fertility, nutrition, environment and management. The number of studies on MOET in camelids remains low and further refinement of existing protocols is required to identify a MOET program that consistently yields an acceptable number of transferable embryos, and is associated with minimal risk of infertility to the elite donor female. Embryos have been yielded on 3 and 4 consecutive MOET programs in the last year, without apparent effect on donor fertility as donor females have readily conceived within 2-4 weeks after their last MOET flush.

Table 1. Single and multiple ovulation embryo transfer results in alpacas in Australia.

	Single-ovulation ET	Multiple-ovulation ET
Donors flushed	340	96
Embryos transferred	235	239
60-day pregnancies	146	181
% pregnancies/donor	43%	189%

Preparation of donors and recipients

Females that are to be used as donors need to be reproductively sound (owners must resist the temptation of preparing females that have been difficult to get pregnant in the past), of superior genetic quality, have good conformation, and be free of all known inherited genetic disorders.

Females that are to be used as embryo recipients must also be reproductively sound in order to optimise the chances of successful embryo implantation and birth of a cria. Demonstrated good mothering ability is an advantage. Females with physical and/or genetic abnormalities (carpal valgus, luxating patellae, fused toes, extra toes, wry face) can be used as recipients since these characteristics will not be transferred to the embryo and gestating foetus.

Attention to detail and thorough preparation of donor and recipient females (and males) is essential for successful embryo transfer. Four factors appear to be important for *all* alpacas participating in an ET program:
- normal fertility.
- body condition score 2.5 to 3 (out of 5).
- access to green grass.
- adequacy of selenium intake.

Most females ovulate one egg after mating, with multiple ovulations occurring in up to 10% of natural matings (Fernandez-Baca, 1993). Ideally, donor females should have their ovaries examined by ultrasonography every second day to monitor follicular wave patterns. When a growing or mature follicle (greater than 7 mm diameter) is present on an ovary, the donor female can be mated to a genetically superior male. If adequate numbers of recipient females are available they should be subjected to ovarian ultrasonography every second day in order to monitor follicle growth. Females with a growing or mature pre-ovulatory follicle on the day the donor female is inseminated should be induced to ovulate.

Both donor and recipient females are induced to ovulate at a similar time so that the uterine environment and circulating progesterone are comparable for donors and recipients. Hormones that can be used to induce ovulation include GnRH analogues such as buserelin (*e.g.* Receptal®). After injection of GnRH, around 90% of females ovulate within 30 hours.

Whilst a second mating 12 hours after the first would not enhance the ovulatory stimulus, additional spermatozoa would potentially be available for fertilisation (Bravo, 1990; Bravo *et al.,* 1990). When a 2-mating protocol was used in superstimulated llamas, the embryo recovery rate was increased by 50%, without a change in ovulation rate (Bourke *et al.,* 1992).

Embryo development in camelids

The embryos of camelids develop faster than in domestic ruminants and morulae have been recovered in the oviducts of llamas as early as 3 days after mating. The faster rate of embryo development in camelids is likely related to early maternal recognition of pregnancy, which needs to occur around Day 8 to 10 after mating to ensure persistence of the corpus luteum of pregnancy (Aba *et al.,* 1997; Del Campo *et al.,* 1995).

Embryos are flushed from donor females a week after mating. Two to three days after ovulation, embryos are found in the oviduct as 2- to 4-cell stage embryos, 8- to 16-cell stage embryos and morulae (Del Campo, 1997). A week after ovulation, embryos have migrated to the uterus and are in the form of a morula, early blastocyst or hatched blastocyst. Embryos are spherical (up to 4 mm in diameter) 8.5 days after ovulation, and thereafter begin to elongate (Adam *et al.,* 1992). Elongation of blastocysts occurs from Day 7 or Day 8 after ovulation and by Day 14, blastocysts can be 10 cm in length. Ninety percent of embryos recovered between 6.0 and 7.5 days after ovulation had hatched from the zona pellucida, the protective coat around the outside of the oocyte and early embryo (Del Campo, 1997; Bourke *et al.,* 1992).

Non-surgical, trans-cervical collection of embryos

This method involves the introduction of a Foley catheter through the cervix and placement of the catheter in the uterus. Medium is flushed through the catheter into the uterus, then allowed to drain, via gravity, into an embryo collection vessel. This method is relatively non-invasive and does not have the attendant risks of abdominal adhesions associated with surgical embryo collection. However, females with a narrow pelvis or excessive fat in their pelvis may not be suitable for non-surgical collection and there is also a risk of rectal trauma with this procedure. The author uses the non-surgical method of embryo collection from alpacas and llamas.

Surgical collection is possible. Using this method, the female is anaesthetised then placed on her back, in a laparoscopy cradle. A small midline incision is made, and the uterus exteriorised. Collection medium is flushed from the ovarian end of the oviduct towards the uterine body, and is collected via a catheter placed near the cervix.

The retrieved fluid is examined under a dissecting microscope for embryos. After collection and washing, single embryos are loaded into small plastic straws similar to those used for artificial insemination and then placed transcervically (non-surgically) into the uterus of the recipient female. In some studies, embryo were placed in the uterine horn on the same side as the CL on the ovary (Bourke *et al.,* 1995, 1991), whilst in other studies embryos were placed at the tip of the left uterine horn, or only females with a CL on the left ovary were used as recipients (Wiepz & Chapman, 1985), as 98% of pregnancies in alpaca are located in the left horn. Pregnancy diagnosis using transrectal ultrasonography can be performed from approximately Day 25 after embryo transfer to assess pregnancy (Parraguez *et al.,* 1997).

Future developments include the continued refinement of multiple ovulation protocols and the freezing of embryos to allow indefinite storage and easy transport of genetic material. Pregnancies have been achieved in camels (Skidmore *et al.,* 2004) and llamas (Aller *et al.,* 2002) following vitrification, thawing and transfer of embryos, but this success has not yet been translated to alpacas.

Conclusion

The development and application of reproductive technologies has been instrumental in the genetic improvement of other domestic species, and this will also be essential for the multiplication and dispersal of genetically superior alpacas within the various camelid industries around the world. Before reproductive technologies become practical in camelids, however, considerable research is required to better understand their unique reproductive physiology and to improve the techniques of collection, processing and preservation of semen and embryos.

References

Aba, M.A., P.W. Bravo, M. Forsberg and H. Kindahl, 1997. Endocrine changes during early pregnancy in the alpaca. Anim Reprod Sci 47: 273-279.

Adam, C.L., D.A. Bourke, C.E. Kyle, P. Young and T.G. McEvoy, 1992. Ovulation and embryo recovery in the llama. Proc. First International Camel Conference, pp. 125-127.

Adams, G.P. and O.J. Ginther, 1989. Reproductive biology of the non-pregnant and early pregnant llama. Proc. Annual Meeting of the Society for Theriogenology, pp. 166-175.

Adams, G.P. and M.H. Ratto, 2001. Reproductive biotechnology in South American camelids. Proc. Revista de Investigaciones Veterinarias, Peru. Vol 1: 134-141.

Aller, J.F., A.K. Cancino, G. Rebuffi and R.H. Alberio, 1999. Inseminacion artificial en llamas en la Puna. Proc. II Congreso Mundial sobre Camelidos, Cusco, Peru, pp. 72.

Aller, J.F., G.E. Rebuffi, A.K. Cancino and R.H. Alberio, 2002. Successful transfer of vitrified Ilama (Lama glama) embryos. Anim Reprod Sci 73: 121-127.

Apaza, N., V. Alarcon, T. Huanca and O. Cardenas, 1999. Avances sobre la inseminacion artificial con semen congelado an alpacas. Proc. II Congreso Mundial sobre Camelidos, Cusco, Peru, pp. 73.

Bourke, D.A., C.L. Adam and C.E. Kyle, 1991. Successful pregnancy following non-surgical embryo transfer in llamas. Vet Rec 128: 68.

Bourke, D.A., C.L. Adam, C.E. Kyle, P. Young and T.G. McEvoy, 1992. Superovulation and embryo transfer in the llama. Proc. First International Camel Conference, pp. 183-185.

Bourke, D.A., C.E. Kyle, T.G. McEvoy, P. Young and C.L. Adam, 1995. Recipient synchronisation and embryo transfer in South American camelids. Theriogenology 43: 171.

Bravo, P.W., 1990. Studies on ovarian dynamics and response to copulation in the South American camelids, *Lama glama* and *Lama pacos.* PhD thesis, University of California-Davis.

Bravo, P.W., 1995. Physiology of reproduction and fertility evaluation in the male alpaca. Proc. University of Sydney Post Graduate Foundation in Veterinary Science, Geelong. Vol 257: 61-66.

Bravo, P.W., 2002. Male reproduction. In: The reproductive process of South American camelids. Bravo Publishing, pp. 49-64.

Bravo, P.W., M. Callo and J. Garnica, 2000a. The effect of enzymes on semen viscosity in Llamas and Alpacas. Small Rum Res 38: 91-95.

Bravo, P.W., D. Flores and C. Ordonez, 1997a. Effect of repeated collection on semen characteristics of alpacas. Biol Reprod 57: 520-524.

Bravo, P.W., U. Flores, J. Garnica and C. Ordonez, 1997b. Collection of semen and artificial insemination of alpacas. Theriogenology 47: 619-626.

Bravo, P.W., M.E. Fowler, G.H. Stabenfeldt and B.L. Lasley, 1990. Endocrine responses in the llama to copulation. Theriogenology 33: 891-899.

Bravo, P.W. and L.W. Johnson, 1994. Reproductive physiology of the male camelid. Vet Clin North Am Food Anim Pract 10: 259-264.

Bravo, P.W., R. Moscoso, V. Alarcon and C. Ordonez, 2002. Ejaculatory process and related semen characteristics. Arch Androl 48: 65-72.

Bravo, P.W., C. Ordonez and V. Alarcon V., 1996. Processing and freezing of semen of alpacas and llamas. Proc. 13th International Congress on Animal Reproduction. Vol 2: 2-3.

Bravo, P.W., J.A. Skidmore and X.X. Zhao, 2000b. Reproductive aspects and storage of semen in camelidae. Anim Reprod Sci 62: 173-193.

Bravo, P.W., G.H. Stabenfeldt, B.L. Lasley and M.E. Fowler, 1991. The effect of ovarian follicle size on pituitary and ovarian responses to copulation in domesticated South American camelids. Biol Reprod 45: 553-559.

Brown, B.W., 2000. A review on reproduction in South American camelids. Anim Reprod Sci 58: 169-195.

Burgel, H., G. Erhardt and M. Gauly, 1999. Cryopreservation of llama (Lama glama) semen. Proc. II Congreso Mundial sobre Camelidos, Cusco, Peru, pp. 82.

Calderon, W., J. Sumar and E. Franco, 1968. Avances en la inseminacion artificial de las alpacas (Lama pacos). Proc. Revista de la Facultad de Medicina Veterinaria, University Nacional Mayor de San Marcos, Lima, Peru. Vol 22: 19-35.

Callo, M., J. Garnica and P.W. Bravo, 1999. Efecto de la fibrinolisina, hialuronidasa, colagenasa y tripsina sobre la viscosidad del semen en alpacas. Proc. II Congreso Mundial sobre Camelidos, Cusco, Peru, pp. 78.

Del Campo, M.R., 1997. Reproductive technologies in South American camelids. In: Youngquist, R.S. (Ed.), Current therapy in large animal theriogenology. WB Saunders, Philadelphia. pp. 836-839.

Del Campo, M.R., C.H. Del Campo, G.P. Adams and R.J. Mapletoft, 1995. The application of new reproductive technologies to South American camelids. Theriogenology 43: 21-30.

Fernandez-Baca, S., 1993. Manipulation of reproductive functions in male and female New World camelids. Anim Reprod Sci 33: 307-323.

Fernandez-Baca, S. and W. Calderon, 1966. Methods of collection of semen in the alpaca. Proc. Revista de la Facultad de Medicina Veterinaria, Universidad Nacional Mayor de San Marcos, Lima, Peru. Vol 18-20: 13.

Fowler, M.E.,1998 Medicine and surgery of South American camelids: llama, alpaca, vicuña, guanaco. Ames, Iowa State University Press.

Garnica, J., R. Achata and P.W. Bravo, 1993. Physical and biochemical characteristics of alpaca semen. Anim Repro Sci 32: 85-90.

Garnica, J., E. Flores and P.W. Bravo, 1995. Citric acid and fructose concentrations in seminal plasma of alpaca. Small Rum Res 18: 95-98.

Gauly, M. and H. Leidinger, 1996. Semen quality, characteristic volume distribution and hypo-osmotic sensitivity of spermatozoa of *Lama glama* and *Lama guanicoe*. Proc. 2nd European Symposium on South American camelids, pp. 235-244.

Hafez, E.S.E.,1993 Artificial insemination, Lea & Febiger, Baltimore.

Huanca, W. and M. Gauly, 2001. Conservacion de semen refrigerado de llamas. Rev Inv Vet Peru 1: 460-461.

Lichtenwalner, A.B., G.L. Woods, J.A. Weber, 1996. Seminal collection, seminal characteristics and pattern of ejaculation in llamas. Theriogenology 46: 293-305.

McEvoy, T.G., C.E. Kyle, D. Slater, C.L. Adam, D.A. Bourke, 1992. Collection, evaluation and cryopreservation of llama semen. J Reprod Fert Supp 9: 48.

Pacheco, C., 1996. Efecto de la tripsina y colagenasa sobre el acrosoma espermatozoide y su relacion con la fertilidad del semen de alpacaMasters thesisUniv Catol Santa Maria

Parraguez, V.H., S. Cortez, F.J. Gazitua, G. Ferrando, V. MacNiven, L.A. Raggi, 1997. Early pregnancy diagnosis in alpaca (Lama pacos) and llama (Lama glama) by ultrasound. Anim Reprod Sci 47: 113-121.

Purohit, G.N., 1999. Biotechnologies in camelid reproduction: current status and future prospectives. Journal of Camel Practice & Research 6: 1-13.

Quispe, G., 1996. Inseminacion artificial en alpacas (Lama pacos) con semen diluido a differentes concentracionesMasters thesisUniv Nac San Antonio Abad.

Ratto, M.H., M. Wolter and M. Berland, 1999. Refrigeration of epididymal sperm from lama with three different extenders. Proc. II Congreso Mundial sobre Camelidos, Cusco, Peru pp. 79.

Raymundo, F., W. Huanca, S. Huertas and M. Gauly, 2000. Influence of different extenders on the motility in alpaca (*Lama pacos*) semen. Proc. 2nd Int Camelid Conference Agroeconomics of Camelid Farming, Almaty, Kazakhstan, pp. 79.

Salamon, S. and W.M.C. Maxwell, 1995. Frozen storage of ram semen I. Processing, freezing, thawing and fertility after cervical insemination. Animal Reproduction Science 37: 185-249.

Simson, A., 2001. Equine reproduction: a clinical perspective. Proc. Australian Embryo Transfer Society p1-8.

Skidmore, J.A., M. Billah and N.M. Loskutoff, 2004. Developmental competence in vitro and in vivo of cryopreserved, hatched blastocysts from the dromedary camel (Camelus dromedarius). Reproduction, Fertility And Development 16: 605-609.

Sumar, J., 1983. Studies on reproductive pathology in alpacasMasters thesisSwedish University of Agricultural Sciences and Universidad Nacional Mayor de San Marcos

Taylor, S., P.J. Taylor, A.N. James and R.A. Godke, 2000. Successful commercial embryo transfer in the llama (Lama glama). Theriogenology 53: 344.

Tibary, A. and M.A. Memon, 1999. Reproduction in the male South American Camelidae. Journal of Camel Practice & Research 6: 235-248.

Vaughan, J.L., D. Hopkins, D. Galloway, Rural Industries Research and Development Corporation, Rural Industries Research and Development Corporation and Rare Natural Fibres (Program), 2003. Artificial insemination in alpacas (Lama pacos): a report for the Rural Industries Research and Development Corporation. Barton ACT.

Vaughan, J.L., K.L. Macmillan and M.J. D'Occhio, 2004. Ovarian follicular wave characteristics in alpacas. Animal Reproduction Science 80: 353-361.

Von Baer, L. and C. Hellemann, 1999. Cryopreservation of llama semen. Reprod Dom Anim 34: 95-96.

Wiepz, D.W. and R.J. Chapman, 1985. Non-surgical embryo transfer and live birth in a llama. Theriogenology 24: 251-257.

Yoshida, M., 2000. Conservation of sperms: current status and new trends. Animal Reproduction Science 60-61: 349-355.

Part I
Sustainable development

Wildlife or livestock? Divergent paths for the vicuña as priorities change in the pursuit of sustainable development

Jerry Laker
Macaulay Institute, Craigiebuckler, Aberdeen AB15 8QH, United Kingdom.

Abstract

The unparalleled success of the international conservation effort of the last 30 years to recover populations of vicuña (*Vicugna vicugna*) from the brink of extinction has resulted in widespread ambitions to derive income from sales of its fibre. Vicuñas are locally abundant in their four main range countries: Peru, Bolivia, Argentina and Chile, to the extent that competition with domestic livestock for grazing resources is an increasingly important issue for pastoral altiplano communities. A condition for the relaxation of international regulations on trade under CITES has been that fibre harvesting should be non-lethal, and this has led to the establishment of a number of different models for exploitation based around the live capture, shearing and release of vicuña. Proyecto MACS, a research initiative with support of the EU INCO programme has been investigating the ecological, economic and social implications of alternative management approaches. Liberalisation has resulted in different strategies emerging in different parts of the altiplano largely as a result of diverse policy priorities in the different countries. This paper reports results from Proyecto MACS to demonstrate some of the implications of these management strategies for the vicuña and its continued conservation.

Resumen

¿Animales silvestres o domésticos? Divergencia en el camino productivo de la vicuña en busca de un manejo sustentable.

El éxito incomparable de los esfuerzos internacionales por conservar la vicuña durante los últimos 30 años y sacarla del borde de la extinción, ha generado una ambición generalizada por la obtención de beneficios económicos a través de la venta de su fibra. Actualmente, las vicuñas presentan una alta abundancia local en su principal área de distribución que involucra a cuatro países (Argentina, Bolivia, Chile y Peru), lo que se manifiesta en una creciente competencia por los recursos forrajeros con ganado domestico en la comunidades altoandinas. Una condición para la relajación de las regulaciones internacionales de comercialización de la fibra de vicuña en CITES, ha sido que la cosecha de la fibra no debe significar la muerte del animal y esto ha conllevado al establecimiento de diferentes modelos de explotación basados en captura de animales vivos, esquila y posterior liberación. El Proyecto MACS, una iniciativa de investigación con financiamiento del programa UE INCO, ha investigado las implicancias ecológicas, económicas y sociales de diferentes alternativas de manejo. Desde la liberalización de CITES diversas estrategias de manejo han emergido a lo largo del altiplano basadas en las propias prioridades políticas de cada país. Este artículo muestra los resultados del Proyecto MACS, los cuales ilustran algunas de las implicancias de las estrategias de manejo y la continuidad en la conservación de la especie.

Introduction

A coordinated international programme of controls on hunting and trade in the vicuña (*Vicugna vicugna*) during a period of some 30 years successfully averted the danger of extinction, and led to a recovery in the population and expansion of range that still continues. The high value of the vicuña's fleece continues, however, to generate interest in its commercial exploitation, and a number of initiatives have been established to enable fibre harvesting, under license to the relevant CITES (Convention on International Trade in Endangered Species of Wild Fauna and Flora) authorities. The revision of conservation policy has generally aimed to balance conservation interests against the pressing need for economic development in rural communities. The widely accepted paradigm has been that:

"..the greater the equity and degree of participation in governance, the greater the likelihood of achieving [biodiversity conservation] for present and future generations" (IUCN Sustainable Use Specialist Group, 2001).

The nature of the systems that have been established since the first relaxation of CITES controls in 1996 has been diverse, with different levels of management "intensity" reflecting different social-cultural realities on the ground. Vicuña have already become livestock in some areas, while they continue to be protected as wildlife in others. This paper explores the nature of this dichotomy, its origins and possible consequences.

Background

The vicuña is found at elevations in excess of 3700m in a range which extends from 9°30'S around Ankash in Peru to 29°30' in the III region (Atacama) of Chile (Novoa & Wheeler, 1984). The vicuña is classified as "lower risk: conservation dependent" in the 1996 Red List of Threatened Animals (IUCN, 1996).

The vicuña has a long history of association with man. Early inhabitants of South America began the process of domesticating the vicuña some 6000 years BP in the Lake Titicaca basin of Peru and Bolivia (Novoa & Wheeler, 1984). By Incan times, management of the wild vicuña was ritualised and followed strict rules, which ensured not only that the fibre was available for the exclusive use of the Incan royal family, but also maintained a pattern of sustainable utilisation of the wildlife resource (Ochoa, 1994a). Hunting of vicuña was prohibited. The harvesting of fibre was a communal activity in organised "chakus", with each population being captured once in every three to four years. By modern standards, these chakus were immense affairs – early reports describe 20-30 thousand people taking part in each chaku, with a catch of 30-40 thousand head (Ochoa, 1994b). All types of animals were shorn, and some of the males were killed for meat for the participants in the chaku. Cloth made from the fibre was highly prized. Garments are reported to have been worn once only by the emperor, thereafter being given away as favours and for burned offerings to the gods (Wheeler, 1995).

This apparently sustainable system of vicuña use broke down completely with the Spanish conquest. It has been estimated that the pre-Colombian vicuña population was in the region of 1.5-2 million (Flores-Ochoa, 1977). Though Incan belief systems persisted in communities of indigenous pastoralists, the introduced Spanish culture along with the firearms that they colonisers brought, recognised no cultural restrictions. With increasing pressure from hunting, vicuña numbers began to fall. Concern about overexploitation was recorded even in the 16th century by Spanish chronicler Pedro Cieza de León, who noted a dramatic decline in the populations of both vicuña and guanaco in Peru following colonisation in 1532 (Flores-Ochoa, 1994). The first conservation legislation was issued by decree in 1777 by the Imperial Court, when it was ruled that it was illegal to kill a vicuña and that it was necessary to have a suitable person, appointed by a magistrate present at captures. Later, at independence in 1825, Simon Bólivar introduced further measures in Peru to prevent hunting of vicuña. Acting against this was the establishment

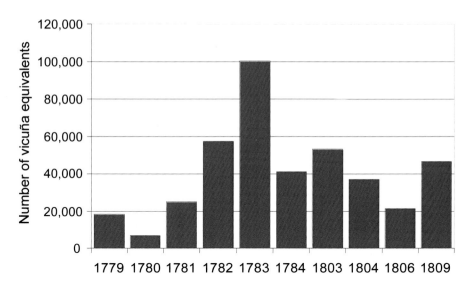

Figure 1. Exports of vicuña fibre from Argentina and Bolivia, 1779 – 1809, calculated assuming an average fleece weight of 250g. After Yaccobaccio et al. (2003).

of new trading links, principally for alpaca fibres, to export markets by British-owned companies based in Arequipa (Orlove, 1977).

Laws to protect the vicuña continued to be introduced. A Supreme Decree in 1920 prohibited trade in vicuña products, and another in 1926 forbade the export of vicuña fibre from Peru. These measures had a limited impact on hunting activities, but in 1933, controls were relaxed again to allow state involvement in licensed vicuña fibre exports. At this time, commercial demand and hence international trade in vicuña skins increased, such that as a result numbers began to crash dramatically. In the period 1937-1965, imports of vicuña fibre to the UK, the principal market, averaged 1,270 kg per year, equivalent to the production from some 5,500 – 6,500 individuals. Over the same period, the vicuña population appears to have fallen from 400,000 in the 1950s to about 10,000 individuals in 1967 (Wheeler & Hoces, 1997). The population estimates at this time are likely not to be particularly accurate, but it seems clear that the surge in pressure on vicuña stocks caused a rapid decline, and exposed the species to a real risk of extinction. Fibre continued to be traded openly (approx. 350 kg/yr) until 1970, when international restrictions on trade were enforced, and conservation measures were agreed multilaterally by the signatories of the first Vicuña Convention.

The vicuña as agent of sustainable development

There is widespread belief that sustainable use of vicuña for its fleece through appropriate management has great potential to contribute both to the long-term conservation of the species and to the economic development of Andean communities sharing the same land (Sumar, 1988; Torres, 1992; IUCN, 1996). This principle is formalised in the 1978 Vicuña Convention, though it was not until 1996 that capture and shearing on a legal commercial basis began. Exploitation of the vicuña is now practised to a greater or lesser extent in all four altiplano countries, though the results in terms of development have been mixed (Lichtenstein *et al.*, 2002). Management practices vary between (and within) the countries, apparently as a result of cultural, political and land tenure differences (Galaz, 1998; Lichtenstein *et al.*, 2002).

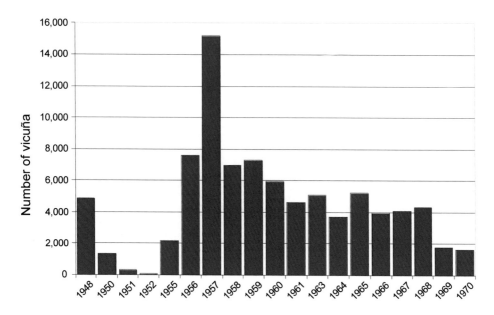

Figure 2. Imports of vicuña fibre by one British company, 1948-1970.

As many of the indigenous communities involved give religious and cultural importance to the vicuña, there is an extra sociological dimension to the dynamics of vicuña ecology (Bernhardson, 1986). The future of vicuña conservation is inextricably linked to future economic and social change in the altiplano. Vicuña may increase in numbers and colonise new areas only if left to do so by local communities. Tolerance, or the lack of it - the trade-off between culturally reinforced positive attitudes towards vicuña and practical concern for their direct impact on forage availability for livestock - may be a highly significant factor influencing vicuña distribution (Cueto *et al.*, 1985). In any case, it is clear that the conservation of vicuñas has in general been successful. Table 1 shows the development of populations in the signatory countries to the Vicuña Convention since protection measures were enforced.

Conservation activities for vicuña were first developed in Pampa Galeras in Peru. In 1972, the reserve received support from the German Federal Government to conduct research, build infrastructure and establish a security system through armed guards patrolling the 6,500 hectare core management zone. The programme proved highly successful. Removal of hunting pressure resulted in initial recruitment rates of 21% per year (Eltringham & Jordan, 1981). However, by the mid 1970's a negative population growth of 11.3% was detected, possibly caused by prolonged droughts and overgrazing (Brack, 1980). In response, a cull was carried out in 1977 (120 head) and 1978 (400 head). This decision involved not only the Peruvian authorities, but also international conservation agencies – IUCN and the World Wildlife Fund – and caused huge controversy at the time (Otte & Hoffmann, 1981). The issue brought into stark focus the differences between conservationists on the one hand and wildlife managers on the other.

In 1980, the National Plan for the Rational Utilization of the Vicuña was introduced in response to the culling controversy. It was recognised that local communities should see some return for their investment in wildlife protection (Brack *et al.*, 1981), and that their involvement in conservation would help to reduce the level of poaching.

Table 1. Convention since protection measures were introduced. Change in the estimated vicuña populations in the 5 signatory counties to the Vicuña.

	1969	1981	1997	2001
Peru	10,000	61,900	102,800	118,700
Bolivia	3,000	4,500	33,800	56,400
Argentina	1,000	8,200	22,100	33,500
Chile	500	8,000	19,800	16,900
Ecuador	0	0	1,600	2,000
Total	14,500	82,600	180,100	227,500

Source: Anon., 1993; Muspratt et al., 1996; CONACS, 1997; D.G.B., 1997; Canedi & Virgili, 2000.

The project established a new set of principles for future management of the species by:
* Local community participation.
* Technology transfer to the Andean *campesino* for effective management of the vicuña.
* Generation and organisation of legal markets for vicuña wool (based on live shearing of vicuña).
* Implementation of housing, health and education programs in the *campesino* communities involved in the project.

Revenue generated by the legal commercialisation of the vicuña wool would, it was hoped, generate additional productive activities for the well-being of the population.

Divergent development paths for vicuña management

The principles established in Peru have underpinned subsequent policy development for vicuña sustainable use throughout the altiplano. In 1991, the law was changed to shift the emphasis of vicuña management from protection to sustainable use (*Ley de promocion de las inversiones en el sector agrario, Decreto Legislativo No. 653*), by transferring the custody of the vicuña to local communities as well as transferring technology and methods for the rational use of vicuña wool as a means of local socio-economic development. At this time, international trade was still heavily restricted under CITES (Convention on International Trade in Endangered Species of Wild Flora and Fauna). The trade ban for cloth made from Peruvian fibre was lifted in 1995.

In the same year, the government of Peru approved a law granting communities the right to manage the land used by the vicuña, and penalizing illegal game practices (Cueto et al., 1985). Local communities began by exporting 2,000 kg of vicuña fibre (produced between 1987 and 1993). The following year, 3,000 kg (produced from 1994 to 1995) were exported. In 1998 the total export was 2,500 kg.

The commercial harvest of vicuña was pushed harder in Peru than in the other three countries because of strong political pressure from local communities to be allowed access to a potentially valuable resource.

In Bolivia, which has the second highest population of vicuña, the approach has been more cautious. There has been a strong emphasis on conservation since the establishment of national parks in 1969. Unlike in Peru, however, a legislative framework for vicuña sustainable use was not introduced until 1997 (*Reglamento para la Conservacion y Manejo de la Vicuña* - D.S. 24.529). Vicuña retain their national heritage status – they belong to the state - and as such may not be kept in enclosures. Rights of use are, however, passed to local altiplano communities who have official approval to undertake vicuña management (Rendon-Burgos, 2000). Three pilot centres were established: Ulla Ulla, Mauri Desaguadero and Sud Lipez, and a programme of capacitation was initiated by the Ministry of Biodiversity and Sustainable Development (DGB) to establish a system of wild capture for the benefit of indigenous communities. The objective of the Bolivian project was clearly stated to involve these communities in decision-making, though unlike in Peru, the government has sought to maintain the conservation of the vicuña as the ultimate objective of management.

In Chile, a conservation programme was initiated in 1970, at which time the national population was estimated at 500 individuals (Cattan & Glade, 1989). Protected areas were established in Region I (Lauca, Tarapaca). The main priority was to stop poaching and illegal traffic of fibre and to apply the recently agreed Vicuña Convention (Miller, 1980; Torres 1992). With the installation of park guards, annual census counts began to rise as the population recovered with the easing of hunting pressure (Rodriguez & Nunez, 1987). By the 1980s, the pressure was beginning to build for sustainable use to be authorised. Several studies were carried out to evaluate fibre quality and ways to distribute benefits of fibre sales (Fernandez & Luxmoore, 1995) and a strategy for the sustainable use of the vicuña was developed (CONAF, 1991). It was expected that in the early 1990s, the vicuña should be in use by local communities (Torres, 1992). However, the sustainable use by indigenous communities has to date never been realised principally because of problems with agreeing a framework for distribution of benefits.

With the successful population recovery in Chile, the reality on the ground is that conservation has to move forward into a sustainable use phase (Bonacic *et al.,* 2002). Both the wild capture and the captive breeding models are being developed simultaneously. A pilot programme for breeding vicuña in enclosures within their habitat was established at Ancara, near the Peruvian border, in 1999, and, following relaxation of the CITES regulations in 2002, further captive management modules have been established at a number of sites in the region (Urrutia, 2004, Muñoz, 2004). Production from captive systems remains low compared to the wild capture modules. For example in 2003, the corrals produced 17 kg fibre, compared to 57kg sheared from captured wild vicuñas.

Argentina has a population of around 23,000 vicuñas (Torres, 1992). The pattern of land use and ownership in the Argentine altiplano is quite distinct from the situation in neighbouring Bolivia. The area is extensively settled by owner-occupier ranchers, with herds of sheep and llamas.

Vicuña distribution in Argentina includes portions of the north-western provinces of Jujuy, the main focus, with vicuña present in Salta, Catamarca, La Rioja and San Juan. The lack of a national census and the scarcity of surveys make it impossible to have reliable data on total vicuña numbers. However most of the researchers in the country agree that some populations have increased their numbers in the last years while others maintained their size (SSN, 2002). Populations from areas that suffered local extinction in the past are slowly repopulating. The distribution of the species is patchy. The attitudes of the local population and the frequency of patrols by wildlife guards appear to be important influences on this, with local abundance of vicuña being associated with communities which have a positive attitude to their presence (Vila, 2002).

Commercial management of wild vicuñas is currently permitted by CITES in Jujuy, however, to date there are no records of this having taken place. Vicuña utilisation in Argentina takes place on farms. The system is promoted by the agricultural extension organisation, the National Institute of Agriculture and Cattle Technology (INTA) Abrapampa, Jujuy. This station donates groups of 12, 24 or 36 vicuñas from their captive herd to individual producers. Young vicuñas plus 10% of their offspring produced under captive conditions have to be returned to INTA station as a compensation for the initial vicuña donation.

Argentine vicuña production created some controversy at the 2002 COP-12 CITES meeting in Chile. The US Fish and Wildlife Service had proposed not to allow Argentine fibre to be imported to the US. Their objection was based in their concerns about the relation between the enclosures and the conservation of the wild populations and the genetic fitness associated with the small of animals in the enclosures. Trade from all producer countries was in the end authorised, on the basis that it would be practically impossible to differentiate traded fibre from different provenance. However, the issue underlines the sensitivity of a major market for the fibre to ethical questions related to animal welfare and conservation.

Discussion

The international conservation efforts brought back the species from the brink of extinction. As a consequence of its success, the vicuña conservation programme became one of the most symbolic projects in Latin America. It is a heartening demonstration that governments, international agencies and local communities can work together to stop species population decline.

As an example of live harvesting of wildlife products, the vicuña is probably unique. As an example of the farming of wildlife for the harvesting of commercially valuable products, the vicuña joins a number of other notable examples worldwide. Farm systems have been established within the last century for the production of other wildlife products, such as bear bile and musk. These predominantly Chinese farms have attracted international criticism on animal welfare grounds. The combination of luxury products with animal abuse is not only ethically questionable, but also disastrous for product image. Both the bear and the deer farms have been the subject of hard-hitting animal rights campaigns (Shrestha, 1998; Homes, 1999). Sustainable use of wildlife is likewise under the spotlight of international concern for both animal welfare and environmental impact (Roe et al., 2002).

So it is that the vicuña producers need to be careful not to establish the type of production that could one day attract such criticism from the animal welfare lobby. The nature of fibre as a product, ensures that its provenance is far more obvious to buyers than for example bile or musk. Consumers are already sensitised to environmental concerns about quality textiles following extensive publicity about shahtoosh fibre, the fine undercoat of the Tibetan antelope or *chiru* (Traffic, 1999). The campaign to increase public awareness of the plight of the chiru has had a significant impact on demand from the US, and should alert vicuña producers to the need to produce fibre within internationally recognised standards of "sustainability".

The impacts of capture, shearing and release are not well known, especially over the longer term (Bonacic et al., 2002). The rigorous and unforgiving climate of the altiplano may cause significant cold stress to animals devoid of the protection of an insulating fleece (Eltringham & Jordan, 1981).

On the other hand, it is also obvious that harvesting systems must be at the same time profitable and practical. With problems being encountered with achieving expected levels of wealth creation, the initial aims of sustainable use defined during the seventies are now being reconsidered. There is still no consensus whether the vicuña should be managed communally as a wild animal or be privatised to be farmed by local communities, or indeed other farmers outwith the altiplano. In

Chile (Galaz, 1998), a series of wild capture-release trials were conducted during the last ten years. In 2000, a programme was initiated of breeding in enclosures on the bofedal habitat of the wild vicuñas. In Peru, which embarked on an ambitious programme of enclosure building in the late 90s, opinion amongst the campesinos appears to be swinging away from fencing towards wild management, as cases of psoroptic mange begin to increase in frequency. Clearly there is a case for improving international collaboration in systems development.

Work within the MACS project (Lichtenstein & d'Arc, 2003) suggests that in Argentina and Bolivia, neither the intensive or extensive management options have so far achieved conservation or local development goals. Management in captivity in Argentina does not provide an incentive towards conservation of vicuñas outside corrals, and the economic benefits are limited. The lack of progress in commercialising vicuña fibre in Bolivia has prevented *campesinos* from realizing economic benefits, and incentives for conservation of vicuñas by local communities remain elusive. However, it does seem clear that while the wild management model does have the potential to bring development benefits, the farm model appears to have neither the capacity to promote conservation of wild vicuña populations outside corrals, or to enhance local livelihoods.

Conclusions

The management of wild vicuñas has genuine potential to augment rural incomes in the Andes, and this potential is being realised in a number of locations where wild vicuña abundance is high, and effective property rights agreements have been reached. Community involvement will probably ensure protection of wild vicuña numbers, at least where such exploitation is seen to bring real economic or community benefits.

However, conservation is more than maintaining populations. The concept includes protection of landscape, animal welfare, genetic diversity, and indeed "wildness". It is important that these secondary benefits of wildlife conservation remain an integral part of the development of commercial exploitation. Farming of vicuñas is not sustainable use. Now that wild vicuñas are out of immediate threat of extinction, the breeding of vicuñas in captivity makes no positive contribution to the conservation of vicuñas in the wild. Enclosures on a larger scale, by restriction of free movement of vicuñas over extended periods have reduced conservation value and create a duty of care for animal health and nutrition, the cost of which has to be met by improved productivity.

The extraction of valuable fleece can be part of an integrated management system for these wild places, enhancing people's lives by sustaining the protection of wild landscapes and traditional culture. The principles for this are already central to international conservation policy but the realisation of such ideals as a recognisable model for vicuña management has yet to be achieved.

Acknowledgements

The authors would like to thank the other members of the Proyecto MACS team.
In particular: Dr Gabriela Lichtenstein, Prof Desmond McNiell, Dr Cristian Bonacic, Prof Iain Gordon, Dr Hugo Yaccobaccio, Vet Med. Pia Bustos. Proyecto MACS receives funding from the European Commission INCO-DEV research programme: ICA4-CT-2001-10044.

References

Anon., 1993. Informe del censo de vicuñas año 1993, Reserva Laguna Blanca. Catamarca, Argentina, Direccion de Ganaderia y Fauna de Catamarca.

Bernhardson, W., 1986. Campesinos and Conservation in the Central Andes - Indigenous Herding and Conservation of the Vicuña. Environ. Conserv. 13: 311-318.

Bonacic, C., D.W. Macdonald, J. Galaz and R.M. Sibly, 2002. Density dependence in the camelid Vicugna vicugna: the recovery of a protected population in Chile. Oryx 36: 118-125.

Brack, A., 1980. Situacion actual de la poblacion de vicuñas en Pampa Galeras y zonas aledanas y recomendaciones para su manejo. Proyecto Especial de Utilizacion Racional de la Vicuña, Ministerio de Agricultura y Alimentacion, Peru, 15 pp.

Brack, E., D. Hoces and J. Sotelo, 1981. Situacion actual de la vicuña en el Peru y acciones a ejecutarse para su manejo durante el ano 1981. In: PEURV, pp. 71.

Canedi, A.A. and R.P. Virgili, 2000. Censo de vicuñas-informe final. Buenos Aires, Argentina, Consejo Federal de Inversiones-Provincia de Catamarca.

Cattan, P.E. and A.A. Glade, 1989. Management of the Vicuña Vicugna vicugna in Chile - Use of a matrix model to assess harvest rates. Biol. Conserv. 49: 131-140.

CONACS, 1997. Censo Nacional de Vicuñas, 1997, Ministerio de Agricultura, Peru. 132 pp.

Cueto, L., C. Ponce, E. Cardiach and M. Rios, 1985. Management of vicuña: its contribution to rural development in the high Andes of Peru. Food and Agriculture Organisation (FAO), Rome. 38 pp.

D.G.B., 1997. Censo Nacional de la Vicuña en Bolivia. La Paz, Bolivia, Dirección Nacional de Conservación de la Biodiversidad, Unidad de Vida Silvestre, 60 pp.

Eltringham, S. and W. Jordan, 1981. The vicuña in Pampa Galeras National Reserve-the conservation issue. In: Problems in management of locally available wild animals. Academic Press, London.

Fernandez, C. and R. Luxmoore, 1995. Commercial ultilization of vicuña in Chile and Peru. In: World Conservation Monitoring Centre, Cambridge.

Flores-Ochoa, J., 1977. Pastores de alpacas de los Andes. In: Pastores de Puna Uywamichiq punarunakuna, Flores-Ochoa, J. (ed.). Instituto de Estudios Andinos, Lima.

Flores-Ochoa, J., 1994. The Andean camelids. Gold of the Andes: the llamas, alpacas, vicuñas and guanacos of South America. J. Martinez, Francis. O. Patthey and sons, Barcelona.

Galaz, J., 1998. El manejo de la vicuña en Chile. In: La conservacion de la fauna nativa chilena: logros y perspectivas. Valverde, V. (ed.), Corporacion Nacional Forestal, Santiago, Chile, pp 178.

Homes, V., 1999. On the scent: conserving musk deer - the uses of musk and Europe's role in its trade. In: TRAFFIC Europe.

IUCN, 1996. The red list of threatened animals. In. International Union for the Conservation of Nature.

IUCN, 2001. Analytic framework for assessing factors that influence sustainability of uses of wild living natural resources. IUCN Sustainable Use Specialist Group.

Lichtenstein, G., F. Oribe, M. Greig-Gran and S. Mazzuchelli, 2002. Manejo comunitario de vicuñas en Peru: Estudio de caso del manejo comunitario de vida silvestre. In: PIE Series No.2 - Evaluating Eden. IIED, London.

Lichtenstein, G. and N. Renaudeau d'Arc, 2003. Vicuña use by Andean communities: a risk or an opportunity? In: The Commons in an Age of Global Transition: Challenges, Risks and Opportunities. Tenth Biennial Conference of the International Association for the Study of Common Property (IASCP), Oaxaca, Mexico, August 9 – 13, 2004. http://www.iascp2004.org.mx/index_eng.html

Miller, S., 1980. Human influences on the distribution and abundance of wild Chilean mammals: prehistoric-present. University of Washington, Seattle.

Muñoz, E., 2004. Aplicación de manejo en cautiverio en la vicuña en la provincia de Parinacota. Wild Camelid Management. Vol II. www.macs.puc.cl.

Muspratt, J., D. Vaysse, et al., 1996. Informe definitivo del censo de vicuñas, 1996, en la reserva Laguna Diamante y Sierra Calalaste. Argentina, Gobierno de Catamarca, Servicio de Ganaderia y Fauna.

Novoa, C. and J. Wheeler, 1984. Lama and alpaca. In: Evolution of domesticated animals, Mason, I. (ed.), Longman, London, pp. 116-128.

Ochoa, J., 1994a. The Andean Camelids. In: Gold of the Andes: the llamas, alpacas, vicuñas and guanacos of South America, Martinez, J. (ed.), Francis. O. Patthey and sons, Barcelona, pp. 22-35.

Ochoa, J., 1994b. Man's relationship with the camelids. In: Gold of the Andes: the llamas, alpacas, vicuñas and guanacos of South America, Martinez, J. (ed.), Francis. O. Patthey and sons, Barcelona, pp. 22-35.

Orlove, B.S., 1977. Alpacas, sheep and men: the wool export economy and regional society in Southern Peru. New York, Academic Press. Studies in anthropology series XX, 270 pp.

Otte, K. and R. Hoffmann, 1981. The debate about the vicuña population in Pampa Galeras reserve. In: Problems in management of locally available wild animals, Hart, D. (ed.), Academic Press, London, pp. 259-275.

Rendon-Burgos, O., 2000. Experiencia Boliviana en el manejo communal de la vicuña. In: Manejo sustentable de la vicuña y guanaco, Iriarte, I. (ed.). SAG - PUC - FIA, Santiago, Chile.

Rodriguez, R. and Nunez E., 1987. El censo de la poblacion de vicuñas. In: Tecnicas para el manejo de la vicuña, IUCN (ed.), pp. 33-57.

Roe, D., T. Mulliken, S. Milledge, J. Mremi, S. Mosha and M. Greig-Gran, 2002. Making a living or making a killing? Wildlife trade, trade controls and rural livelihoods. In. TRAFFIC/ IIED.

Shrestha, M.N., 1998. Animal welfare in the musk deer. Applied Animal Behaviour Science 59: 245-250.

SSN, 2002. Vicuña conservation in Argentina. In. Species Survival Network. Information for COP-12 CITES.

Sumar, J., 1988. Present and potential role of South American camelids in the high Andes. Outlook on Agriculture 17: 23-29.

Torres, H., 1992. Camelidos Silvestres Sudamericanos. Un plan de Accion para su Conservacion. IUCN/ CSE South American Camelid Specialist Group, 58 pp.

Traffic, 1999. Fashion statement spells death for Tibetan antelope. IUCN.

Urrutia, J.L., 2004. Proyecto Vicuña FIA-CONAF. Wild Camelid Management. Vol II. www.macs.puc.cl.

Vila, B.L., 2002. La silvestría de las vicuñas, una característica esencial para su conservación y manejo. Ecología Austral 12: 79-82.

Wheeler, J.C., 1995. Evolution and Present Situation of the South-American Camelidae. Biol. J. Linnean Soc. 54: 271-295.

Wheeler, J.C. and D. Hoces, 1997. Community participation, sustainable use, and vicuña conservation in Peru. Mountain Research and Development 17: 283-287.

Yaccobaccio, H.D., L. Killian and B.L. Vilá, 2003. Explotacion de la vicuña durante el período colonial (1535-1810). Proceedings of the. III Taller Internacional de Zooarqueología de Camélidos Sudamericanos, (GZC – ICAZ): Manejo de los Camélidos Sudamericanos.

Mixed camelids-sheep herds, management practices and viability analysis: some considerations for a sustainability framework of Andean pastoral systems

D. Genin[1] and M. Tichit[2]
[1]*IRD, Laboratoire Population, Environnement, Développement, UMR IRD-Université de Provence 151, Centre St-Charles, Case 10, 13331 Marseille cedex 3, France; didier.genin@ up.univ-mrs.fr*
[2]*UMR INAPG INRA-SADAPT, 16 rue Claude Bernard, 75231 Paris cedex 05, France.*

Abstract

In this communication we aim at documenting how herd management strategies, particularly those related to herd composition and to breeding and off-take practices, contribute to satisfying basic household needs and influencing the sustainability of the mixed llamas-sheep pastoral systems of Bolivia. An extensive survey to assess factors affecting herd composition, and herd management monitoring were performed, focusing particularly on breeding and off-take practices. A model was built, using the viability theory, in order to assess the incidence of these management practices along with climatic uncertainty on long-term household survival. Results show that llamas and sheep play complimentary roles in the functioning of the overall system, due to their differentiated biological characteristics and management. Raising these two species together is viewed as a sound strategy to limit various kinds of risks. The sustainability of the system depends on the balance between the two species, managed with a variety of practices, and the wealth of the household. These considerations lead to the proposal of a conceptual framework based on the links between pastoral family wealth and the range of practices they can mobilise in order to face environmental limitations as well as endogenous uncertainty related to household decisions.

Resumen

En esta ponencia se muestra como las estrategias de manejo de rebaños, particularmente las que conciernen con la composiciòn del rebaño (proporciòn entre llamas y ovinos) y las pràcticas de reproducciòn y descarte, contribuyen a satisfecer los requerimientos domésticos e influyen sobre la sostenibilidad a largo plazo de los sistemas de crianza mixta llamas-ovinos del altiplano àrido boliviano. Se realizò una encuesta extensiva para caracterizar factores que afectan la composiciòn del rebaño y un seguimiento de las pràcticas ganaderas en 14 unidades de producciòn representativas. Se construyò un modelo matemàtico mediante la teorìa de la viabilidad, poniendo en perspectiva la incidencia de la practicas ganaderas sobre la perenidad a largo plazo de los sistemas pastoriles en un contexto de incertidumbre climàtica. Los resultados muestran que las llamas y los ovinos juegan papeles complementarios en el funcionamiento del sistema en su conjunto, debido a caracterìsticas biològicas y de manejo diferenciadas. La crianza mixta es en este sentido una buena estrategia para enfrentar riesgos de varios tipos. La sostenibilidad del sistema depende del balance entre las dos especies, manejadas mediante pràcticas diversas, y del nivel de riqueza de la unidad de producciòn. Estas consideraciones llevaron a proponer un marco conceptual para entender mejor las lògicas de las pràcticas que pueden mobilizar los criadores del altiplano para enfrentar limitaciones ambientales y llevar a cabo los proyectos familiares.

Keywords: Andes, pastoral systems, mixed herds, management practices, viability

Introduction

Pastoral systems are often found in variable environments, characterized by highly limiting factors for crop and livestock production. In the arid Andes, altitude, frost along with low and highly variable rainfall induce pastoral practices and strategies largely influenced by risk management and long term survival concepts (Kervyn, 1988; Genin *et al.*, 1995).

These systems present common features with those found in other parts of the world, such as very extensive production systems, an almost exclusive use of heterogeneous and highly variable natural forage resources, a strong self-regulation of the functioning of the systems, and a highly developed endogenous culture (Landais & Balent, 1993). Others features are proper to the Andean region such as the livestock species reared (South-American camelids), community social organization, the Andean pastoral cosmogony (Flores Ochoa, 1988). Mixed herds of camelids (mainly llamas) and sheep are a general feature in the western highlands of Bolivia and constitute an interesting example due to their significance in relation to the long-term household survival in an extremely harsh environment (Tichit & Genin, 1997). Due to their feeding and spatial complementarity (Genin *et al.*, 1994), mixed herds of camelids and sheep make better use of the ecological diversity of the available rangelands. Moreover, raising various species of livestock is viewed as a sound strategy to limit various kinds of risk (drought periods, disease outbreaks, possibilities of choice for animal off-take in relation to reproductive capacities, market price fluctuation). However, camelids and sheep have specific biological characteristics and constraints which can be partially modulated and adapted by management practices in order to cope with socio-economic needs of pastoral households and to ensure their sustainability. Particularly, sheep are more productive than llamas due to higher breeding rates. Consequently, sheep flocks have a higher potential off take rate than llamas, but seem to be more sensitive to climatic conditions and require more care in their management.

Household production decisions in these low input livestock farming systems can then be modulated in order to cope with the specific constraints each pastoral unit faces. These decisions are principally linked with 1) traditional knowledge and perceptions of the environment, 2) the contemporary structure of means of production (Sieff, 1997), 3) the necessity to cover the immediate domestic needs whatever happens, and to face eventual risks resulting from a large spectrum of causes such as climatic uncertainty, socio-economic constraints, health problems, *etc.*, which can occur in a wider scale of time, and 4) the individual projects of different people, which also contribute to lengthening the time scale involved in this process.

In this paper we document how herd management strategies, particularly those related to herd composition and breeding and off-take practices, can contribute to satisfying basic household needs and influencing the sustainability of pastoral systems. These considerations lead to the proposal of a conceptual framework for sustainability assessment based on a new approach for analysing the significances of livestock management practices.

Materials and methods

Research was conducted at Turco in a pastoral area in the western part of the Department of Oruro in Bolivia during the 1990s decade. Elevation ranges from 3,800 to 4,500 m, and climate is characterised by mean annual rainfall of about 330 mm and almost 300 days of frost per year, which largely hinder crop production. Extensive rangelands constitute the almost exclusive forage resources. They can be categorised in four types *Pajonales* which occupy extensive areas and are dominated by tall, coarse bunchgrasses of *Festuca orthophylla*; *Tholares*, located in the plain and formed by shrubs of genera *Parastrephia; Bofedales,* swamp areas with perennially green, tundra-like pasture where the mat-like plant community of forbs, sedges and reeds is irrigated throughout the year by melt water from ice fields; and finally *Gramadales* with short low-growing grasses which develop over small areas of sedimentary soils in plain and piedmont. Using a NOAA satellite imagery of the Turco area, four types of landscape units were defined, combining topographical units with plant communities: mountains, piedmont, pajonal plain and tholar plain.

On the basis of a previous survey conducted to characterise herd composition in 93 pastoral households (Tichit & Genin, 1997), a monitoring of 14 differentiated households was implemented for two years on a monthly basis, in order to assess the effect of management practices on the performances of livestock systems (Tichit, 1998). It involved accurate herd recording of a population of 800 llama females and 1,500 ewes which included the record of zootechnical parameters concerning female careers and reproduction (fecundity, mortality, intervals between parturition), as well as young survival and body growth. Livestock management practices analyses were primarily concerned by the ones related to animal reproduction, caretaking and off take. A quantification of the economic requirements of families was carried out, it represented the annual cash value necessary to cover expenses (Morlon *et al.*, 1992). The survey also involved retrospective interviews on careers of females in relation to climatic conditions.

All these data were used to identify key variables and estimate parameters that influence herders' strategies, and therefore their capacity to ensure the reproduction of the livestock system on the long-term. Herding strategy is defined here as the combination of decisions on herd composition and management practices. To analyse long-term herd dynamics under climate uncertainty and herding strategies, we relied on the viability theory (Aubin, 1991). This mathematical framework deals with uncertain dynamic systems under state and control constraints. The viability of a dynamic system is related to the maintenance of some conditions to be met for the survival of the system. This approach differs from optimization techniques (Mace & Houston, 1989) by focusing on the viability constraints themselves, and not on the maximisation of a given objective function. Structure of the model relies on a state variable – *i.e.* total household wealth – influenced by climatic uncertainty (*i.e.* the natural growth rate of animal species depends on unpredictable climatic variable). Total household's wealth is composed of the numbers of breeding females in each flock, it is expressed in monetary units by weighting the number of females of each species by its sale price. Herder's decisions on female harvest rate and herd composition account for control variables that represent management practices and strategies. The harvest rate of females ensures a minimum income C to the family when all male offspring have been sold, and herder can manipulate herd composition to different degrees by selling llama females in order to buy ewes. These management decisions closely interact with the herd's endogenous growth rate influenced by climate variability. A full description of the model and its differential equations can be found in Tichit *et al.* (2004).

Results and discussion

Herd Composition

Mixed herds of llamas and sheep were found in the most of the pastoral units surveyed, while monospecific herds of llamas or sheep remained rare. We defined herd composition RA, as the ratio between numbers of sheep and camelids, expressed in ovine units (OU). According to Tichit and Genin (1997), one adult llama is 2 OU, one adult alpaca is 1.5 OU and one adult sheep is 1 OU. Figure 1 illustrates that there is a relatively strong relationship between herd composition and the dominant landscape unit on the farm. Big flocks of llamas are found in mountain (average of 147 llamas including males and 77 sheep) and pajonal plain (average of 148 llamas including males and 108 sheep), while mixed and more reduced herds dominated by sheep are more frequently observed in piedmont (average of 48 llama females and 80 sheep) and in lesser extend in tholar plain (average of 89 llamas females and 87 sheep). However, data showed that family herd composition in all types of landscape, also depended upon socio-economic variables such as labour force availability and cash needs of the family (Tichit & Genin, 1997). This diversity agrees with Norman's assertion that constraints households face lead to the most important differences in farming systems (Norman, 1992), and with Birbuet (1989) who argues that differences in herd size and composition among households are as much a function of households' socioeconomic strategies as they are a factor of families' access to pastures. These considerations led to focus investigations on the functioning of livestock systems.

Diversity of management practices

Our monitoring showed that, among the numerous practices implemented in these livestock systems, breeding and male off take practices varied greatly between both species and among flocks for a given species. We define breeding practices as the combination of mating and caretaking practices (herder's intervention on the mother - young relationship at parturition). Differences in mating practices were observed exclusively in llama flocks. Some herders bred all adult females (> 2 years) each year. We define such flocks as "uncontrolled". Others selected for mating two-year-old females, females that did not produce any offspring in previous years and females that have lost their offspring within their first weeks of life. We call them "controlled flocks". No particular trend was observed for the mating practices among sheep flocks (rams

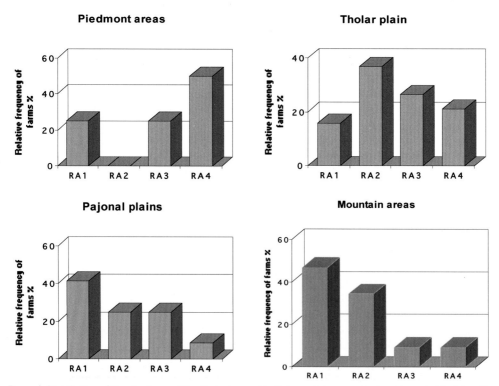

Figure 1. Distribution of the sheep-camelid ratio (expressed in Ovine units) among pastoral households depending upon the localisation of the farm. RA1: <0.25; RA2<0.5; RA3<0.8; RA4>0.8.

were always introduced into the ewe flock in June where they remained until November), but two types of flock caretaking were observed for this species. In this arid environment ewes tend to abandon their young just after lambing. This behaviour is likely to be related to a lower parental care in this species than in llamas. Some herders reinforced the mother - young relationship by reducing the need for the mother to travel. The presence of the shepherd induced the ewes to remain near their young to facilitate the formation of bonds. We define such flocks as "high care". Others had low flock supervision at lambing and many lambs were orphaned a few hours after birth, principally due to forage scarcity for ewes. Then, in order to control the mortality of orphaned lambs, herders bottle-fed them. We call these "low care flocks". These particular caretaking practices were not observed for llama flocks because females never abandon their young, thus herders do not have to act over the mother - offspring relationship.

For both species, male offspring off take practices occurred at two stages: males were slaughtered or sold just after weaning (6 and 12 months for sheep and llamas, respectively), or later at maturity (18 and 36 months for wethers and llamas, respectively). During the annual production cycle, we analysed the way breeding practices were combined with the off take practices of young males. In controlled llama flocks and low-care sheep flocks, male off take was usually delayed until males reached maturity, whereas in uncontrolled llama flocks and high-care sheep flocks, male off take occurred earlier (at 6 and 12 months respectively for lambs and llamas). Delayed male off take appeared to be a response strategy for compensating a lower number of individuals (related to controlled breeding in llamas or low lamb survival attributable to lack of care of the mother - young pair) with a higher individual live weight. The efficiency of these practices was evaluated through annual flock performances.

Flock numerical productivity as an indicator to evaluate management practices

Breeding practices were evaluated using numerical productivity index. It expresses the number of offspring weaned per year as a percentage of the number of breeding females. It is a synthesis of various parameters linked with animal breeding process.

Uncontrolled llama flocks reached higher annual numerical productivity than controlled ones (70% and 44%, respectively). This difference is significant, indicating that uncontrolled breeding practices were more efficient than controlled ones. Due to a lower rate of fecundity (47% versus 80%), controlled llama flocks produced 37% less offspring each year than uncontrolled ones. In the case of sheep, numerical productivity at 6 months of high-care flocks was higher than that of low-care ones (83% and 69%, respectively). This difference is not significant and it was not possible to conclude whether high-care practices were more efficient in increasing numbers than low-care ones. High- and low-care sheep flocks started with the same fecundity rate and low-care flocks produced 17% less offspring than high-care ones at weaning. The higher decrease in numerical productivity during the first month after lambing in low-care flocks illustrated that bottle-feeding was insufficient to control lamb mortality. At the level of the annual production cycle, the endogenous flock growth rate (Upton, 1993) is influenced by the tradeoffs between replacement (in this case 50% of the numerical productivity at weaning as all young female are kept) and female culling rate according to household subsistence requirements. However, this is not the only determinant of herd change over time in arid areas. Numerical productivity is also influenced by an environment that is radically unpredictable and lead to high interannual variations in the natural growth rates of the number of females. This situation highlighted that efficiency of management practices, on the basis of their annual outcomes, only provided a partial assessment, which was inadequate to understand why some pastoralists chose to reduce their annual production. The short-term restraint illustrated by controlled breeding practices may be better understood in relation to their role in terms of anticipating uncertainty. Flock growth rates in bad years were investigated through interviews with pastoralists who reported flock performances during 1983 and 1992 drastic droughts. Herders involved in controlling the breeding rate of their llamas, even in good years, stated that this practice aims at preserving females during bad years, which resulted in an average growth rate higher than in uncontrolled flocks (estimated respectively at 5% versus 2%). In uncontrolled flocks, herders reported that lactating females showed reduced survival during drought periods. Conversely, there were no reports that breeding practices could influence the rate of growth of sheep. Sheep were liable to heavy loss during bad years, as their average growth rate was always lower than that of llamas, and sometimes negative (-2%). This highlighted the importance to assess management practices on a longer-term perspective.

Viability analysis of mixed herds

The model focuses on the long-term interactions between household subsistence requirements and inter-annual variability of mixed herd performances. It represents household wealth that grows or declines as a result of both the endogenous herd growth rate (influenced by the environment and breeding practices) and the effect of the direct management decisions (take-off). Changes in wealth (herd size) are mainly determined by the balance between household needs (minimum income) and each of the species female growth rates. Analytical results showed that thanks to the complementary growth rates of both species, a mixed herd allowed the herder to take advantage of a wider range of environmental situations. They also indicated that management practices consisting in controlling the breeding rate of llama flock strengthened the evolution of the mixed herd only when the herd was large enough so that relatively low off take rate satisfied the minimum income.

Results are illustrated here with two simulations based on the same climatic scenario. By dividing his total wealth between both species (with 60% llamas), the herder secured long-term herd evolution. Figure 2 shows that the herder who did not control the breeding rate of his llama flock secured a minimum income at each time period with an initial herd composed of 36 llamas and 76 sheep (i – Figure 2). Conversely, the herder who chose to control the breeding rate of his llama flock even in good years, needed a larger mixed herd (27% larger in wealth) in order to secure herd viability: 46 llamas and 97 sheep heads (iii – Figure 3). All trajectories starting below these thresholds (a wealth of respectively 3,630 and 4,030 US$) are not viable in finite time (ii – Figure 2 & iv – Figure 3). Given that the culling rate depends on herd size, it decreased as herd size increased. Thus, to satisfy the minimum income, rich herders adopted a lower culling rate than poor ones. The only deterministic situation (only one decision is viable) appeared at critical thresholds *i.e.* when herd size was on the discriminating boundary and adversity was such that herd yield only permitted to harvest a minimum income. In this case, the herder's decisions were reduced to exchange all his wealth for the species with the highest growth rate. This decision has been described for various pastoral systems; it is referred to as "upstocking" (Mace & Houston, 1989). However, even in this extreme case, we can formulate doubts on the chance of an exchange of wealth from one species to another to occur in the systems we are concerned with, mainly because the market for females is nearly non-existent even on good years and because socio-cultural aspects of camelid rearing still take a great importance on their functioning. Both simulations also showed non-viable wealth evolution (dotted line) of a mixed herd composed of 10% of llamas. This means that for this particular scenario in which between years 2 and 7, the climatic scenario is unfavourable to the dominant species (sheep), the off take rate necessary to satisfy the minimum income induced an irreversible de-capitalisation. The contrary would also be true for a mixed herd with 90% llamas in another climatic scenario. In earlier work, it was shown that a mixed herd with a close proportion of each species allows to

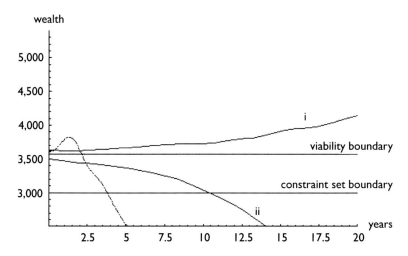

Figure 2. Wealth evolution in a mixed herd with an uncontrolled breeding rate of the llama flock. i) Strongly viable wealth evolution: initial wealth is 3,630 US$, herd structure is made up of 60% of llamas and offtake rate on females satisfies a minimum income at every point in time. The viability kernel boundary is 3,571 US$ and the constraint set boundary is 3,000 US$. ii) Non viable wealth evolution. Dotted line: non viable wealth evolution of a mixed herd composed of 10% llamas.

overcome any possible climatic scenario and that a monospecific herd can also be viable but at higher herd size (Tichit *et al.*, 2004).

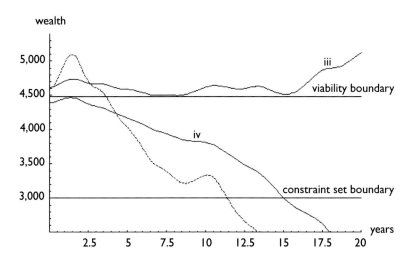

Figure 3. Wealth evolution in a mixed herd with a controlled breeding rate of the llama flock. iii). Strongly viable wealth evolution: initial viable wealth is x_0 = 4,618 US$, herd structure is made up of 60% of llamas and off take rate on females satisfies a minimum income at every point in time. The viability kernel boundary is 4,478 US$ and the constraint set boundary is 3,000 US$. iv) Non viable wealth evolution. Dotted line: non viable wealth evolution of a mixed herd composed of 10% llamas.

Toward a conceptual framework for sustainability assessment based on the analysis of the significance of livestock management practices

Our results show that herd management practices reflect household's strategies involved in the maintenance and dynamics of pastoral systems. There are a variety of such practices, finding their logic only in the context of the specific opportunities and constraints each pastoral unit faces. Results of the model showed that there is a minimum size for the herd to have a chance to be viable. This minimum viable wealth is dependant upon herd composition, meaning that animal species are not interchangeable. There is obviously a minimum level of wealth below which households are caught in a declining cycle, and we can empirically define wealth thresholds above which some practices are possible and under which they are impossible. This is illustrated by the "controlled" llama flocks, in which pastoralists slow down the growth rate of the flock in order to preserve females during bad years. Such a practice appears to be difficult to implement in the case of poor herders who find it difficult to satisfy basic household needs with small flocks, and so always tend to maximise herd growth rate, even if it is at higher risks. To assess the interaction between biological and decision processes, herd viability is an essential point to take into account, notably in systems where sustainability is the core of the production project. Most herd dynamic models estimate long term production along with herd demography. However, they consider that herd size has to remain stable from one year to the next. Models, testing breeding and off take practices, that consider herd size as a simulation output remain rare (Lebenhauer & Oltjen, 1998).

Livestock management practices take their rationality within a large spectrum of considerations, and usually have to be evaluated at a larger temporal scale than their immediate outcomes. In this sense, practices are usually described in terms of their modalities and assessed on the basis of their short-term technical efficiency. We argue that they also should be viewed in terms of their significance within the overall functioning of the farming system. From this point of view, they are parts of the elements involved in the functional integrity approach developed by Thompson (1997) to assess sustainability in livestock systems.

In Figure 4 we propose a conceptual framework on the representation of the links between total wealth of pastoral families and the range of practices they can mobilize in order to face exogenous (environmental limitations) and endogenous uncertainty (related to herders' decisions). A variety of practices can be implemented in order to meet objectives as different as:
- Covering immediate subsistence needs (*subsistence practices*).
- Anticipating unpredictable events which could eventually put the pastoral system at risk, such as drought, changes in the meat market, *etc.* (*"reactive" anticipatory practices*).
- Anticipating a future event who has been wished for or planned, and which requires cash mobilisation (building a house, scholarisation of children, social charge[1], *etc.*) or necessitates the adaptation of production means (less labour availability in future, changes in types of products, *etc.*) (*"proactive" anticipatory practices*).
- Accumulating richness, for an increase of the global pastoral patrimony (*accumulation practices*).

---

[1] In the pastoral zone of the bolivian highlands, the Head of the community (*Jilakata*) is chosen by the community members for one year. This function constitutes a fundamental part of the identity sign of peasants who have to occupy it at least once during their life. However, it also implies an accumulation of wealth in order to personally face problems that could arise to any member of the community. This is the reason why this nomination is made seven years before the responsibility is assumed.

At a given time, pastoralists will implement some particular practices which will depend on the range of practices they can mobilise -*i.e.* linked with the level of wealth-, but also on their psychology. The combination of practices involved in herd management should constitute a compromise between maximising yield, perception and anticipation of risk, and projects. Functional characterisation of this combination could be helpful to better evaluate the overall functioning of the system and constitute an interesting decision-making tool for development projects. Moreover, this approach should be placed within a longer temporal scale, and should characterise "trajectories" of functioning of pastoral systems. It also requires that attention be paid to the "significance of practices" since the trajectory of a foreseen project can move due to unexpected events and lead to the implementation of *substitution* and *opportunity practices*, relating to changes in production conditions and/or projects.

Conclusion

The choice of which animal species to breed and their management assume great importance in the functioning of Andean pastoral systems since llama and sheep are not interchangeable species and have effects both on the short-term domestic economy and on the long-term sustainability of the enterprise. A mixed herd enables herders to take advantage of opposed species-specific traits. Llamas behave as a stabilising component owing to their ability to thrive during perturbation, whereas sheep can promote a far higher wealth dynamic but at higher risk for the long-term viability of the system. Wealth level of wealth and attitude toward risk and opportunities will largely contribute to differentiate households and reveal different strategies more or less sensitive to the long-term capacity of a pastoral system to perdure. In this sense, practices constitute the basic elements of strategies adopted by pastoralists and should be better formalised to characterise and assess the sustainability of pastoral systems.

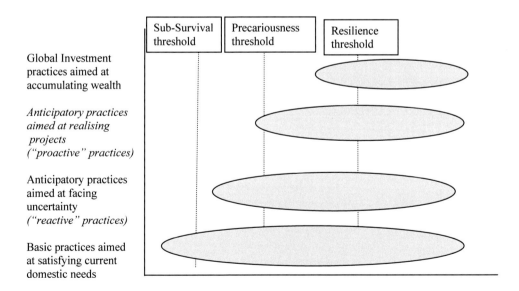

Level of wealth

Figure 4. A preliminary conceptual framework on the representation of the links between total wealth of pastoral families and the range of practices they can mobilize in order to face environmental limitations and endogenous uncertainty related to their own decisions.

Part I

References

Aubin, J.P., 1991. Viability Theory. Birhaüser, Boston.

Birbuet, D.G., 1989. La ganaderia campesina en Pacajes (peasant livestock rearing in Pacajes). SEMTA, La Paz, Bolivia, 156 pp.

Flores Ochoa, J. (Ed.), 1988. llamichos y Paqocheros, pastores de llamas y alpacas. (Herders of llamas and alpacas). CONCYTEC, CEAC, Cuzco, Perù, 318 pp.

Genin, D., H.J. Picht, R. Lizarazu and T. Rodriguez (Eds), 1995. Waira Pampa, un sistema pastoril camélidos-ovinos del altiplano arido boliviano (Waira Pampa, a mixed camelid-sheep pastoral system of the arid highlands of Bolivia). ORSTOM-CONPAC-IBTA, La Paz, Bolivia. 299 pp.

Genin, D., Z. Villca and P. Abasto, 1994. Diet selection and utilization by llamas and sheep in a high altitud –arid rangeland of Bolivia. Journal of Range Management 47: 245-248.

Kervyn, B., 1988. La economia campesina en el Perù, teorias y politicas. (Peasant economy in Peru, Theories and Politics). Centro de Estudios rurales Andinos «Bartolome de las Casas», Cuzcu, Perù, 92 pp.

Landais, E. and G. Balent, 1993. Introduction à l'étude des systèmes d'élevage extensif. In: Landais, E. (Ed.), Pratiques d'élevage extensif (Practices in extensive livestock rearing), INRA Editions, Paris, France, pp. 13-34.

Lehenbauer, T.W. and J.W. Oltjen, 1998. Dairy cow culling strategies: making economical culling decisions. J. Dairy Sci. 81, 264-271.

Mace, R. and A. Houston, 1989. Pastoralists strategies for survival in unpredictable environments: a model of herd composition that maximise household viability. Agricultural Systems 31: 185-204.

Morlon, P., B. Montoya and S. Channer, 1992. Dix ans dans la vie de paysans des rives du Titicaca. In: Morlon, P. (Ed.), Comprendre l'Agriculture Paysanne dans les Andes Centrales,(Understanding peasant agriculture in the central Andes) INRA, Paris, pp. 331-364.

Norman, D.W., 1992. Household economics and community dynamics. In: Valdivia, C. (Ed.), Sustainable Crop Livestock Systems for the Bolivian Highlands, Columbia: University of Missouri-Columbia, pp. 39-69.

Sieff, D.F., 1997. Herding strategies of the Datoga pastoralists of Tanzania: is household labor a limiting factor? Human Ecology 25: 519-544.

Tichit, M., 1998. Cheptels multi-espèces et stratégies d'élevage en milieu aride: analyse de viabilité des systèmes pastoraux camélidés - ovins sur les hauts plateaux boliviens.(Multispecies herds and livestock management strategies in arid environment: mixed camelid-sheep pastoral systems analysis in the arid Puna of Bolivia). Thèse de doctorat, Institut National Agronomique Paris - Grignon, Paris, 283 pp.

Tichit, M. and D. Genin, 1997. Factors affecting herd structure in a mixed camelid-sheep pastoral system in the arid Puna of Bolivia. Journal of Arid Environments 36:167-180.

Tichit, M., B. Hubert, L. Doyen and D. Genin, 2004. A viability model to assess sustainability of mixed herds under climatic uncertainty. Animal Research 53: 405-417.

Thompson, P.B., 1997. The varieties of sustainability in livestock farming. In: Proceedings of 4th International Symposium on Livestock Farming Systems, Foulum, Denmark, EAAP Publication no 89, Wageningen Pers.

Upton, M., 1993. Livestock productivity assessment and modelling. Agricultural Systems 43: 459-472.

Linking community aims with vicuña conservation: A Bolivian case study

Nadine Renaudeau d'Arc
School of Development Studies, University of East Anglia, Norwich NR4 7TJ, United Kingdom;
n.d-arc@uea.ac.uk

Abstract

Community-based wildlife management through sustainable use has been widely promoted in the past few decades as an appropriate strategy to link community aims with wildlife conservation. In the case of the vicuña, *Vicugna vicugna,* a wild camelid found in the high Andean region, conservation policies have shifted away from strict protection to allowing sustainable use by Andean communities. The general argument behind this move is that the generation and distribution of benefits derived from the commercial use of fibre at the community-level is likely to encourage local conservation of the vicuña. In Bolivia, the state grants custodianship and exclusive use rights to local communities. The unit of custodianship and use is the 'communal management area' that is within the control of one or more communities. This paper analyses the key factors affecting community involvement in vicuña capture and shearing events, based on data collected in San Andres de Machaqa (Province of Ingavi, Department of La Paz). The analysis focused on the first period of implementation of the programme (1997-2002) in the absence of commercialisation of fibre. The paper identifies two groups of factors affecting communities involvement: those related to size and boundaries of communal management areas; and those related to the internal dynamics of local communities.

Resumen

El manejo comunitario de la vida silvestre a través del uso sustentable ha sido promovida, en las últimas décadas, como una estrategia apropiada para unir los intereses de las comunidades con la conservación de la vida silvestre. En el caso de la vicuña, Vicugna vicuña, un camélido silvestre en la región Andina, las políticas de conservación han pasado de la protección estricta al uso sustentable por comunidades andinas. El argumento general detrás de estos cambios es que la generación y distribución de beneficios a partir del uso comercial de la fibra a nivel comunal puede contribuir a la conservación local de la vicuña. En Bolivia, el Estado otorga custodia y derecho exclusivo de aprovechamiento de la vicuña a las comunidades que conviven con dicha especie. La unidad de custodia y aprovechamiento es el área de manejo comunal que puede estar integrada por una o más comunidades. Este trabajo analiza los factores que afectan la participación de las comunidades en los eventos de captura y esquila de la vicuña, basado en datos obtenidos en San Andrés de Machaca (Provincia Ingavi, Departamento de La Paz). El análisis se basa en el primer período de implementación del programa (1997-2002), sin comercialización de fibra. El trabajo identifica dos grupos de factores afectando la participación de las comunidades: aquellos relacionados con el tamaño y límites de las áreas de manejo comunal; y aquellos vinculados con la dinámica interna de las comunidades.

Keywords: vicuña, conservation, communities, collective action, Bolivia

Introduction

Wildlife conservation perspectives have shifted away from strict-protection to sustainable use, where benefits from wildlife utilisation are expected to provide incentives for community-based conservation (Adams & Hulme, 2001). The community-based strategy relies on a number of assumptions, one being that the target community is a willing participant in collective action (Hutton & Leader-Williams, 2003). Collective action refers to the joint collaboration or involvement of a group of people to achieve a common goal or interest. This paper re-examines this assumption by exploring the factors community-based management of vicuña Vicugna vicugna, a wild South American camelid which has a limit range in the high Andean region of five countries (Argentina, Bolivia, Chile, Peru, and Ecuador).

Due to the success of vicuña conservation and the high commercial value of its fibre, conservation policies have shifted away from strict protection to allowing sustainable use of the species. This is done under specific commitments signed at regional level (Vicuña Convention[1], 1979) and conditions established by the Convention for the International Trade in Endangered Species of Wild Fauna and Flora (CITES). Under this global policy framework, different exploitation systems have been developed to shear vicuñas, with diverse shortcomings for both vicuña conservation and providing meaningful benefits to local people (Lichtenstein, 2004).

Vicuña management and property rights in Bolivia

In Bolivia, the actions of local communities has brought about the success of vicuña conservation[2]. In 1997, three target populations of vicuña, located in Apolobamba, Mauri-Desaguadero, and Lipez-Chichas were passed by CITES from strict-protection (Appendix I) to allowing commercial use of fibre obtained from live shorn vicuñas (Appendix II). From 1997-2002, technical, logistical and financial support[3] has been provided by the State through two government agencies: the General Biodiversity Bureau (DGB) in Mauri-Desaguadero and Lipez-Chichas, and the National Service of Protected Areas (SERNAP) in Apolobamba (DNCB, 1997).

Under this new international policy framework, the government formulated the Vicuña National Regulation[4] supporting community-based management of vicuña in the wild. Vicuña is property of the state, but the government grants custodianship to local communities and exclusive use rights over those wild vicuña populations living in their communal lands. The unit of custodianship and shearing activities is the 'communal management areas' (CMA). The general assumption is that CMAs designed by communities themselves, will fit their territorial and social organisation, and facilitate collective action in vicuña management. Bolivia provides a pertinent case to identify the factors affecting community involvement because, in spite of the lack of direct economic[5] benefits from shearing activities, the number of communal areas for vicuña management (CMA) has been increasing from 3 CMAs in 1998 to 32 CMAs in 2002 (DGB, 2003). In 2000, five Regional Associations for Vicuña Management (ARMV) were created to group together communal management areas for the future distribution of benefits from fibre commercialisation.

---

[1] The Bolivian government is one of the first (together with Peru) to sign the Convention for the Conservation of Vicuña (Tratado de La Paz, 1969) and also to agree, ten years later, to the inclusion of the concept of 'sustainable use' in the new Convention for the Conservation and Management of Vicuña (1979).

[2] The term 'community' or 'ayllu' within the Andean context, refers to a group of families sharing control of a territory and is the social unit used by Quechua and Aymara speaking Andean people for the defence of their rights.

[3] The establishment of partnerships between government and international aid agencies (e.g. SERNAP-AECI in Apolobamba) or projects and programmes (e.g. DGB-Proquipo in Sud Lipez) have played a key role during this period, often attaching conditions to their funding.

[4] The Vicuña National Regulation is divided in VII Titles, XIV Chapters and 64 Articles; authorised by Supreme Decree in May 1997 (DS 24.529)

[5] The fibre sheared from 1998-2002 is stocked by government for future commercialisation.

Study area and research methods

This paper uses primary and secondary data collected in the Regional Association[6] Machaqa. The area is situated at (approximately) 120 km South West from La Paz city, corresponding to the new Municipality of San Andres de Machaqa in the north part of the Mauri-Desaguadero pilot area for vicuña management. It covers around 150,000 hectares surrounded by the river Desaguadero in the North; the frontier with Peru in the west and the Province of Pacajes in the East and South as shown in Figure 1.

The area of Machaqa corresponds to the traditional Marka San Andres de Machaqa. The term Marka refers to the patrimony of minor units represented by kinship jurisdictions called Ayllus[4], surrounding an administrative or ceremonial center (Ticona Alejo & Albo Corrons, 1997). Figure 1 shows how the Marka San Andres de Machaqa is divided into six Ayllus: Collana, Levita, Choque, Alto Achacana, Bajo Achacana and Yaru. These converge in a town called San Andres de Machaqa, which is situated at their geographical centre. These Ayllus are divided into minor 'communities' that group between 30-60 families.

Vicuña capture and shearing events started in the year 2000, with technical, financial and logistical support from by the DGB. The main role of DGB (as the CITES authority in Bolivia) is to control and certify that fibre has been obtained from live sheared animals. But, during the first three years of implementation (2000-2002), DGB also provided broad support for the communities.

Research methods combined participative observation with other research techniques such as semi-structured interviews and group discussions before and during vicuña capture and shearing events. Information was triangulated with secondary data obtained from review of documents and official records.

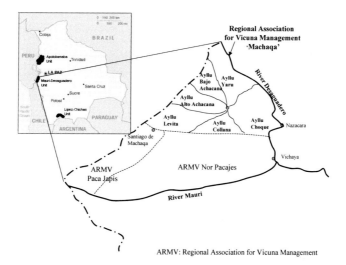

Figure 1. Mauri-Desaguadero area for vicuña management.

---

[6] In 2000, five Regional Associations for Vicuña Management (ARMV) were created to group communal management areas for the future distribution of benefits from fibre commercialisation. These are distributed in Potosi department (Sud Lipez) and La Paz department (Apolobamba; Machaqa; Paca Japis; and Nor Pacajes).

Results and discussion

Vicuña capture and shearing events take place in each of the communal management areas that are willing to participate in the activities, in joint collaboration with two DGB technicians and 16 wildlife wardens[7] from the Mauri-Desaguadero area. The two-day events are divided in a number of stages and organisational processes as seen in Table 1.

The first day of the event consists of the construction of enclosure, while the second day consists of the round-up, capture, and shearing. These different stages and the organisational processes involved are described below.

Construction of the capture enclosure

The capture enclosure (*manga de captura*) is a structure installed forming a V-shaped design to permit the round-up of vicuñas to a corral at its end. The structure is constructed with posts 3 m long and 10cm in diameter that are joined by fish netting (type "Raschell") 200 m in length and 2 m wide. Netting and posts are shared between communities. The DGB supports the transport of nets, but relies on the pre-planning and coordination of communities for the transport of posts to the capture site.

Round-up and capture of vicuña

People are first grouped in different points far away from the corral (depending on the design of the capture enclosure) in order not to be seen by vicuñas that are grazing inside the V shaped enclosure. With the use of walkie-talkies the different groups are coordinated to gradually walk towards the corral, slowly reducing the distance from each other until they are joined in the

Table 1. Organisational processes during vicuña capture and shearing events.

Stages	Organisational processes		
	Coordination and planning	Decision-making arrangements	Collective action arrangements
Construction of capture enclosure	Date of the event	Criteria for selection of capture site and design of enclosure	Human labour
	Site of the event		
	Transport of posts		
Round-up and capture of vicuña	Coordination of people	Round-up strategy	Human labour
	Social contract (Ayni)		
Shearing and certification	Coordination and division of labour	Shearing technique	Skills and capabilities

Source: Participative observation during vicuña capture and shearing events in Mauri-Desaguadero (2002)

---

[7] Wildlife wardens are community members employed by DGB agency to monitor and control vicuña all year round as well as produce a monthly census report informing on the status of vicuña populations.

same line until the vicuñas are enclosed in the corral. Once inside, the vicuña's legs are tied up (following the same traditional method as domestic animals) and laid down on the floor.

Shearing and certification of vicuña

Animals are divided by sex, offspring younger than two years old are not sheared and are kept to one side; females are first sheared and then released together with their offspring. Once all vicuñas have been released into the wild, the fibre sheared from each animal is put into a plastic bag and the weight recorded using an electronic balance provided by DGB. This number is registered in the Community Minutes[8] and the DGB agency files; and stored in DGB central offices in La Paz with the name of the communal management area awaiting future commercialisation.

Table 2 introduces the main characteristics of the two main vicuña management units: Protected Areas and Communal Management Areas.

Table 2 shows that each Ayllu within Machaqa coincides with the Vicuña Protected Area established by government for monitoring and control of vicuña populations (DNCB, 1996). Within each Ayllu the distribution of vicuña is not homogeneous, depending on the dynamics of vicuña populations and the habitat availability that can also vary from one season to the next (Renaudeau d'Arc *et al.*, 2000). Ayllu Choque, for example, has (approximately) 1,672 vicuñas

Table 2. Characteristics of the main vicuña management units in San Andrés de Machaqa.

Protected Areas (PA)	Area (Ha)	N° of Vicuñas	N° of Wildlife wardens	Communal Management Area (CMA)	N° of communities involved	Community involvement in vicuña capture and shearing events		
						2000	2001	2002
Ayllu Choque	20,112	1,672	3	San Antonio	1		-	✓
				Kanapata	1	-	✓	✓
				Nazacara	1	✓	✓	✓
				J.deManquiri	1	✓	✓	✓
				Huallaquiri	1	-	-	✓
				Pachamaya	1	-	✓	-
Ayllu Yaru	13,104	500	1	Conchacollo	3	-	✓	✓
Ayllu Collana	21,141	955	1	Collana	6	-	-	✓
Ayllu Levita	41,186	1,996	2	Laquinamaya	8	-	✓	✓
Ayllu Bajo Achacana	23,163	1,050	0	Chijipucara	1	-	-	✓
				Chuncarcota	1	✓	✓	-
Ayllu Alto Achacana	30,200	1,314	2	Antaquirani	1	-	-	✓

Sources: Integration of primary (Field work) and secondary sources (DGB, 2003; Prefectura La Paz, 2002)

---

[8] The Community Minutes (Actas de la Comunidad) also records the names of the participants and their community of origin.

distributed in 20,112 hectares patrolled and monitored through periodic census by three wildlife wardens[9]. Each of the six communities in Choque decided to manage vicuña at the community level without grouping their territories. But within Ayllu Choque, community involvement has not been continuous and while new communities get involved (*e.g.* Huallaquiri), others decide not to get involved (*e.g.* Pachamaya).

The analysis of past experiences reveals that the different patterns of community involvement observed depend on a compromise between size and boundaries of communal management areas, and the internal dynamics of communities as discussed below.

In the case of CMA Huallaquiri, the community authority first expressed her reserves towards participating in the shearing event that year (2002) arguing that it was not good for the capture because it was on a slope, there was low density of vicuña and there were too few people. Her argument was based on the perception that her community did not fulfil the three criteria for the selection of a capture site (high density of vicuñas, water availability and accessibility to the area). This perception was influenced by observations during vicuña capture and shearing events in CMA Nazacara, a communal management area with a high density of vicuñas (probably related to the good local habitat conditions) as well as good topography for the placement of enclosure. Communal management areas integrated by one community only can provide 1 to 3 (in the best of cases) appropriate sites where to capture vicuña. Where the density of vicuña is low, varying spatially and over time (probably related to the poor habitat quality), the availability of vicuña at time of need (day of capture) is unpredictable.

While the unpredictable availability of vicuña at time of need is a common problem observed in the literature of common pool resources (Ostrom, 1990), the analysis suggests that communal management areas integrated by more than one community can better address this problem than those areas integrated by only one community by increasing the number of appropriate sites to capture vicuña. For example, in 2002, the shared territories of the eight communities in CMA Laquinamaya provided two appropriate sites to capture and shear vicuña.

In reporting Pachamaya community decisions to get involved or not in vicuña capture and shearing events, views are strongly shaped by past experiences. During a group discussion amongst wildlife wardens (one belonging to the Pachamaya community) they revealed that Pachamaya did not want to participate because half of them were in discord between being involved or not. The underlying reason for this disagreement was related to past experiences that members from Pachamaya community had faced during their participation in the shearing event in 2001 (the previous year). In this regard, the wildlife warden from Pachamaya community remarked that the previous year, when they captured vicuñas, people started to yell that they were vicuñas from CMA Jesus de Manquiri, and asked for them to be released again.

This difficulty of defining boundaries of communal management areas is related to open debates about how boundaries and borders should be conceptualised in common property regimes (Geisler *et al.*, 1997; Sturgeon, 2004). The demarcation of boundaries in San Andrés de Machaqa is based on social norms (informal laws) frequently unknown except to community members (Astvaldsson, 1997; Plata Quispe *et al.*, 2002; Ticona Alejo & Albo Corrons, 1997). But, other processes are influencing communities perceptions towards common property management as a tenure strategy (Albó, 2002). And, this is reflected in the divisions and fusions of community boundaries for vicuña management. This illustrated by the case in CMA Laquinamaya.

---

[9] Wildlife wardens are both community members and DGB employees. The three wildlife wardens in Ayllu Choque, for example, are community members from Nazacara, Huallaquiri and Pachamaya.

In the case of CMA Laquinamaya, the eight communities decisions to manage vicuña in joint collaboration may be based on their past successful experience in managing water resources[10]. But, three communities from the Ayllu Levita are not included in this joint collaboration. One reason of this is because these three communities want to divide their boundaries from the rest of the Ayllu and create their own canton. This process of division of communities through land titling (Land Reform INRA 1996) has also been observed in other parts of Machaqa area such as the case of Chuncarcota in Ayllu Bajo Achacana (Plata Quispe *et al.*, 2002).

Vicuña capture and shearing events relies heavily on collective action institutions already in place in the Andean communites, that also reduces other costs associated with the two-day events, such as the access to tools, infrastructure or transport of posts (for example, posts are shared by each of the three Ayllus in Machaqa). The social bond to collaborate in the activities is already established in those communal management areas formed by more than one community. However, those communal management areas formed by one community only depend on the establishment of these social contracts to compensate the lack of human labour.

The size of communal management areas plays a key role since they represent a strategy to compensate the effects of out-migration on the human labour available. For example, in the case of CMA Huallaquiri, the community authority's willingness to participate in the event (supported by community members) was strong enough and the problem of lack of human labour was resolved through the establishment of a social contract or mutual reciprocity agreement with Jesus de Manquiri and Nazacara called Ayni.

In spite of low expectations (such as the school teacher from Jesus de Manquiri who told me he really doubted Jesus de Manquiri would give much support), on the day of the event there were approximately 20 community members amongst which approximately half came from Jesus de Manquiri and from Nazacara. The community involvement in these cases was related to the legitimacy to the social contract made between communities (Ayni). This social contract is a key element in the organisational process during the round-up and capture of vicuña for those CMA integrated by one community only. The number of people available during the capture is a key factor affecting vicuña events. In those CMAs where social contracts have not been established (*e.g.* CMA Chijipucara) the lack of labour affected the vicuña capture event, which resulted in a small number of vicuña being captured.

This type of institutions for mutual aid, reciprocity and collective work (such as Ayni) depends heavily on the characteristics of community members, and, in the future, it is difficult to predict whether the internal dynamics of communities is changing towards reinforcing or on the contrary, weakening these links.

Conclusions

While the immediate decisions to join the programme may be influenced by past experiences, the analysis in this paper highlighted that issues within communities is a complex affair that must be considered when exploring local institutions for vicuña management. These are the size and boundaries of communal management areas, and the collective action arrangements during vicuña capture and shearing events, both happening within a changing socio-political context in which communities exist today.

---

[10] This development intervention has been supported by the Misión Alianza Noruega and is still functioning.

Acknowledgements

I would first like to give a special thank to the communities visited; the wildlife wardens from Mauri-Desaguadero and to A. Velazco Coronel and A. Orosco Mita for their support in the field. I would also like to thank G. Lichtenstein, M-C Badjeck, I. Gordon and J. Laker for their valuable comments and revisions on the final article. The results presented in this paper forms part of a PhD research that is undertaken within the Project MACS, funded by the INCO-DEV programme of the European Union, Contract ICA 4-2000-10229.

References

Adams, W. and D. Hulme, 2001. Conservation and Community. Changing narratives, policies and practices in African conservation. In: African Wildlife and Livelihoods, D. Hulme and M. Murphree (eds.), James Currey Ltd, Oxford, pp. 336.

Albó, X., 2002. Pueblo indios en la política. *Cuadernos de Investigación CIPCA N° 55*, La Paz.

Astvaldsson, A., 1997. Jesus de Machaqa: La marka rebelde. Volumen 4: Las Voces de los Wak'a. Cuadernos de Investigación CIPCA N° 54, La Paz.

DGB, 2003. Informe a la XXII Reunion Ordinaria Comision Tecnico Administradora del Convenio de la Vicuña. Dirección General de Biodiversidad, Ministerio de Desarrollo Sostenible, Republica de Bolivia, La Paz.

DNCB, 1996. Informe Nacional Gestion 1996. In: XVI Reunion de la Comision Tecnico-Administradora del Convenio de la Vicuña, La Paz, Bolivia.

DNCB, 1997. Censo Nacional de la Vicuña en Bolivia Gestion 1996. In: Producciones Cima, La Paz, Bolivia, pp. 60.

Geisler, C., R. Warne and A. Barton, 1997. The wandering commons: a conservation conundrum in the Dominican Republic. Agriculture and Human values 14: 325-335.

Hutton, J.M. and N. Leader-Williams, 2003. Sustainable use and incentive-driven conservation: realigning human and conservation interests. Oryx 37: 215-226.

Lichtenstein, G., 2004. The Paradigm of Sustainable Use and INTA breeding ranches. Presentation at 4th International Seminar of South American Camelids and 2nd International Seminar of Decama Project, Facultad de Ciencias Agropecuarias de la Universidad Católica de Córdoba, Argentina.

Ostrom, E., 1990. Governing the commons: the evolution of institutions for collective action. Cambridge University press, Cambridge.

Plata Quispe, W.C., Colque Fernandez, G., & Calle Pairumani, N. (2002) Visiones de desarrollo en comunidades aymaras. Un estudio en Jesus, San Andres y Santiago de Machaca. In Programa de Investigacion Estrategica en Bolivia, pp. 21, La Paz.

Prefectura La Paz, 2002. Manejo y Aprovechamiento de la vicuña en el Departamento de La Paz. Dirección Departamental de Recursos Naturales y Medio Ambiente, Republica de Bolivia., La Paz.

Renaudeau d'Arc, N., M.H. Cassini and B.L. Vila, 2000. Habitat use by vicuñas Vicugna vicugna in the Laguna Blanca Reserve (Catamarca, Argentina). Journal of Arid Environments 46: 107-115.

Sturgeon, J.C., 2004. Border, Practices, Boundaries, and the Control of Resource Access: A case from China, Thailand and Burma. Development and Change 35: 463-484.

Ticona Alejo, E. and X. Albo Corrons, 1997. Jesus de Machaqa: La marka rebelde. Volumen 3: La Lucha por el poder comunal. Cuadernos de Investigación CIPCA N° 47, La Paz.

Raising camelids up the Andes: Aymara indians animal and vegetable farming complementarities in Chile

I.M. Madaleno

Department of Natural Sciences, Tropical Institute, Apartado 3014, 1349-017 Lisbon, Portugal;
isabel-madaleno@clix.pt

Abstract

A joint Portuguese-Chilean team field researched Aymara Indian communities in extreme northern Chile, a millenary ethnic group still stubbornly inhabiting hard to reach adobe houses settlements, located up the Andes and high oasis valleys, persevering against all odds their nature respectful farming traditions. Results revealed that ancestral solidarity chains were always vital for Indian camelid herders' survival because very few vegetable farming can be practised 3,800 metres above sea level, quinua being a rare exception. Multiethnic occupation of the Andes and low valleys was closely intertwined with socio-economic complementarities from step to step, a sort of archipelago exploitation of the territory based in diverse ecological niches, characteristic of arid mountainous South American environments. Aymaras preferred to live in the high plateau though (average 4,000m) where along the years they built many villages in support of their essential activity, pastoralism, customarily developed in the wettest areas of the heights termed *bofedal*. Camel family species domesticated long ago were llamas and alpacas, resistant to the cold and lack of oxygen, bred for meat, milk and wool. The paper portrays traditional cultural aspects related to camelids' husbandry as well as their current status in Chile.

Resumen

Un equipo luso-Chileno ha investigado las comunidades Aymaras del extremo norte de Chile, un grupo étnico milenario que aún y siempre vive en los Andes, valles altos y oasis, en sus casitas de adobe, persistiendo en prácticas agrícolas amigas del ambiente. Los resultados de esa pesquisa revelaron que para los pastores de camélidos son vitales las cadenas de solidaridad establecidas entre los indios, porque muy pocas especies vegetales superviven a más de 3,800 metros del nivel del mar, con la rara excepción de la quinua. La ocupación demográfica de los Andes y valles bajos siempre estuvo agregada a complementariedades socio-económicas entre etnias y comunidades indígenas, organizada de escalón en escalón, cordillera abajo, en distintos nichos ecológicos que conforman una especie de explotación en archipiélago del territorio en análisis, típica de ambientes áridos de América del Sur. Los indios Aymaras han demandado aficionadamente el altiplano (altitud media de 4.000 metros) donde han edificado muchos pueblos de amparo a la actividad ganadera, buscando sobre todo los pastos húmedos o *bofedales*. La crianza está sustentada, desde tiempos inmemoriales, en llamas y alpacas, especies muy resistentes al frío y escasez de oxígeno, creadas por la carne, la leche y la lana. El trabajo que se presenta reporta los aspectos culturales relacionados con la producción de camélidos, revelando su importancia y estadio de desarrollo actual en Chile.

Keywords: Aymara indians, agro-ecological levels

Introduction

In Chile, the Andes extend over the horizon, proud of their summits greatness, full of prolific glaciers, lakes, lagoons and crystalline-streamed valleys, fruit of a remarkable geological history and extreme climates. Dazzling Andean peaks spine western South America, in a north-south orderly fashion, reaching 6,340 metres high Parinacota volcano in the north of the country, where crests are knit together by a far-reaching high plateau located about 4,000 metres above sea level. Then the mountains descend into a series of interior plane deserts, here and there river oasis snaked, fed by icy water from the heights, sliding in a toboggan movement that takes less than 200 km to plunge westwards, toward the Pacific Ocean (Charrier & Muñoz, 1993; Oyarzún, 2000). Fragile ecosystems prevail over Chile's northern territories, from Peruvian border till about 27° south of Equator, along the Capricorn high pressures zone, closely associated to precipitation meagreness. Scarce moisture is either sea breeze supported (*camanchaca*) or summer rainfall related, mostly following the annual behaviour of the Inter-Tropical Convergence that brings moisture from the northeast (easterlies), starting in December and finishing in March, phenomenon improperly known as Bolivian winter.

Annual precipitation of 400 mm is registered in the extreme northern high plateau but lowers as the latitude increases and the altitude diminishes, whereas average maximum temperatures reach 16.8° in April and October in the city of Putre (3,500 m) and minimum temperatures frequently lower –10° Celsius in July (Aceituno, 1993; Mendez, 1993). High thermal amplitudes are registered year round, freezing nights and winters predominate, dry air and intense solar radiation prevail, oxygen scarcity and vegetation rarity are all strong limiting factors experienced in the heights, making life harder for both animals and humans, demanding physiological adaptations from the first and slow movement activities and special cares from the second. Nevertheless, Chilean Andes have an interesting number of species able to survive under extreme climatic and nutritional conditions, some of them endemic and with great biological value. Birds and mammals are less stressed by cold than reptiles and amphibious, nesting places being a primary concern and food a scanty resource.

The biggest wild mammals are spitting guanacos (*Lama guanicoe*) spread over an extensive habitat existent from 3,000 to 3,600 m above sea level, all yellowish herbivorous on the verge of extinction, and vicuñas (*Vicugna vicugna*) that dominate the high plateau, sharing out extremely chill (2.5° C mean annual temperature), air scarce planes with Suris (*Pterocnemia pennata*), a sort of South American ostrich, in areas located from about 3,800 m and upwards where only ground-level plant species, forming small rosettes and grasses like llareta (*Azorella compacta*) have freezing tolerance mechanisms to the cold (Arroyo *et al.*, 1993). However, the richest endemic fauna and flora is associated to lakes like Chungará, where one can meet three different flamingo species: the Chilean (*Phoenicopterus chilensis*), the Andean (*Phoenicoparrus andinus*) and James (*Phoenicoparrus jamesi*), the last being quite rare (Raggi, 1993b).

In spite of superseding water scarcity and all environmental constraints there are appreciable natural resources though, sometimes depredated yet never depreciated. Lakes, lagoons, plateaus and river valleys have legitimated refuge habitats for humans, no more than subsistence niches located at different altitudinal steps along the Andes mountains slopes. Activities rang from animal husbandry through mixed animal and vegetable farming, into irrigated agriculture and fishing, all connected through proficient trade networks and cemented by means of good neighbourhood relationships, encouraged by survival needs and not imposed by force. Small Indian communities developed distinctive activities in each agro-ecological level from immemorial times, supported on reciprocity and complementarity relations, linked to strong ethnic and even multiethnic alliances, interdependencies and cultural bonds in a sort of archipelago human geography. Study object will be shrinking spaces existent in Northern Chilean Andes and oasis valleys, increasingly menaced with external cost benefit directives, growing tourism and secluded mineral exploitation, object of depredation by higher and quicker return activities and target of state despise. The only omnipresent ethnic group still stubbornly fighting for recognition of their identity are Aymara Indians, which culture as well as renowned Incas, was originated in the high plateau around Titicaca Lake (Barros, 1987; Millar, 2000).

Methodology

A joint Portuguese Chilean team, directed by Portuguese Tropical Research Institute, has been researching Aymara culture and farming practises in Northern Chile and Bolivia. The multidisciplinary team includes geography, agronomy, veterinary, architecture and regional planning experts. Fieldwork included photo, film records, enquiries to animal and vegetable farmers, sample research in pueblos and riverine settlements, following itineraries over the First Region of Chile, known as Tarapacá, and neighbouring Bolivian villages and cities, so far covering an extension of 6,362 kilometres. Interviews to representative community leaders, Aymara Indian associations, local and regional authorities, planning cabinets, indigenous population commissions (CONADI) were added in order to permit wider data collection and full documentation of official policies and ancestral practises archetypical of the ethnic group under examination. Research results are thus far quite rich and linkage with history and anthropology documentation, as well as water and land tenure regulations versus ancestral customs have given course to reflection and nurtured current analysis. Even though a geographer wrote the paper, it results from multinational and interdisciplinary debates over Indigenous populations in Chile and Aymaras, in particular.

Aymara indians habitats

According to research results, agro-ecological levels managed by Indian communities currently include: (1) The high Andean plateau, sometimes termed *Puna*, located above 3,800 metres where llamas, alpacas and sheep are raised using enlarged family work; (2) The Andean slopes, better known as *Pre-Andes* (3,000 to 3,800 m), where mixed animal and vegetable farming dominate; (3) Upper river valley oasis (mostly situated above 2,000 metres) with small livestock, horticulture and fruit culture; (4) Peri-urban lower valleys, maize, olives and fresh vegetables dominant, where modern agriculture prevails. These days the Indians are increasingly settling in port cities and abandoning traditional activities in favour of small business or qualified jobs, because education is sought, whilst public subsidies to housing benefit urban areas. It's controversial though to include urban centres in the agro-ecological framework. Even so, and because two thirds of Aymaras live in the cities at the present time, they will be focused as an Indian habitat in this paper.

The high plateau

Aymara communities were historically integrated in *ayllus*, a stratified system based in subordination to regional caciques, in the incorporation of Indian peoples in *markas* (villages), in the institution of human mobility from step to step down the Andes Mountains, mere reflex of vertical economic models, based in solidarity alliances between llama and alpaca herders and Andes slopes or valley oasis farmers, which has been object of a political disruption some authors date back the end of the eighteen-century (Rivera, 1994). In matter of fact, along the years colonization waves introduced new settlers and drove to new customs, first the haciendas and then the mining fields, especially nitrate companies, contributing to break ancestral traditions and loyalties. The less affected area is indeed the Andean high plateau, for it's far too distant from main urban centres, away from paved roads, having an impossible climate, icy in *Thayapacha* (Winter), unbearably rainy in *Jallupacha* (Austral Summer).

In the heights of the Andes one can find a few remaining communities of animal farmers, relying on enlarged family work and sharing usufruct rights over pastures. Flocks of South American camels pop out from the wilderness, deserving special mention domestic species like llamas (*Lama glama*) and alpacas (*Lama pacos*), resistant to the cold and lack of oxygen, bred for meat, milk and wool. Llamas raised in the First Region of Chile (Tarapacá) correspond to 90.12% and alpacas to 89.09% total animals registered in the country (see Table 1). Livestock management under extreme conditions requires wide extensions with distinctive plant species:

1. The *hok'o (bofedal* in Spanish*)*, which includes herbs, algae and several aquatic genera, dominates all wetter surfaces above 3,200 metres. Plant growth is interrupted during the coldest months, from May to July, and gets its peak from December to February. Botanic species consumed as fodder are *Oxychloe andina, Distichia muscoides, Festuca rigescens*, and genera *Deyeuxia, Werneria, Lemma, Carex*.
2. The *waña*, (*tolar* in Spanish), can be found at less than 4,000 m above sea level, and has both small bushes like *Parastrephia, Adesmia, Bacharis, Senecio* and *Fabiana* and annual herbs, *Munroa, Eragrostis, Aristida* and *Euphorbia* species.
3. Finally there is the *zuni*, no more than perennial herbs that include multipurpose *llareta (Azorella compacta)* and constitute the highest pastures in the Andes. Predominant species belong to genera *Stipa, Festuca* again, and *Calamagrostis* (Arroyo *et al.*, 1993; Villagrán *et al.*, 1999).

Llamas can be fed in the last two natural associations because they are bigger and cope better with aridity and biodiversity but alpacas are quite selective and prefer the small grasses of the *bofedal*, here and there punctuated by small water ponds where they refresh themselves and which water they eagerly consume. The animals' paws are quite small and light, giving them agility and an elegant walk, which minimizes scarce pastures deterioration, in addition to having the unparalleled ability to cut the vegetation with their teeth without jerking out or destroying grazing fields as other ruminants do (Raggi, 1993a).

Aymara females customarily execute daily shepherd tasks, particularly sheep care for these are less adapted animals to the extreme conditions experienced in the high plateau. Animal possession is individual, young boys and girls being given from birth and along their youth the assets that will permit them easier survival in adulthood. Families cooperate in animal farming sharing refuges or temporary settlements were more than a few members make use of informal land rights and pastures – the *estancias* – transmitted from father to son, meaning male lineage, no matter whom they legally belong to in official Chilean property records. The estancias can also enclose a small number of isolated permanent houses that punctuate the short food sources. Hereditary rights over fodder supplies are recognised by Chile to any family member (again only males through Indian law), in theory, but usufruct of *hok'o* (wet grazing lands or *bofedales*) is informally established and respected, in spite of the frequent land and water disputes (Gundermann, 1986; Castro, 1993). In matter of fact, land tenure has long been a contending issue for Chilean state never registered communal land, and a 1979 regulation further wrecked Aymaras traditions, because private ownership was imposed. Last but not least, the 1981 Water Code separated land from water rights and gave way to disruption of *Suma Qamaña* millenary practises, seen as the sum of material growth, spiritual equilibrium and fair ecosystem management (Comisión Verdad y Nuevo Trato, 2003).

Andes slopes and terraces

In the surroundings of most villages visited in the Andean slopes the team examined vestiges of many ancient terraces built up the mountains and not excavated, as elsewhere in Indian land, for according to Aymara cosmic vision one cannot hurt the motherland, locally known as *la pachamama* (Beach *et al.*, 2002). Terracing is still a common practise and a manifestation of intensive agriculture, even though population is no longer as large as it was in old days for migration to less extreme environments has increased in the last half of the 20th century. Terraces occur in perpendicular position to gradient inclination, intended to facilitate water channelling

Table 1. Livestock raised in Extreme Northern Chile.

Municipalities and PROVINCES	Cows	Sheep	Goats	Llamas	Alpacas
Putre	1,140	9,053	772	30,613	23,624
General Lagos	14	7,401	3	18,904	11,760
PARINACOTA	1,154	16,454	775	49,517	35,384
Arica	1,964	5,767	718	114	1,405
Camarones	822	6,667	881	1,277	327
ARICA	2,786	12,434	1,599	1,391	1,732
Iquique	1	231	43	35	5
Huara	295	2,228	475	1,198	38
Camiña	346	2,624	125	913	354
Colchane	4	3,334	28	14,767	2,421
Pica	1	1,071	10	2,916	340
Pozo Almonte	31	7,629	7,777	787	67
IQUIQUE	678	17,117	8,458	20,616	3,225
Total I REGION of TARAPACA	4,618	46,005	10,832	71,524	40,341

Source: INE, 1997.

step by step, to slow rain water runoff, to built up planting surfaces in steep slopes, to avoid soil erosion and to maximize soil moisture. Fertilization uses livestock excreta and guano brought from lower valleys and littoral areas, whereas irrigation channels are a classic landscape element, which construction was further encouraged by ruling Chile, following this country's victory over Peru and Bolivia in the War of the Pacific (1879).

Water and land rights were theoretically recognised to Indian communities before and after 1879. In matter of fact, either Spanish viceroy or republican rulers from Peru and Chile legislated in respect with Aymara's consuetudinary laws yet always guided their jurisprudence by sheer entrepreneurial rationality (Dougnac, 1975; Van Kessel, 1985). As a result communal landownership and water rights were soon considered impossible to register and so most Indian properties were individualised whenever and wherever the Aymaras were educated and diligent enough to descend to the port cities and sign down their property rights, whilst unclaimed territories were nationalised. Regarding the water, drama is nitrate, copper and other mining companies continuously pressure natural resources and easily get usufruct over the precious liquid, using concessions granted under 1981 Water Code, as mentioned. Three main consequences are the outcome: 1) land accumulation in the hands of a few Indian families as well as the State of Chile (public properties); 2) slow death of high plateau and pre-Andes *bofedales*, together with irrigation water scarcity in terraced slopes used for subsistence agriculture, and extermination of formerly productive oases; 3) finally the growing destruction of ancestral alliances, communal cooperation and reciprocity relations, typical of Andean societies.

In the old days, animals and herders from the high plateau came down to lower altitudes during the winter, whist the Aymaras sought to trade animals and wool for vegetables. Disarticulation of millenary agro-ecological levels being steadfast, it has given path to enforced reorganization of communal activities and landownership, Indians using more and more paid workforce instead of family cooperation (Castro, 1993). Actually in the second agro-ecological step Aymara Indians mainly grow potatoes (*qhathi*) brought from Titicaca Lake margins where it was consumed at least since Tiwanaku golden era (500-1200 A.D). Long-established sustainable agriculture practises and culinary traditions associated potatoes with quinoa (*Chenopodium quinoa*) and oca (*Oxalis tuberosa*), also grown in slopes orientated to the North, but nowadays maize outnumbers any other crop, albeit it is not clear whether it was brought from Central America after Spanish colonisation or in earlier times from the Amazon basin and eastern tropical jungles, for together with llamas and alpacas, these were mentioned to be the real Inca treasures around 1590 (Kalinovsky, 1993). Alfalfa occupies considerable terraced plots of land and elder Aymaras, particularly women, raise cattle in extensive regime (cows, horses, sheep, plus llamas and alpacas), flocks being seen and heard in the still numerous Pre-Andes settlements either in the very early mornings and late summer evenings, when it's time for the animals to return to the stables.

The astonishingly green valleys

Rivers and streams hinder steps comprised between 2,000 and 3,000 meters to be desolate tracts. In fact, rocky brownish landscapes rule amid those valleys, being the elected habitat for mimetic desert foxes, the arid soil scarcely punctuated by strange lines of candelabra *(Browningia candelaris)*, cacti that grow in Equator oriented slopes, nurtured by rare summer rains. Codpa, pueblo located along Victor waterway, is the most outstanding example of Aymara ingenious work and continuous struggle for sustainable agriculture practises in Arica province. Figs, sweet oranges, cherimoyas, guavas, avocados, tunas and grapes are actually the main bets for trade, wine being produced with utmost care, widely appreciated in the region. Horticulture produces, maize, potatoes are cropped for subsistence purposes, in close association with goat and sheep farming. In Iquique province, Camiña (2,400 metres) is an emblematic example of Indian ingenious and strenuous work, whilst dispersed populations crop maize and raise cattle along Tana and Camiña valleys. As researched few young Indians inhabit the upper valleys these days, except for tourism related activities because oasis gorges are stunning, river waters incredibly fresh, clean and clear, as mild climates attract urban dwellers and seasonal visitors.

Peri-urban agriculture oasis

Evidence suggests irrigation water channelling to be a millenary practise used either by Aymara Indians and the Incas, recognised master builders. A remarkable example of such ingenious work is Azapa valley, a beautiful oasis basin located near Arica, the northernmost port city in Chile. The stream was initially quite meagre and the Indians only terraced the slopes; then underground water was added to the scarce local reserves till Chilean public investment finally built a dam up in Chapiquiña (Pre-Andes level) making use of River Lauca waters, the longest in the study area and quite voluminous, fed by glaciers and high plateau rains, which in turn was connected through a wide duct to the valley (Rivera, 1985; Hidalgo, 1985). In matter of fact, projects to irrigate lowlands using highland sources of water were long a menace to extreme northern territories, making *Puna* peoples twice as misfortunate for not only the precious liquid was taken away from them but they were also required as workforce for construction purposes until finally being eradicated from the small pueblos or dispersed pasture refuges, where subsistence animal and vegetable farming dominated, to littoral highly modernised and market orientated realms, like the ones found in this step.

Several ethnic groups inhabited lower Azapa basin through times, as numerous and valuable archaeological data confirm, including Aymara Indians, attracted by soil fertility and water availability (Lehuede *et al.*, 1990). Nowadays Azapa supplies the whole region with its diverse horticulture produces, corn is omnipresent and the valley is famed for mangoes and particularly the appetizing olives, trees and technology having been introduced by Italian and Spanish immigrants, so bountiful that Azapa currently places 3rd in national ranking of olive producers. Drop irrigation is adamant, organic fertilisers combined with chemicals are common inputs in Azapa and Lluta valleys, both located in the outskirts of Arica, for modern agriculture practises are the obvious choice. The difficult task is to locate Aymara workers and landowners because most of them adopted Spanish names, very few speak the traditional language and even fewer are proud of their Indian identity.

In Iquique province, the southern territories in the study area, rain is even scarcer than in Arica and rare oases, such as Matilla or Pica, depend exclusively on irrigation. Ancestral Aymara techniques were the *canchones*, based on superficial water retention, following evaporation during hotter months, juicy melons and watermelons being nurtured in the moisturized soil, after salty layer extraction. Spaniards further developed water captivity and distribution techniques, building extensive underground water galleries from river sources and Andean lakes to lower valleys and oases, termed *socavones* (Figueroa, 2003). As underground water towels are shrinking and sinking, due both to unacceptable demographic pressure (see Pica and Iquique's populations in Table 2) and excessive resources depredation, mostly resulting from uncontrolled mining companies water concessions, these Atacama Desert fringes become less and less productive.

Port cities

Modern cities like Arica and Iquique were settled on top of ancient Indian villages essentially devoted to fishing. Reportedly just before Spanish colonization, multi-ethnicity predominated along present Chilean coast (Hidalgo & Focacci, 1986). After Peruvian independence from Spain (1821) most of what is today the Extreme Northern Chilean Territory, especially Arica, remained part of Peru for more than 5 decades. Definitive and quite disputed possession of the northernmost port was attributed to Chile in 1929. In our days not only multi-culture prevails,

Table 2. Human population in Extreme Northern Chile.

PROVINCE and municipality	1992	2002	1992-2002 (%)
Iquique	151,677	215,233	41.9
Camiña	1,422	1,268	-10.8
Colchane	1,555	1,460	-6.1
Huara	1,972	2,593	31.5
Pica	2,512	6,185	146.2
Pozo Almonte	6,322	10,801	70.8
IQUIQUE	165,460	237,540	43.6
Arica	169,456	184,134	8.7
Camarones	848	1,203	41.9
ARICA	170,304	185,337	8.8
Putre	2,803	2,179	-22.3
General Lagos	1,012	1,295	28.0
PARINACOTA	3,815	3,474	-8.9
Tarapacá or I Region	339,579	426,351	25.6

Source: INE (2002).

also economy is greatly diversified in both littoral cities, widely stimulated by manufacture sector growth and intense port activity prevalence in Arica, and booming import export trade by means of a tax paradise institution (*Zona Franca*) in Iquique (1976). Rural exodus was adamant for Aymaras throughout *chilenization* process (1890-1950), meaning the forced integration of former local residents in Chile, Indians persevering the migratory movement after the Second World War because higher education, subsidies to house acquisition and employment remain sited on the coast.

It's statistically difficult to know how many Indians moved into port cities through times, because the central bureau neglected that question frequently and Indians of any ethnic group are not keen to admit their origin anyway. According to the last census exactly 48,501 Aymaras subsist in Chile (against 48,477 in 1992), but only one third inhabits the Extreme North, remainder preferring big cities, especially Santiago (Grebe, 1998). As to the study area city pull is crystal clear if one analyses demographic evolution in the last decade (Table 2). High Plateau municipalities such as Colchane and Putre have been loosing considerable portion of residents in benefit of valleys (Camarones) and oasis (Pica), but mostly to the cities. A visit to Arica's markets, to central bus station, particularly to the numerous taxi bureaus, a booming activity engaged in daily public transportation to and from Tacna (Peru), give us the picture of business the less wealthy and undereducated Aymaras engage these days: trade of clothes and food, chiefly performed by women – *warmis* – meaning elder than 18 and less than 60 years of age, sometimes caring for their young kids all together, children easily noticed hanging on their backs by typically coloured cloths (Vega, 2002). Educated Aymaras are chiefly doctors, university professors, teachers, engineers, nurses, administrative officers and computer technicians, having notable dominance the scientific and technical professions, on account of their proficiency in mathematics supposedly due to the ancestral idiom and prevailing cosmic vision.

Conclusion

Chile has a narrow territory sandwiched between the Andes and Pacific Ocean but so dazzling and biodiverse that in itself is a Geography lesson. In 2003-2004 a joint Portuguese-Chilean team chose to study Aymara Indians culture, architecture and economy and travelled throughout the Extreme North of the country, mostly a dry desert area here and there interrupted by waterways escorted by exquisite fauna and flora. In this paper dominant focus were remainder Aymara settlements, ancestral farming practises, and agro-ecological niches, ranging from the Andean high plateau, a surface lying beneath numerous volcano peaks where camel family species feed under scarce human supervision, through breath-taking oasis valleys tree covered and painfully managed by the elderly, till littoral urban centres the new generations are eager to call home whilst denying their roots.

Even though the team acknowledged that intensively cultivated terraces and oasis valleys still point out the landscape on the way up the Andes; exquisite blinding white Indian agglomerations are recognizable along the slopes; highland pastures continue displaying flocks of llamas, alpacas and vicuñas, whereas virtually inhabited settlements enclose enchanted adobe churches; truth is ancestral respect for the motherland (la pachamama), the usage of sustainable agriculture practises, and the demonstration of traditional environmental concerns are deeds from few remaining Indians. Pastoralism is the most noticeable traditional activity in the study area, lesser supported on vegetable farming complementarities these days than it was in the past. In spite of all, the paper tells survival stories: the fauna; the flora; the ethnic group. Refuges of resistance reported and illustrated are not only sanctuaries of Indian ways of life but struggle sites where elder Aymaras keep millenary customs alive, because they know culture is a tax free heritage and the most difficult asset to colonize.

Acknowledgements

I wish to thank Prof. Luis Riveros, Rector of the University of Chile, in Santiago, and Prof. Raggi, Director of the International Centre for Andean Studies, in Putre, for their support to the 2003 high plateau scientific mission, as well as to Prof. Alberto Gurovich Weisman, Department of Architecture and Urban Planning, for the magnificent photos. Thanks are due to Prof. José Delatorre, University of Arturo Prat, in Iquique, for his assistance with agronomic data in the year 2004.

References

Aceituno, P., 1993. Aspectos Generales del Clima en el Altiplano Sudamericano. El Altiplano, Ciencia y Conciencia en los Andes, pp. 63-70.

Arroyo, M.T.K., F.A. Squeo, H. Veit, L. Cavieres, P. Leon and E. Belmonte, 1993. Flora and Vegetation of Northern Chilean Andes. El Altiplano, Ciencia y Conciencia en los Andes, pp. 167-178.

Barros, H.L., 1987. Etnogeografía de Chile. Instituto Geográfico Militar, Santiago.

Beach, T., S. Luzzadder-Beach, N. Dunning, J. Hageman and J. Lohse, 2002. Upland Agriculture in the Maya Lowlands: Ancient Maya Soil Conservation in Northwestern Belize. Geographical Review 92: 372-399.

Castro, M.L., 1993. El Campesinado Altoandino del Norte de Chile. El Altiplano, Ciencia y Conciencia en los Andes, pp. 243-253.

Comisión Verdad Y Nuevo Trato, 2003. Informe de la Comisión Verdad y Nuevo Trato de los Indígenas. Gobierno de la República Chilena, Santiago.

Charrier, R. and N. Muñoz, 1993. Geología y Tectónica del Altiplano Chileno. El Altiplano. Ciencia y Conciencia en los Andes, pp. 23-25.

Dougnac, R.F., 1975. La Legislación aplicable a los indígenas del Norte Grande chileno. Norte Grande, I (3/4): 437-446.

Figueroa, C., 2003. Galerías Filtrantes en el Oasis de Pica: tecnología y conflicto social, siglos XVII-XVIII. Congreso Americanista, Valparaíso.

Grebe, M.E.V.,1998. Culturas indígenas de Chile: Un estudio preliminar. Pehuén, Santiago.

Gundermann, H.K., 1986. Comunidades ganaderas, mercado y diferenciación interna en el altiplano chileno. Revista Chungará 16-17: 233-250.

Hidalgo, J. and G. Focacci, 1986. Multietnicidad en Arica, S.XVI. Evidencias etnohistóricas y arqueológicas. Revista Chungará 16-17: 137-147.

Hidalgo, J., 1985. Proyectos coloniales inéditos de riego del desierto: Azapa (Cabildo de Arica, 1619); Pampa Iluga (O'Brien, 1765) y Tarapacá (Mendizábal, 1807). Revista Chungará 14: 183-222.

INE, 1997. Censo Nacional Agropecuario. Instituto Nacional de Estadísticas, Santiago.

INE, 2002. Censo de Población. Instituto Nacional de Estadísticas, Santiago.

Kalinovsky, J.S., 1993. Evolución y Desarrollo de la Ganadería Camélida en el Altiplano de Latinoamérica. El Altiplano, Ciencia y Conciencia en los Andes, pp. 211-221.

Lehuede, J.H., M.M. Cardozo, C.S. Vargas and R.C. Castro, 1990. Compraventa de una hacienda en el valle de Azapa por Gaspar de Oviedo, 1661. Documento inédito del Archivo General de la Nación. Lima, Peru. Dialogo Andino 9: 83-106.

Mendez, C.S., 1993. Hidrologia del Sector Altiplanico Chileno. El Altiplano, Ciencia y Conciencia en los Andes, pp. 71-77.

Millar, W., 2000. Historia de Chile. Zig-Zag, Santiago.

Oyarzún, G., 2000. Andes, Chile. Kactus, Santiago.

Raggi, L.A.S., 1993a. Características Fisiológicas y Productivas de los Camélidos Sudamericanos Domésticos. El Altiplano, Ciencia y Conciencia en los Andes, pp. 223-225.

Raggi, L.A.S., 1993b. La Fauna Altiplanica. El Altiplano, Ciencia y Conciencia en los Andes, pp. 199-202.

Rivera, M.A.D., 1994. Hacia la Complejidad Social y Política: El desarrollo Alto Ramírez del Norte de Chile. Dialogo Andino,13: 9-37.

Rivera, S.A., 1985. Una visión del lago Chungará. Revista Chungará 14: 131-134.

Van Kessel, J., 1985. La lucha por el agua de Tarapacá: la visión andina. Revista Chungará 14: 141-155.

Vega, V.G., 2002. Seeking for Life: Towards a theory on Aymara gender labor division. Revista Chungará 34: 101-117.

Villagran, C., V. Castro, G. Sanchez, F. Hinojosa and C. Latorre, 1999. La Tradición Altiplánica: Estudio Etnobotánico en los Andes de Iquique, Primera Región, Chile. Revista Chungará 31: 81-186.

The European Endangered Species Programme (EEP) for vicuñas

C.R. Schmidt

Zoo Frankfurt, Alfred-Brehm-Platz 16, D-60316 Frankfurt am Main, Germany; christian.schmidt.zoo@stadt-frankfurt.de

Abstract

The International Studbook for the Vicuña was started in 1969. It includes all pure bred Vicuñas outside the four native countries since 1945. Between 1946 and 1971 28 Vicuñas have been imported from Argentina, Bolivia and Peru. Additional 41 specimens specimens were imported from unknown origin. The increase from 11 specimens in 1945 to 59 specimens in 1985 is mainly due to these importations. The natural growth rate of the population from 1957 until 1985 was only 0.2% annually. 576 births have been recorded since 1945. In 1985 the European Endangered species Programme (EEP) was established and the one for the Vicuña was one of the initial programmes. Thanks to this coordinated breeding programme the number increased to 163 Vicuñas in 34 European and one North American collections in the year 2003. The annual growth rate of the populaion in this period was 5.9%. The whole present population belongs to the Southern subspecies *Lama v. vicugna*. Since the number of active founders is only four males and eight females, the EEP urgently needs some unrelated specimens. To imitate natural social behaviour, groups of one male and two to three females are created. As a consequence there is a surplus of males. In the wild they form large bachelor groups. This proved to be impossible in enclosures of up to 4,000 sqm. Research is needed in this context to enable us to form long standing bachelor groups. Other aspects of reproduction and behaviour resemble the conditions in the wild. The main birth season is August to October, a six months transition from the Southern to the Northern hemisphere. Time of birth remains as 79% occur in the morning even in the sixth zoo generation. The breeding male chases out of the group male and female offsprings at an age of 4 to 14 months.

Keywords: vicuña, European Endangered species Programme (EEP), zoo management, population growth, biology in zoos

Introduction

The Vicuña was scientifically described in 1782 by Molina. The type specimen obviously belongs to the larger, somewhat lighter-yellowish-coloured Southern subspecies *Lama v. vicugna* (Figure 1) from Northern Argentina and Chile with the whitish belly-hair coming up the flanks and with the short bib of off-white hair hanging from the chest. The Peruvian Vicuña (*Lama v. mensalis* Thomas, 1917)(Figure 2) is smaller, darker-reddish-coloured and with a long bib of off-white hair hanging from the chest. One specimen from Bolivia (Figure 3) looks rather intermediate. So far subspecies´ borders have never been published. Therefore research is needed to establish whether there are two valid subspecies or a cline with extreme populations easily to distinguish.

In precolumbian times, the Vicuña fleece was reserved for the priests for the service of the temples of the sun. Inspite of protection laws uncontrolled hunting started after the end of the Inca empire. Poaching by shooting and poisoning drinking pools reduced the number of Vicuñas to some 12,000 specimens 35 years ago (Grimwood, 1969). Through protection the number of Vicuñas is again up to some 200,000 specimens.

Figure 2. Peruvian Vicuñas in Pampa Galeras.

Figure 1. Male Southern Vicuña Zurich 35 at an age of four years.

Figure 3. Southern Vicuña Bolivia 1 at an age of 15 years with four months old daughter Antwerp 52.

Material and methods

With the establishment of the International Society for the Conservation of the European Wisent in 1923 by the former Director of Frankfurt Zoo, Kurt Priemel, the first International Studbook was established, followed 1956 by the second for the Przewalski Horse. In the meantime there are over 150 International Studbooks, 140 European Studbooks and many Studbooks in other regions. A studbook is a pedigree registration for a wild species in zoos.

The author started the International Vicuña Studbook in 1969. It includes all pure bred Vicuñas outside the four native countries since 1945. Between 1946 and 1971 22 Vicuñas have been imported from Argentina, 1 from Bolivia and 5 from Peru (Figure 4). Additional 41 Vicuñas have been imported from unknown origin (Schmidt, 1972a, 1973). 576 Vicuña births have been recorded since 1945 (Schmidt, 1986-2004).

To improve a genetically sound breeding management North American zoos started the Species Survival Plans (SSP) and the zoos in British Islands the Joint Management of Species Group (JMSG). In 1985 six colleagues and the author started the European Endangered species Programme (EEP). The EEP-Commission so far appointed Coordinators for 150 programmes. The Vicuña EEP was one of the initial programmes. The Coordinator is supported by an elected Species Committee. The aim of an EEP is to retain 90% of the genetic variability over the next 100 years. In reaching this aim we try to start with at least 15 unrelated founders (representing 95% of the taxon´s genetic variability), to avoid inbreeding, to use the same number of males and females for breeding if socially advisable, to produce approximately the same number of offsprings per breeder. In addition it is important to offer the correct social environment and a keeping system of high standard.

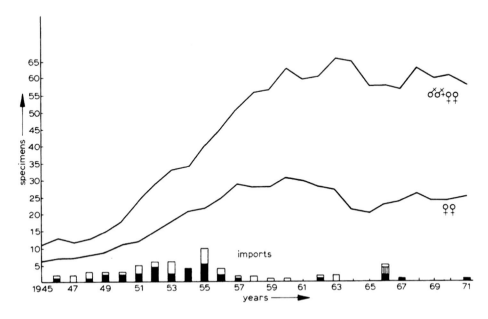

Figure 4. Total number of Vicuña (upper curve) and the number of female Vicuña (lower curve) kept in between 1945 and 1971. Blocks show the minimum number of Vicuña imported from South America (white = males, black = females, hatched = unsexed specimens).

Results

Before the EEP-management started in 1985 the breeding results were very poor. 11 Vicuñas survived World War II. The increase to 57 specimens in 1971 was mainly due to the importations (Figure 4). The average annual growth rate by breeding from 1957 until 1985 was only 0.2% (lambda 0.89–1.13, Figure 5a). On average 27% (11–42%, Figure 6a) of the females gave birth every year between 1957 and 1970. Juvenile mortality was 32% from 1945 until 1971. Due to this poor breeding performance there was almost no population growth from 1971 until 1985 (from 57 to 59 specimens, Figure 7a). It is quite interesting that between 1957 and 1971 65% males and only 35% females were born (n = 97), an additional reason for a very slow growth rate. There were inbreeding coefficients of up to 0.4375 (Table 1). Table 2 shows that inbred specimens have a low breeding performance.

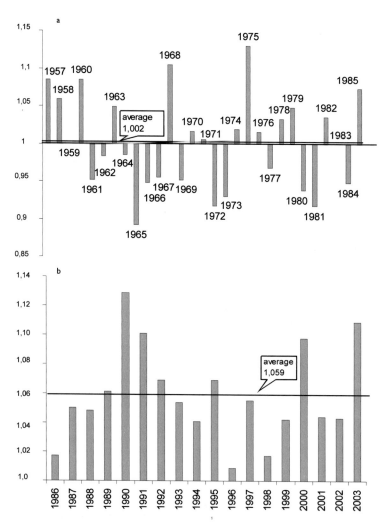

Figure 5. Growth of the Vicuña population in (a) the EEP-period 1957-1985 and (b) the EEP-period 1986-2003.

Figure 6. Percentage of female Vicuñas giving birth annually in (a) the pre-EEP-period 1957-1971 and (b) the EEP-period 1985-2003.

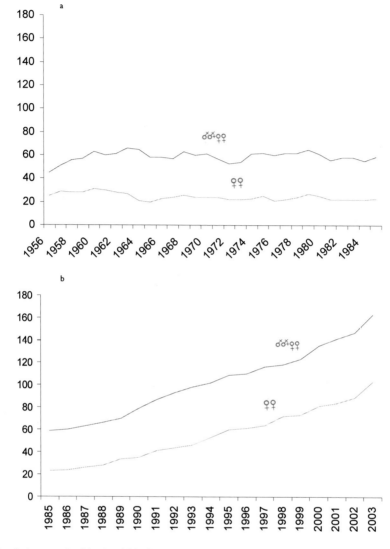

Figure 7. Population growth of Vicuñas (a) before and (b) after establishment of the Vicuña-EEP in 1985.

Table 1. Inbreeding coefficients of possible combinations of all Vicuñas living on 31. December 1986. The underlined coefficients are those of realized combinations.

	Pol 1	Amsterdam 12	Amsterdam 19	Antwerpen 46	Antwerpen 49	Antwerpen 52	Antwerpen 53	Brasschaat 11	Hannover 2	Hannover 4	Hannover 9	Hannover 14	London 2	London 4	Munich 7	Munich 12	Pretoria 6	Zurich 32	Zurich 38	Zurich 52	Zurich 53	Zurich 59	Zurich 61	Zurich 63
Amsterdam 11	0							0																
Amsterdam 15	0							0																
Amsterdam 17	0							0																
Amsterdam 18	0							0																
Amsterdam 20	0							0																
(Antwerp 18)	0							0																
Antwerp 31	0								0	0	0				0	0								
Antwerp 36	0								0	0	0				0	0								
Antwerp 38	0							0	0	0	0				0	0								
Antwerp 40	0								0	0	0				0	0								
Antwerp 41	0								0	0	0				0	0								
Antwerp 45	0.25	0.1875						0	0	0	0				0	0	0.3475							
Antwerp 50	0.25								0	0	0				0	0								
Antwerp 51	0		0.0927734		0.06641			0	0	0	0				0	0								
Brasschaat 19	0		0																					
Brasschaat 21	0		0																					
Frankfurt 9	0																							
Frankfurt 11	0																0.0410156							
Frankfurt 14	0											0					0							
Hannover 11	0			0	0	0	0	0	0	0	0	0	0	0									0	
London 3	0							0							0	0				0				
London 5	0			0	0	0	0	0					0	0	0	0	0	0	0	0	0	0		
Munich 2	0				0	0	0	0							(0.375)	(0.375)								
Munich 13	0			0	0	0	0	0						0	(0.4375)	(0.4375)	0	0	0	0	0	0	0	
Munich 18	0				0	0	0	0						0	0	0	0	0	0	0	0	0	0	
Pretoria 5	0							0									0			0	0			
Zurich 26	0							0	0	0	0				0	0								
Zurich 27	0							0	0	0	0				0	0								
Zurich 31	0							0	0	0	0				0	0								
Zurich 34	0.09375	0.09375						0	0	0	0		0.09375		0	0		(0.25)	(0.25)					
Zurich 44	0							0	0	0	0				0	0								
Zurich 49	0							0	0	0	0				0	0								
(Zurich 55)	0							0	0	0	0				0	0								
Zurich 56	0							0	0	0	0				0	0								
Zurich 58	0							0	0	0	0				0	0								
Zurich 64	0							0	0	0	0				0	0								

Column-head coefficients: Zurich 63: 0.0351563 0.017570; Zurich 61: 0.0620313; Zurich 59: 0.0703125; Zurich 53: 0.0703125; Zurich 38: 0.0703125; Zurich 32: 0.0703125.

Table 2. Effects of inbreeding in captive-bred Vicuñas up to 1985.

Inbreeding coefficient	Mortality within 5d	Mortality 6d to 6m	Surviving 6m	n
0	12 = 11.5%	15 = 14.4%	77 = 74.0%	104
-0.25	17 = 18.7%	15 = 16.5%	59 = 64.8%	91
-0.4375	8 = 23.5%	5 = 14.7%	21 = 61.8%	34

Inbreeding coefficient	Males breeding	Males not breeding	Females breeding	Females not breeding
0	18 = 64.3%	10 = 35.7%	14 = 73.7%	5 = 26.3%
-0.4375	9 = 32.1%	19 = 67.9%	15 = 71.4%	6 = 28.6%

With the EEP-management inaugurated in 1985 a sound breeding policy was started resulting in a good population growth from 59 to 163 Southern Vicuñas (*Lama v. vicugna*) in 34 European and one North American collections in 2003 (Figure 7b). The average annual population growth was 5.9% (lambda 1.01 – 1.13, Figure 5b). On average 37% (22–46%, Figure 6b) of the females gave birth every year. Juvenile mortality decreased to 24%. The longevity record of 31 years and 7 months was achieved by the male 140 Antwerp 18. The sex ratio of the neonates changed to only 45% males and 55% females (n ‒ 316), helping the fast growth rate. Inbreeding coefficiants of young ones born in 2003 ranged from 0 to 0.230. Since the number of active founders is only four males and eight females (Figure 8) the EEP urgently needs some unrelated specimens.

Husbandry and patholgy is reported elsewhere (Schmidt, 1972b). Groups of one male and two to three females are created to imitate natural social behaviour (Koford, 1957). First the females have to be put together before the male joins the group, because most males do not tolerate new group members. The male leader chases out of the group male offsprings at an average age of 8 months (n = 20, 4–13 months), female offsprings at an average age of 9 months (n = 16, 4–14 months). Some of the leaders seem to be more tolerant than others and chase the offsprings at a

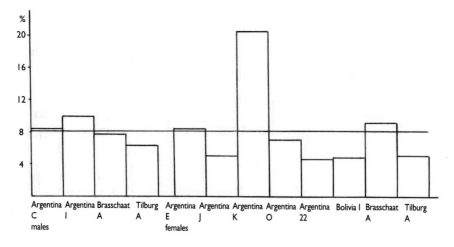

Figure 8. Representation of 4.8 foundermembers in all 36.23 captive Vicuñas living on 31/12/86.

higher age. Rarely young ones still suckle at an age of 8 months, however most are weaned at 6 months. As observed in the wild (Franklin, 1974, 1978), zoo groups of Vicuñas cannot grow. Because of the harem structure of the breeding groups, there is a surplus of males. In the wild they form bachelor groups of 7 to 75 members (Koford, 1957). This proved to be impossible for a longer time period in enclosures of up to 4,000 sqm, whereus bachelor groups in enclosures of several hectars could be kept together over several years.

The main birth season is August to October (Figure 9), a six months transition from the Southern wet season (Koford, 1957) to the Northern hemisphere. Pregnancies in male young ones lasted on average 350 days (n = 11, 329–359 days), in female young ones 355 days (n = 8, 330–369 days). 80% of 79 observed births occured between 07:00 and 14:00 o´clock (Figure 10). These morning births are observed in the wild too: In Peru during wet season it is raining almost every afternoon. The female Vicuña does not lick dry her infant. If this is born in the afternoon it stays wet and dies of pneumonia. Surviving male neonates weigh on the average 7.0 kg (n = 24, 5.9–8.2 kg), surviving females 6.8 kg (n = 15, 5.0–8.1 kg).

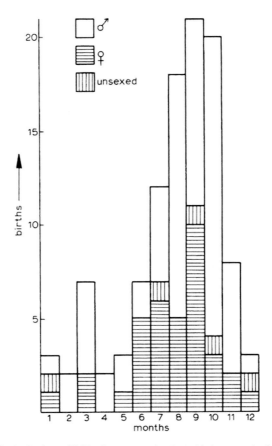

Figure 9. Distribution of Vicuña births in 11 Northern zoos, showing a birth season in the period July – October.

Figure 10. Female Southern Vicuña Zurich 18 at an age of nine years giving birth to a male offspring.

Discussion and conclusion

In 1972 (Schmidt) my opinion was *"Indeed, we must admit that that the existing management of captive Vicuña is a total failure... In my opinion, further distribution of captive Vicuña should no longer depend on dealers alone, but should be arranged in agreement with an international committee.... I am fully aware that these are very rigorous recommendations and that the present level of international cooperation between zoos is still one of the weakest points, but perhaps the future will bring the necessary changes."*

I am indeed happy that I was able to personally contribute to these changes with the formation of the EEP. The EEP management for the Vicuña was very successful in every respect. All measured parameters improved. Since the inception of the Vicuña EEP there is a flourishing population in European Zoos with the possibility to supply North American and Japanese Zoos with the species. The EEP is not only a success story for the Vicuña, but so far all EEP species survived in European Zoos.

Research has to be undertaken on the geographical distribution of the two different subspecies of Vicuñas and on how to form long-standing bachelor groups in zoos.

Acknowledgements

I would like to thank all participants in the International Studbook and in the EEP of the Vicuña for the annual contributions. In addition I am very grateful to Annett Wagner for preparing the manuscript, to Roland Gramling for arranging the graphs, my son Fabian Schmidt for compiling the 2003 edition of the International Vicuña Studbook and for many fruitful discussions. All photographs were taken by Christian R. Schmidt.

References

Franklin, W.L., 1974. The social behavior of the vicuña. In: The Behaviour of Ungulates and its relation to management, V. Geist and F. Walther (eds.), IUCN Publ. 24, pp. 477-487.

Franklin, W.L., 1978. Socioecology of the vicuña. Utah State Univ., Logan.

Grimwood, I.R., 1969. Notes on the distribution and status of some Peruvian mammals, 1968. Amer. Committee Int. Wild Life Protection & N.Y. Zool. Soc. Spec. Publ. 21.

Koford, C.B., 1957. The vicina and the puna. Ecol. Monogr. 27:153-219.

Molina, I., 1782. Saggio Storia. Nat. del Chili 1:317.

Scnmidt, C.R. 1972° International Studbook for the Vicina, *Lama vicugna.* Int. Zoo.Yearb. 12: 147-150.

Schmidt, C.R., 1972b. Captive breeding of the Vicuña. In: Breeding Endangered Species as an Aid to their Survival, pp. 271-283.

Schmidt, C.R., 1973. Breeding seasons and notes on some other aspects of reproduction in captive camelids. Int. Zoo Yearb. 13: 387-390.

Schmidt, C.R., 1986-2004. International Vicuña Studbook.

Part II
Breeding and genetics

Genetic parameters for coat characteristics in Bolivian llamas

M. Wurzinger[1]*, J. Delgado[2], M. Nürnberg[2], A. Valle Zárate[2], A. Stemmer[3], G. Ugarte[4] and J. Sölkner[1]

[1]*BOKU-University of Natural Resources and Applied Life Sciences, Vienna, Department of sustainable agricultural systems, Gregor-Mendel-strasse 33, A-1180 Vienna, Austria.*
[2]*University Hohenheim, Institute of livestock production in the tropics and subtropics, Garbenstrasse 17, D-70599 Stuttgart, Germany.*
[3]*Universidad Mayor de San Simón, Cochabamba, Bolivia.*
[4]*Asociación de Servicios Artesanales y Rurales Asar, Cochabamba, Bolivia.*

Abstract

Llamas display a great variability of fibre traits that determine the quality of the fleece as raw material for textiles. Little research has been conducted on the extent of this variability, although it is important for optimal use of natural resources in the Andean region. Fibre samples of 1,869 llamas were analysed with the optical fibre diameter analyser (OFDA). The following traits were considered: Mean fibre diameter (MFD), standard deviation (SD), diameter of fine fibre (DFF), proportion of fine fibre (PFF), proportion of kemp (PK) and proportion of medullated fibre (PMF). The effects of type of llama, age, sex and coat colour were studied. The type of llama influences all traits showing that Th´ampulli (fibre type) is better than Kh´ara (meat type). With increasing age of the animal MFD, SD, DFF and PK increased whereas PFF decreased. Comparing the two sexes, females showed better fibre quality. Light coat colours tend to be of better quality than darker ones. Heritabilities and genetic correlations were estimated using animal model procedures where all information came from mother-offspring relationships. Heritability estimates were 0.33, 0.28, 0.36, 0.32 and 0.25 for MFD, SD, DFF, PFF and PK, indicating potential for genetic selection.

Resumen

Efectos geneticos y no-geneticos que influyen en la calidad de la fibra de llamas en Bolivia
Existe una gran variabilitad en las características de la fibra de llamas, estas son determinantes para la calidad del vellón como materia prima para los textiles. Pocas investigaciones se han realizado respecto a la magnitud de esta variabilidad; no obstante su importancia para el uso óptimo de los recursos naturales en la región andina. Muestras de fibra de 1,869 llamas fueron analizadas con la metodología OFDA (optical fibre diameter analyser). Las siguientes caracteristicas fueron evaluadas: diámetro total de fibras (DT), desviación estandar del diámetro total de fibras (DE), diámetro de fibras finas (DFF), proporción de fibras finas (PFF), proporción de kemps (PK) y proporción de fibras meduladas (PFM). Los efectos del tipo, edad, sexo y color del animal fueron estudiados. El tipo del animal influye sobre todos los parámetros de calidad: Th`ampullis (tipo para fibra) son superiores a Kh`aras (tipo para carne). Con el aumento de la edad del animal, el diámetro total de fibras, desviación estandar del diámetro total de fibras, diámetro de fibras finas y proporción de kemps se incrementaron mientras que la proporción de fibras finas disminuyó. Comparando los sexos, las hembras mostraron mejor calidad de la fibra que los machos. Los colores claros tienden a tener mejor calidad que los colores oscuros. Heredabilidades y correlaciones genéticas fueron estimadas mediante el modelo animal involucrando información de pares madre-progenie. Las heredabilidades de DT, DE, DFF, PFF y PK fueron estimadas en 0.33, 0.28, 0.36, 0.32 y 0.25 respectivamente, por tanto con potencial para selección genética.

Keywords: llama, genetic parameter, fibre

Introduction

In Bolivia 2.4 mio llamas are kept in the marginal areas of the High Andes (UNEPCA, 1999). Llamas are very well adapted to the harsh environment and provide the farm households with a varity of products such as meat, fibre, dung and are used as pack animals. A large part of the products is consumed in the household and sale of meat and fibre is irregular. Absence of infrastructure and unsteady quality and quantity of fibre are main factors preventing a more market oriented llama husbandry. Llama fibre generally has a bad reputation for being coarse and is therefore not used in the textile industry. However Delgado (2003) and Iñiguez et al. (1998) showed that there are subpopulations in Bolivia with a high fibre quality.

Material and methods

The study was carried out in 4 indigenous communities of the eastern slope of the Andes in Bolivia, in the Province of Ayopaya, Department Cochabamba (66°30`S, 17°00`W). The communities are located between 3,800 and 4,300 m above sea level.

Data collection took place between November 1998 and January 2001. A total number of 2,533 fibre samples were taken from the left side of the animals. The weight of each sample was between 5 and 10g. Multi-coloured samples and samples from animals older than 7 years were not used for analysis. Samples were classified in 6 colours: white, beige, light fawn, brown, grey and black.

Two types of llamas were distinguished. Th`ampullis are regarded as "fibre-type" with finer fibre and higher fleece weight. Kh`aras are a more meat oriented type. Both types are kept together in herds and mating between these two types occurs frequently. Classification of types was done by farmers. Age of the animals was identified by denting for young ones and relied on information from farmers for older animals. Distribution of sexes and types according to colour is shown in Table 1.

Laboratory analysis

Analysis of fleece characteristics were conducted in the wool laboratory of University Hohenheim, Stuttgart, Germany. Every raw sample was blended, scoured, dried and conditioned in a standard atmosphere (20°C ± 2°C and 60% ± 2% rH). Snippets of 2 mm length were cut with a guillotine and spread on an open glass slide using an automatic spreader. Then the slides were analysed by OFDA (Optical Fibre Diameter Analyser) according to IWTO-47-95 standard (IWTO, 1995). 3,000 to 6,000 measurements per sample were carried out.

Table 1.Total numbers (n) and distribution (in percent) of types and sexes by colours. Only single–coloured animals are considered.

	n	white	beige	light fawn	brown	grey	black
Kh`ara female	156	9.6	20.5	1.9	56.4	7.7	3.8
Th`ampulli female	1,249	10.0	17.5	4.9	52.8	8.8	5.9
Kh`ara male	33	9.2	24.3	3.0	51.5	6.0	6.0
Th`ampulli male	431	11.4	16.5	7.4	47.3	10.0	7.4
total	1,869	10.3	17.7	5.2	52.0	8.9	6.0

The following traits were analysed: Mean fibre diameter (MFD) which is the most important trait for fibre quality. Standard deviation of the mean fibre diameter (SD) is an indicator for the homogenity of the fleece. Diameter of fine fibre (DFF) and proportion of fine fibre (PFF) as indicators for the quality of the under coat. Proportion of medullated fibre (PMF) is defined by the OFDA as the proportion of fibres with an opacity of $\geq 80\%$. Only data from white samples were used for this trait, because OFDA seems to overestimate this proportion in coloured fibres (Delgado *et al.*, 1999). Proportion of kemps (PK) is defined by the OFDA as the proportion of fibres with an opacity of $\geq 94\%$ and diameter larger than 25 µm. Kemp is a technical term referring to a fraction of medullated fibres.

Traits were tested for normal distribution and transformed when necessary. Only DFF showed a normal distribution. A logarithmic transformation ($y = \log x$) was performed for MFD, SD was transformed with a trigonometric function ($y = \text{atan } x$), PFF with a logarithmic function ($y = \log(x/(100-x))$), PMF with the square root function ($y = \sqrt{x/100}$) and PK with the function of Box and Cox ($y = ((x^{0.2}-1)/0.2)-1$) (Delgado, 2003).

Statistical analysis

The following statistical models were used:

Model 1
$$Y_{ijklmn} = \mu + F_i + A_j + S_k + T_l + C_m + YS_n + (A*S)_{jk} + e_{ijklmn}$$
Y_{ijklmn} = transform of mean fibre diameter, diameter of fine fibre (<30µm)

Model 2
$$Y_{ijklmn} = \mu + F_i + A_j + S_k + T_l + C_m + YS_n + e_{ijklmn}$$
Y_{ijklmn} = transform of standard diviation of mean fibre diameter

Model 3
$$Y_{ijklmn} = \mu + F_i + A_j + S_k + T_l + C_m + YS_n + (A*S)_{jk} + (A*T)_{jl} + e_{ijklmn}$$
Y_{ijklmn} = transform of proportion of fine fibre

Model 4
$$Y_{ijklmn} = \mu + F_i + A_j + S_k + T_l + C_m + YS_n + (A*T)_{jl} + e_{ijklmn}$$
Y_{ijklmn} = transform of proportion of kemp

Model 5
$$Y_{ijklm} = \mu + F_i + A_j + S_k + T_l + YS_{ml} + e_{ijklm}$$
Y_{ijklm} = transform of medullated fibre

with
μ_i = constant common to all individuals;
F_{ij} = fixed effect of farmer j, j = 1,65;
A_j = fixed effect of age j, j = 1,5 (j = $\leq$ 0.5, > 0.5 - $\leq$ 1, >1 - $\leq$ 3, >3 - $\leq$ 5, > 5 - $\leq$ 7);
S_k = fixed effect of sex k, k = 1,2 (male, female);
T_l = fixed effect of type l, l = 1,2 (Th`ampulli, Kh`ara);
C_m = fixed effect of colour m, m =1,6 (white, brown, black, light fawn, beige, grey);
YS_n = fixed effect of year-season n, n = 1,3 (Nov. 1998, Oct. 1999 to Feb. 2000, Oct. 2000 to Jan. 2001);
$(A*S)_{jk}$ = effect of interaction between age and sex;
$(A*T)_{jl}$ = effect of interaction between age and type;

e_{ijklmn} = random residual effect;

Estimation of fixed effects was carried out with the GLM procedure of SAS (SAS, 1999). The effect of the interval between sampling and last shearing was tested and showed no significant influence. Tukey`s test was performed for pairwise comparison of group means. The values presented in Table 2 are least squares means of the untransformed data, the attached probabilities and tests relate to the transforms. This procedure was followed as backtransformation of least squares means of transformed values is mathematically inappropriate.

To estimate genetic parameters, all traits were joined in a mulitvariate model and a random additive genetic animal effect was added. The relationships included were all of the type mother-offspring. The estimation was carried out with the VCE program (Neumaier and Groeneveld, 1998).

Results and discussion

Comparison of results from literature with the results presented here is difficult. Different methods such as air flow, microprojection or OFDA are used in laboratories for determing fibre traits. In addition it is not always clear if published figures come from complete or dehaired fleeces and definitions for some traits such as kemp or medullated fibre vary. Table 2 provides an overview of the fixed effects.

Table 2. LSMeans of different fibre traits.

	n	MFD (µm)	SD (µm)	DFF (µm)	PFF (%)	PK (%)	n	PMF(%)
Kh`ara	189	22.78[a]	8.09[a]	20.77[a]	88.67[a]	0.75[a]	18	34.35[a]
Th`ampulli	1680	21.71[b]	6.59[b]	19.90[b]	89.26[b]	0.10[b]	175	14.58[b]
female	1405	21.96[a]	7.37[a]	20.14	89.94[a]	0.45[a]	141	23.89
male	464	22.54[b]	7.32[b]	20.53	88.00[b]	0.40[b]	52	25.03
age								
0 - <=0.5	130	20.68[a]	7.12[a]	19.08[a]	90.44[a]	0.15[a]	18	32.48
>0.5 - <=1	501	20.19[a]	6.66[b]	18.97[a]	94.03[b]	0.38[b]	54	23.72
>1-<=3	489	22.08[b]	7.42[c]	20.30[b]	90.59[a]	0.50[c]	44	23.82
>3-<=5	444	23.81[c]	7.62[c]	21.45[c]	85.15[c]	0.54[c]	48	19.73
>5-<=7	305	24.48[c]	7.90[d]	21.87[c]	84.62[c]	0.55[c]	29	22.56
colour								
white	193	21.99[a]	7.15[a]	20.19[a]	89.54[ab]	0.40[bc]		
beige	330	21.87[a]	7.41[ab]	20.07[a]	90.13[a]	0.45[ab]		
light fawn	97	21.96[a]	7.27[ab]	20.12[ab]	89.34[abc]	0.45[abc]		
brown	968	22.63[b]	7.43[b]	20.63[c]	88.33[bc]	0.44[a]		
grey	167	22.14[ab]	7.41[ab]	20.25[ab]	89.17[abc]	0.43[abc]		
black	114	22.89[b]	7.40[ab]	20.74[cb]	87.27[c]	0.37[c]		
year x season								
JS = 1	743	22.80[a]	7.57[a]	20.61[a]	87.18[a]	0.42	83	23.98
JS = 2	897	21.43[b]	7.18[b]	19.80[b]	91.11[b]	0.41	86	24.84
JS = 3	229	22.52[a]	7.29[ab]	20.59[a]	88.61[c]	0.45	24	24.57

Table 2. Continued.

	n	MFD (µm)	SD (µm)	DFF (µm)	PFF (%)	PK (%)	n	PMF(%)
Interaction age x sex[A]								
0 - <=0.5 x female	64	20.54		18.98	91.24			
0 - <=0.5 x male	66	20.82		19.18	89.64			
>0.5 - <=1 x female	261	20.23		19.00	93.92			
>0.5 - <=1 x male	240	20.14		18.94	94.14			
>1-<=3 x female	364	22.25		20.41	90.23			
>1-<=3 x male	125	21.92		20.19	90.95			
>3-<=5 x female	421	23.03[a]		20.98[a]	87.81[a]			
>3-<=5 x male	23	24.59[b]		21.91[b]	82.50[b]			
>5-<=7 x female	295	23.75		21.33[a]	86.48[a]			
>5-<=7 x male	10	25.22		22.41[b]	82.75[b]			
Interaction age x type[B]								
0 - <=0.5 x Kh`ara						88.14	0.23	
0 - <=0.5 x Th`ampulli						92.74	0.07	
>0.5 - <=1 x Kh`ara						93.84	0.70[a]	
>0.5 - <=1 x Th`ampulli						94.22	0.06[b]	
>1-<=3 x Kh`ara						90.80	0.90[a]	
>1-<=3 x Th`ampulli						90.38	0.10[b]	
>3-<=5 x Kh`ara						84.72	0.98[a]	
>3-<=5 x Th`ampulli						85.59	0.10[b]	
>5-<=7 x Kh`ara						85.86[a]	0.96[a]	
>5-<=7 x Th`ampulli						83.38[b]	0.15[b]	
R[2]		0.41	0.31	0.42	0.35	0.39		0.47
se		0.10	1.17	1.46	0.59	1.08		0.14

MFD = mean fibre diameter; SD = standard diviation of MFD; DFF = diameter of fine fibre; PFF = proportion of fine fibre; PK = proportion of kemp; PMF = proportion of medullated fibre.
[a, b, c] = different letters indicate significant differences at P=0.05.
[A] = tests of significans only between the two sexes within age group; [B] = test of significances only between the two sexes within types.

Types

Th`ampullis are superior to Kh`aras in all fibre traits. Th`ampullis showed a significantly lower MFD than Kh`aras and a significantly more homogenous fleece. Iñiguez et al. (1998) recorded the lowest MFD of 21.1 µm in a population in the South of Bolivia, but without differences between types. Parra (1999) could not find differences between both types regarding MFD. Whereas Marti et al. (2000), Súmar (1991) and Maquera (1991) reported differences between types.

Diameter of fine fibre was significantly lower in Th`ampullis, but with similar proportion of fine fibre in both types. In contrast Ayala (1999) found similar DFF, but a significant higher proportion of fine fibre in Th`ampullis. The proportion of Kemp for Kh`aras is 0.75% and for Th`ampullis 0.10%. Both Iñiguez et al. (1998) and Parra (1999) give much higher values, but this is due to different definitions of the trait and differences in laboratory analysis.

Sex

Differences between sexes are significant for MFD, DFF and PFF. Males have coarser fibre (MFD: 22.54 ± 7.32 μm) than females (MFD: 21.96 ± 7.37 μm) and a lower PFF. These results correspond with the findings of Chávez (1991). Whereas Ayala (1999), Martinez *et al.* (1997) and Iñiguez *et al.* (1998) could not find differences. For the traits PK and PMF no differences could be found. These results are in agreement with Iñiguez *et al.* (1998), but contrary to other studies (Ayala, 1999; Martinez, 1997; Marti *et al.*, 2000).

Age

With aging MFD, SD, DFF and PK increase and at the same time PFF decreases. Animals between 1 and 3 years have fine fibre with 22.08 μm with a standard deviation of 7.42 μm. The first shearing is recommended at the age of 2 years and should be organized in similar age groups (Martinez *et al.*, 1997). These results are in accordance with reports from other authors (Chávez, 1991; Iñiguez *et al.*, 1998; Marti *et al.*, 2000; Martinez *et al.*, 1997; Parra, 1999).

Colour

MFD and DFF is lower and at the same time PFF is higher in light colours than dark ones. SD is smallest for the colour beige and highest for black. These results are contradictory to the findings of Iñiguez *et al.* (1998) where white fibres were thicker than coloured ones. Parra (1999) and Martinez *et al.* (1997) did not find differences between colours.

Heritabilities and genetic correlations

Estimates of heritabilites and genetic correlations for different fibre traits are shown in Table 3.

In llamas data of heritabilities for different fibre traits are rare. Results presented here are in line with Frank *et al.* (1996) who reported for fibre diameter a heritability of 0.29 and for standard diviation of 0.23. The highest positiv genetic correlation was found between MDF and DFF. Selection for a lower MFD will change DFF in the same direction. At the same time the PFF will increase and the SD will decline. By improving the fibre diameter the homogenity of the fleece will therefore also be better.

Table 3. Estimates of heritabilities (diagonal), genetic correlations (above diagonal) and correlation of residuals (below diagonal) of different fibre traits.

	MFD	SD	DFF	PFF	PK
MFD	0.33 ± 0.05	0.62 ± 0.07	0.96 ± 0.008	-0.94 ± 0.02	0.37 ± 0.11
SD	0.71 ± 0.05	0.28 ± 0.05	0.44 ± 0.09	-0.72 ± 0.06	0.72 ± 0.07
DFF	0.97 ± 0.05	0.59 ± 0.05	0.36 ± 0.05	-0.82 ± 0.04	0.33 ± 0.11
PFF	-0.92 ± 0.06	-0.72 ± 0.06	-0.83 ± 0.06	0.32 ± 0.06	-0.25 ± 0.13
PK	0.47 ± 0.05	0.60 ± 0.05	0.42 ± 0.05	-0.35 ± 0.05	0.25 ± 0.05

MFD = mean fibre diameter; SD = standard diviation of MFD; DFF = diameter of fine fibre; PFF = proportion of fine fibre; PK = proportion of kemp; PMF = proportion of medullated fibre.

Conclusions

The llama population in Ayopaya shows a high genetic potential for high quality fibre production. Fleece weight was missing to make more general conclusions. Improvements in the management and breeding strategies should be carried out at the same time. Improvements of management include shortening of shearing intervall, separation of wool according to colour and fibre diameter.

References

Ayala, C., 1999. Características físicas de la fibra de llams jóvenes (Physical characteristics of fibre from young llamas). Progress in South American camelids research EAAP publication No. 105, pp. 27-34.

Chávez, J.F., 1991. Mejoramiento genético de alpacas y llamas. (Breeding of Alpacas and Llamas.) In: S. Fernández-Baca (ed.), Avances y perspectivas del conocimiento de los camelidos sudamericanos, Santiago, Chile, pp. 149-190.

Delgado, J., 2003. Perspectivas de la producción de fibra de llama en bolivia. (Prospects of llama fibre production in Bolivia). Dissertation, Universität Hohenheim, Germany.

Delgado, J., A. Valle Zaráte and C. Mamani, 1999. Fibre quality of a Bolivian meat-oriented llama population. Progress in South American Camelids Research EAAP Publication No. 105, pp. 101–109.

Frank, E.N., M.H.V. Hick, H.E. Lamas and V.E. Whebe, 1996. A demographic study on commercial characeristics of fleece in argentine domestic camelids (cad) flocks. In: M. Gerken and C. Renieri (eds.). Proc of the 2nd European Symposium on South American Camelids. 30.8. – 2.9.1995, Università degli Studi di Camerino, Camerino, Italy.

Iñiguez, L.C., R. Alem, A. Wauer and J. Mueller, 1998. Fleece types, fiber characteristics and production system of an outstanding llama population from southern Bolivia. Small Ruminant Research 30: 57–65.

IWTO, 1995. Measurement of the mean and distribution of fibre diameter of wool using an optical fibre diameter analyser (OFDA). International Wool Textile Organisation (IWTO), London, UK, pp. 24.

Maquera, F., 1991. Caracterización y persistencia fenotípica en llamas karas y lanudas del centro experimental La Raya-Puno. (Phenotypic characterisation of Kara and Lanuda llamas of the experimental station La Raya-Puno). Tesis Ing. Agr., Universidad Nacional Agraria La Molina, Lima, Peru.

Marti, S.B., M. Kreuzer and M.R.L. Scheeder, 2000. Analyse von einflussfaktoren auf die faserqualität bei neuweltkameliden mit dem OFDA-verfahren. (Analysis of of influencing factors on fibre quality in New World Camels with OFDA). Züchtungskunde 72, 389–400.

Martinez, Z., L.C. Iñiguez and T. Rodríguez, 1997. Influence of effects on quality traits and relationships between traits of the llama fleece. Small Ruminant Research 24, 203-212.

Neumaier, A. and E. Groeneveld, 1998. Restricted maximum likelihood estimation of covariances in sparse linear models. Genetic Selection Evolution 30, 3-26.

Parra, G., 1999. Evaluación del potencial productivo de la llama en la quinta seccion municipal Charaña. (Evaluation of production of llama in the fifth section of Charaña). Tesis Ing. Agr., Universidad Mayor de San Simón, Cochabamba, Bolivia.

SAS Institute Inc., 1999. SAS/STAT Software Release 8.0, SAS Institute Inc., Cary, NC.

Súmar, J., 1991. Caracteristicas de las poblaciones de llamas y alpacas en la sierra sur del Peru. (Characteristics of llama and alpaca populations in the South of Peru). In: Informe de las mesa ronda sobre camelidos sudamericanos. Lima, Peru, pp. 17-80.

UNEPCA, 1999. Censo nacional de llamas y alpacas. (National Census of llamas and alpacas). Oruro, Bolivia.

Genetic diversity and management implications for vicuña populations in Peru

C.S. Dodd[1], J. Rodriguez[2], D. Hoces[2], R. Rosadio[2], J.C. Wheeler[2] and M.W. Bruford[1].
[1]Cardiff School of Biosciences, Cardiff University, Park Place, Cardiff, CF10 3TL, United Kingdom.
[2]CONOPA – Coordinadora de Investigacion y Desarrollo de Camelidos Sudamericanos, Lima, Peru.

Abstract

The aims of this study were to elucidate the recent evolutionary history and current genetic diversity of wild Peruvian vicuña populations with the intention of identifying demographically independent 'management units' within these populations and to assess the likely genetic effects of past and future management strategies. Twelve populations were sampled throughout the range of habitat and reserve coverage in Peru since they were thought to have had relatively long histories of demographic isolation and were not thought to have been influenced by recent translocations of animals from the Pampa Galeras reserve. Blood or skin samples were collected and analysed with eleven previously published South American camelid (SAC) microsatellite DNA markers (Lang *et al.*, 1996; Penedo *et al.*, 1998). These markers proved highly polymorphic and informative in Peruvian vicuña, with mean expected heterozygosity values over all loci varying between 0.377 (Tarmatambo) and 0.586 (Lucanas 2). A total of 20 private alleles were found which may be explained by the relatively low levels of within population compared with among population diversity and indicates some level of local isolation and genetic drift. Therefore, vicuña populations in Peru seem to possess several interesting genetic features, which are a result of biology, habitat occupancy, evolutionary history and management by people in the recent past. The implications of these results for the future management of the Peruvian vicuña are that there should be four demographic management units North-western Junín, Southern Junín, Central Andes and Puno. Translocations of animals should only be carried out within the same, but not between different management units. Free movement of individuals within localities must be ensured to minimise further inbreeding and genetic drift.

Resumen

El propósito de este estudio fue elucidar la reciente historia evolutiva y la actual diversidad genética en poblaciones de vicuñas silvestres en el Peru, con la intención de identificar unidades de manejo demográficamente independientes así como evaluar los probables efectos genéticos de pasadas y futuras estrategias de manejo. Doce poblaciones fueron muestreadas a lo largo del rango de habitat en el Peru, pues se creía que estas poblaciones tenían una larga historia de aislamiento demográfico y que no han sido influenciadas por recientes movimiento de animales desde la reserva de Pampa Galeras. Muestras de sangre o piel fueron colectadas y analizadas con 11 marcadores de ADN microsatélite para camélidos sudamericanos (CSA) previamente publicados (Lang *et al.*, 1996; Penedo *et al.*, 1998). Estos marcadores probaron ser altamente polimórficos e informativos en la vicuña peruana, con un valor promedio de heterocigocidad esperada, para todos los loci, que vario entre 0.377 (Tarmatambo) y 0.586 (Lucanas 2). Un total de 20 alelos privados fueron encontrados lo cual puede ser explicado por los niveles relativamente bajos de diversidad dentro de poblaciones comparados con los niveles de diversidad entre poblaciones e indican algún nivel de aislamiento local y deriva genética. Por lo tanto, las poblaciones de vicuñas en el Peru parecen poseer varias características genéticas interesantes, las cuales son el resultado de su biología, habitat de ocupación, historia evolutiva y por el manejo humano en el pasado reciente. Las implicancias de estos resultados para el futuro manejo de la vicuña peruana son que deben existir cuatro unidades de manejo demográficas Junín Noroeste, Junín Sur, Andes Centrales y Puno. Movimientos de animales solo deben ser realizados dentro de la misma unidad de manejo, pero no entre ellas. Libre movimiento de los individuos dentro de las localidades debe ser asegurado para minimizar adicionalmente la endogamia y la deriva genética.

Introduction

Vicuña are one of the two native wild South American camelid species. Their range extends from 9° 30' to 29° 00' south, in Ecuador, Bolivia, Peru, Chile and Argentina, in the Puna regions of the high Andes between 3,000 to 4,600 meters. Based on recent population censuses by CONACS (2001 and 2003), the Peruvian population of vicuña is the largest in South America, numbering 141,500 animals, with an estimated total South American population of 276,462 animals. However, prior to the Spanish conquest of Peru, vicuña populations were estimated to number as high as two million (Wheeler, 1995). More recently, commercial interest in vicuñas led to their near extinction from illegal hunting to obtain their high value fleece. This had a drastic effect on vicuña populations so that by 1966 only and estimated 5,000-10,000 animals remained. In 1975 the vicuña was listed on Appendix I of CITES and all trade in vicuña fibre and products was banned. By 1987 the Peruvian population had increased to 63,000 and populations in Ayacucho, Junin, Puno and Arequipa were transferred to Appendix II. Trade in fibre from live-shorn animals was reopened and by 1995 the entire Peruvian population was transferred to Appendix II. However, the Andean Vicuña Convention and Peruvian Law prohibit the exportation of live animals. Vicuña populations are now managed by the communities on whose land they live, either as free ranging populations that are periodically captured in chaccus or as captive populations contained in large corrals.

Considering the history of vicuña in Peru, this study uses molecular markers to elucidate the recent evolutionary history of wild Peruvian vicuña populations and evaluates present-day genetic diversity and its partitioning in those populations. This will enable the identification of demographically independent management units (MUs) within these populations for future management and will enable the assessment of the likely genetic effects of past and future management strategies, including the likely consequences of sustainable utilisation practices.

Methods

Sampling rational and strategy

Blood or skin samples were collected from 488 individual vicuñas from 12 populations within Peru (Figure 1). The geographic locations of the particular populations covered the habitat and reserve range of vicuñas in Peru and had relatively long histories of demographic isolation.

DNA extraction and amplification

Total genomic DNA was isolated from blood or skin samples using phenol and phenol/chloroform once cellular material had been digested with proteinase K. DNA was precipitated in 100% ethanol following the method of Bruford et al. (1998) and resuspended in TE buffer (10 mM Tris-HCl, 1 mM EDTA, pH 8.0) prior to analysis.

Eleven microsatellite loci were amplified (YWLL08, YWLL29, YWLL36, YWLL38, YWLL40, YWLL43, YWLL44, YWLL46, LCA5, LCA19, LCA22). These primers were originally designed from llama DNA (Penedo et al., 1998; Lang et al., 1996) but were demonstrated to be polymorphic and informative for all South American camelids (Kadwell et al., 1999).

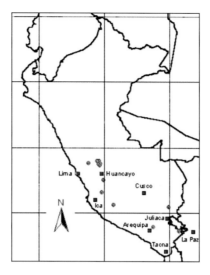

Locality, Department	Grid reference (Latitude, longitude)	Samples collected
1. **Tinco Cancha**, Junín	75°38" W, 11°02" S	36
2. **Villa Junín**, Junín	75°52" W, 11°05" S	30
3. **Yantac**, Junín	76°18"W, 11°20" S	35
4. **Tingo Paccha**, Junín	75°27" W, 11°25" S	30
5. **Tarmatambo**, Junín	75°43" W, 11°30" S	27
6. Sto. Domingo de **Cachi Cachi**, Junín	75°33" W, 11°38" S	21
7. San Pedro de **Huacarpana**, Ica	75°04" W, 12°50" S	20
8. **Ayavi**-Tambo-Huaytará, Huancavelica	75°15" W, 13°42" S	15
9a. **Lucanas** (Pampa Galeras), Ayacucho (1)	74°24" W, 14°39" S	20
9b. **Lucanas** (Pampa Galeras), Ayacucho (2)	74°24" W, 14°39" S	78
9c. **Lucanas** (Pampa Galeras), Ayacucho (3)	74°24" W, 14°39" S	111[a]
10. **Toccra** (Aguada Blanca), Arequipa	71°20" W, 16°10" S	3
11. S.A.I.S. **Picotani**, Puno	70°00" W, 14°55" S	38
12. **Ingenio**, Puno	69°20" W, 16°40" S	24

[a] Samples collected as skin from animals culled between 1977-1983.

Figure 1. Location, numbers of specimens and coordinates (latitude and longitude) for vicuña sample sites within Peru. The name by which the population will be referred to in the text is highlighted in bold.

Data analysis

Within population genetic diversity was measured by allelic diversity (A), observed heterozygosity (H_O) and expected heterozygosity (H_E) (Table 1) and population differentiation was investigated using pair-wise F_{ST} (Table 2), calculated in GENEPOP v3.1 (Raymond & Rousset, 1995). Gene flow between populations, as the number of migrants per generation, was estimated using Nm, which was calculated from F_{ST} using the formula:

$$Nm = (1-F_{ST})/4F_{ST}.$$

A distance tree (Figure 2) was derived using neighbour-joining (NJ) analysis of Nei's 1972 genetic distance (Nei, 1987; Saitou & Nei, 1987) in the computer program MEGA (Kumar et al., 1993) to give an indication of the genealogical relationships between populations. The relationships among the 12 populations were explored using 2-dimensional factorial correspondence analysis (2D-FCA) (Figure 3) in GENETIX v4.0 (Belkhir, 1999).

Results and discussion

Genetic diversity

For the 11 microsatellite loci allelic diversity ranged from 2.6 to 6.8 for the 12 populations (Table 1). YWLL08 was exceptionally informative, possessing 31 alleles with nearly 80% of individuals being heterozygotes. Although some loci possessed a higher number of alleles per locus in this vicuña data set compared with llamas, for which the microsatellites were developed (Lang et al., 1996; Penedo et al., 1998), the overall levels of heterozygosity are very much lower in this study compared with the original papers. We would normally expect wild populations to be more diverse than those that are domestic, since the process of domestication normally results in reduced variability of the captive population. Alternatively this reduced variability could be the result of ascertainment bias, which would result in decreased allelic polymorphism in a species for which the microsatellites were not originally derived. Mean expected heterozygosity values (Table 1) over all loci vary between 0.377 (Tarmatambo) and 0.586 (Lucanas 2) and are lower than those commonly found in continental mammal populations, which usually do not fall below 0.5. Observed heterozygosity is lower than expected heterozygosity in all populations except one (Toccra, which suffers from a small sample size and may be biased) and probably reflects a recent history of population isolation, with correlated genetic drift and localised inbreeding or selection against heterozygotes. Furthermore, there is a significant excess of homozygosity in all populations except Huacarpana, Lucanas 1, Cachi Cachi, Toccra and Tarmatambo, which suggests that inbreeding and/or genetic drift is a fairly common occurrence. In addition, the proportion of population-specific alleles is also higher than expected for a continental population sample and is further suggestive of some level of local isolation and genetic drift and is especially evident in the northernmost population of Yantac.

Finally, although it is clear that the allelic diversity of the entire population is high in comparison to the domestic populations chosen by Lang et al. (1996) and Penedo et al. (1998), the mean number of alleles per locus within populations is lower than often observed in other mammals, and it is only the Lucanas 2 and 3 that show the relatively high allelic diversity often associated with such studies.

Population differentiation

Pairwise comparisons between populations showed significant F_{ST} values ranging between 0.000 (Lucanas 2 vs. Lucanas 3) and 0.313 (Toccra vs. Tarmatambo) (Table 2). Significant differentiation was shown between all pairs of populations except two at the $p < 0.05$ level and all except eight at the $p < 0.01$ level. However, the values broadly reflect the known demographic relationships among the populations, with populations that are the closest geographically (e.g. Lucanas 2 vs. Lucanas 3) being the least differentiated and populations such as Toccra and Tarmatambo being highly differentiated in accordance with their geographic separation. Such population groups are demographically dependent according to these data. The relatively low levels of within vs. between population genetic diversity seen in Peruvian vicuña is characteristic

Table 1. Genetic diversity of vicuña populations in Peru calculated as expected heterozygosity (H_E), observed heterozygosity (H_O), allelic diversity (A) and the number of private alleles per population.

	H_E	H_O	Mean # alleles per locus (A)	# Private Alleles
1. Tinco Cancha	0.538	0.492	5.0	2
2. Villa Junín	0.450	0.381	4.7	2
3. Yantac	0.447	0.373	4.6	4
4. Tambo Paccha	0.391	0.385	3.3	0
5. Tarmatambo	0.377	0.364	2.7	0
6. Cachi Cachi	0.419	0.416	3.4	1
7. Huacarpana	0.481	0.468	4.8	2
8. Ayavi	0.471	0.448	3.9	1
9a. Lucanas (1)	0.538	0.536	5.2	2
9b. Lucanas (2)	0.586	0.531	6.8	0
9c. Lucanas (3)	0.580	0.528	6.4	1
10. Toccra	0.444	0.455	2.6	0
11. Picotani	0.509	0.507	4.5	2
12. Ingenio	0.523	0.485	5.4	3

of patterns commonly observed in other threatened species with formerly large ranges which have become isolated from each other, perhaps in National Parks where they have suffered drastic demographic contraction in recent generations, *e.g.* in South African buffalo (O'Ryan *et al.*, 1998) and red squirrels (Barratt *et al.*, 1999).

Demographic dependence or independence can also be inferred from the Nm values (Table 2) although the study populations vary in size considerably and because many are known to have

Table 2. Pairwise F_{ST} comparisons (above diagonal) and Nm values (below diagonal) between populations of vicuña in Peru.

	Tinco	Villa	Yantac	Tingo	Tarma	Cachi	Huaca	Ayavi	Luc1	Luc2	Luc3	Toccra	Pico	Ingen
Tinco	-	0.114	0.097	0.114	0.164	0.138	0.096	0.145	0.057	0.059	0.067	0.128	0.119	0.079
Villa	1.94	-	0.103	0.114	0.157	0.159	0.127	0.233	0.099	0.080	0.078	0.196	0.146	0.131
Yantac	2.33	2.19	-	0.146	0.217	0.183	0.127	0.241	0.099	0.107	0.100	0.195	0.159	0.121
Tingo	1.95	1.95	1.46	-	0.042	0.114	0.143	0.265	0.125	0.101	0.098	0.263	0.214	0.189
Tarma	1.28	1.34	0.90	5.77	-	0.105	0.183	0.282	0.158	0.106	0.132	0.313	0.267	0.212
Cachi	1.56	1.32	1.12	1.95	2.14	-	0.154	0.193	0.106	0.110	0.117	0.214	0.213	0.161
Huacar	2.38	1.71	1.71	1.50	1.11	1.38	-	0.149	0.025	0.038	0.032	0.046	0.134	0.120
Ayavi	1.48	0.83	0.79	0.69	0.64	1.04	1.43	-	0.127	0.099	0.119	0.139	0.241	0.187
Luc1	4.14	2.28	2.27	1.75	1.33	2.11	9.62	1.72	-	0.017	0.015	0.064	0.088	0.069
Luc2	3.98	2.88	2.09	2.22	1.77	2.02	6.26	2.27	14.33	-	0.000	0.066	0.103	0.066
Luc3	3.39	2.97	2.25	2.30	1.65	1.88	7.42	1.86	16.21	9999	-	0.049	0.085	0.072
Toccra	1.7	1.02	1.03	0.7	0.55	0.92	5.21	1.55	3.67	3.55	4.81	-	0.159	0.145
Picotani	1.86	1.47	1.51	0.92	0.69	0.92	1.62	0.79	2.58	2.19	2.71	1.32	-	0.079
Ingenio	2.93	1.66	1.82	1.07	0.93	1.31	1.83	1.09	3.4	3.57	3.22	1.47	2.93	-

been demographically isolated for hundreds of generations, such values need to be interpreted in the context of known population history. For example we would expect the Yantac population to exchange a higher number of migrants per generation with the population in Cachi Cachi, which is closer geographically, than with Lucanas 1 whereas the analyses suggest the opposite.

Tree topology based on Nei's 1972 genetic distance also infers relationships between populations that make sense biogeographically and demographically (Figure 2). For example the three southern Junín populations cluster closely together, as do two of the three northern Junín populations and the two Puno populations. However, less structure is apparent for the populations in Lucanas, Huancavelica and Arequipa and is probably indicative of a loose demographic affiliation between these populations, which is also suggested by the FCA analysis.

Within the FCA plots many individuals within each population cluster tightly together in 2-dimensional space and form four broad geographic groups (Figure 3a-d). The populations in northern Junín (Yantac, Villa Junín and to a lesser extent Tinco Cancha) form a group occupying (x/y) 0.0-0.5/0.2-1.5 (Figure 3a). Of these, Yantac, located at the north western extreme, is clearly the most different from the remainder of the populations in this region (Figure 3e). The populations in southern Junín cluster tightly, occupying the (x/y) vector coordinate range of 0.0-0.5/-0.5-0.0 (Figure 3b). This coordinate space is unique among the Peruvian vicuña populations sampled and is strongly indicative of an independent demographic history for this region. The population of Picotani, situated in Puno north of Lake Titicaca and not far from the Bolivian border, occupies its own 2D space (x/y = -1.5 to –0.5/-0.8 to 0.2) (Figure 3c) and forms a distinct geographic group, which requires further investigation. The remaining populations form a large central group (Figure 3d) comprising the populations of Lucanas, Ayavi, Huacarpana, Toccra and to a lesser extent Ingenio in Puno (x/y range –0.5 to 0.2/-0.8 to 0.5). These results recapitulate the relationships indicated by the genetic distance dendrogram in Figure 2, and suggest that the populations from Huancavelica to Ayacucho and Arequipa form a single demographically linked group of subpopulations, with some linkage to the Ingenio population south of Lake

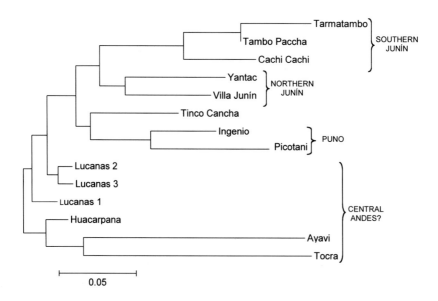

Figure 2. Neighbour-joining tree of Nei's (1972) genetic distances between Peruvian vicuña populations.

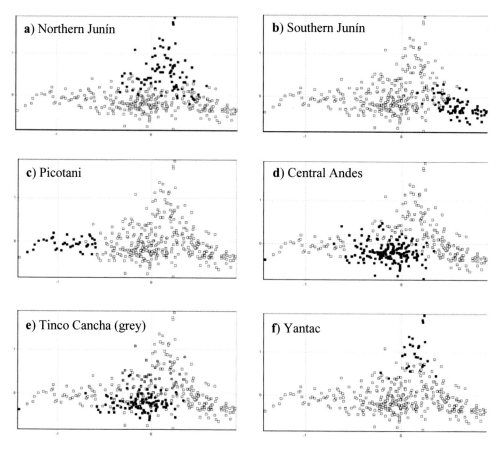

Figure 3. 2-D FCA plots for Peruvian populations of vicuña showing the four main demographic groups and significantly distinct populations: a) populations in Northern Junin (Yantac, Villa Junin and Tinco Cancha); b) populations in Southern Junin (Tarmatambo, Tambo Paccha and Cachi Cachi); c) Picotani (Puno); d) Central Andean populations (Lucanas 1, 2 and 3, Huacapana, Ayavi and Toccra); e) Yantac; f) Tinco Cancha population (grey) in comparison with Lucanas 1, 2 and 3 populations (black).

Titicaca in Puno. The population in Tinco Cancha in northern Junín genetically resembles this central Peruvian group, in particular the population in Lucanas, more closely than the northern Junín group from where it originates (Figure 3f). A similar effect is evident in Villa Junín and individuals from the population in Huacarpana (Ica) are also genetically more similar to the Lucanas populations than would be expected from its location. Transfers of animals from the Pampa Galeras Reserve to other populations between 1979-1981 (1 and 3 in Table 3) are the most likely explanation for these results. These translocations have severely affected the genetic structure of these populations and future management strategies should take this into account. The influence of transfers 2, 4 and 6 (Table 3) on the recipient populations remains to be studied. However, Andean animal populations are clearly occupying one of the most extreme and variable topographical environments in the world, and one where climate change, geological upheaval and changing colonisation routes must have a crucial role in shaping the genetic relationships among populations. Thus it may be reasonable to expect many vicuña populations to have been naturally isolated, then perhaps reconnected, and then possibly isolated again throughout

Table 3. Vicuña translocations from Pampa Galeras Reserve to other vicuña populations in Peru during the past 35 years.

Vicuña transfers from Pampa Galeras to:	Number of animals translocated and date
1. Laccho (Huancavelica) (74°55"W, 12°37"S)	121 in 1979; 601 in 1980
2. Cañaguas (Parque Nacional Aguada Blanca-Arequipa) (71°20"W, 16°10"S)	40 in 1979
3. Atoxaico (Junín) (76°7"W, 11°7"S)	395 in 1980; 617 in 1981
4. Yanganuco (Parque Nacional Huascarán-Ancash) (77°25"W, 9°15"S)	108 in 1980; 100 in 1997 to a community outside the park
5. Cooperativa Atahualpa (Cajamarca) (78°30"W, 7°9"S)[a]	25 in 1994; 170 in 1997
6. Toccra (Parque Nacional Aguada Blanca-Arequipa) (71°20"W, 16°10"S)	95 in 1997

[a] There are no native vicuña populations in Cajamarca.

much of their history. These effects, coupled with the recent anthropogenic influence on vicuña populations, make it difficult to assess how much of the genetic pattern seen here is a result of recent human exploitation and how much is a natural consequence of evolutionary history. A conservative approach to future management, however, would be to pay attention to the current patterns of genetic diversity described here, and to attempt to avoid future losses in diversity and structure.

Management recommendations

In respect of these results, it is suggested that vicuña populations in Peru should be conservatively managed in four demographic units (North-western Junín (except Tinco Cancha), Southern Junín, Central Andes and Puno). Individuals should be moved between populations within the same management unit (MU) to prevent further loss of genetic diversity, but there should be no further translocations from Pampa Galeras to populations outside the Central Andes MU due to the profound effect this may have on the genetic structure of the recipient population. Within populations measures are required to minimise further inbreeding and genetic drift and this must be considered in management practices for fibre production. Hence the free movement of individuals within localities must be ensured. If potential barriers to free movement (*e.g.* fences) exist, large gaps should be kept open at all times, and these should only be closed for intensive management (*i.e.* chaccu). Further, since any fencing will act as a barrier to movement, methods of intensive management should be found which would lead to the elimination of fencing in the medium term.

It is especially important that these should take account of the future genetic diversity of the populations and that the reproduction of individuals (especially males) be monitored and managed to maximise genetic diversity. This precludes, for example, the exclusive use of a few productive males or of active selection programs to increase fibre productivity in wild populations, since they would be likely to lead to further inbreeding and genetic drift, which may have deleterious effects on the health of individuals in the population and such an approach is unlikely to result in increased fibre quality, since domestication in other fibre producing species has universally led to a significant decrease in quality.

Conclusions

Vicuña populations in Peru seem to possess several interesting genetic features, which are a result of biology, habitat occupancy, evolutionary history and management by people in the recent past. The genetic diversity of Peruvian vicuña is characterised by relatively low levels of genetic diversity within populations, but high levels of genetic differentiation between populations. Previous range contraction and isolation of subpopulations has likely accounted for the patterns of diversity seen today in Peruvian vicuña and should be acknowledged in future management strategies including those for fibre production, which should be designed to minimise further loss of genetic diversity within individual vicuña populations.

There is clear biogeographical structure of genetic diversity in Peruvian vicuña populations, which enables the distinction of four demographic groups. Such groups could conceivably form separate MUs for future demographic augmentation within and between populations. Of particular note is the extreme northwestern population of Yantac, which is genetically unique in this sample set and may represent a northern evolutionary group requiring special attention and further research. Also of note is the distinctiveness of the Puno population of Picotani, which conceivably represents a genetic group, linked to the Bolivian population. It is important, therefore, to establish the genetic relationships between the Picotani population and those in Bolivia and to assess both how far this genetic group penetrates into Peru and how much it differs from the population of Ingenio in the south of Puno. The populations in southern Junín form a clearly independent demographic unit worthy of special consideration, and the large central Peruvian population, which includes the highly diverse population at Lucanas, appears to form a large demographic management unit. However, it should be noted that the Toccra sample analysed here is not large enough to form firm conclusions. Therefore, the proposed management regime divides these populations into four demographic units (North-western Junín (except Tinco Cancha), Southern Junín, Central Andes and Puno). These groups should be managed to prevent further loss of genetic diversity and inbreeding such that individuals should only be exchanged between populations within the same MU. In addition, there should be no further translocations from Pampa Galeras to populations outside the Central Andes MU due to the profound effect this may have on the genetic structure of the recipient population.

Acknowledgements

This work was funded by a Darwin Initiative for the Survival of Species grant (N251). We would like to acknowledge the efforts of Dr. Helen F. Stanley at the Institute of Zoology for help with sampling and project design. We would also like to thank Valerie Richardson and Maria Stevens at the Darwin Initiative Secretariat, DETR London for much flexibility and help with project management. At IVITA we would like to acknowledge the help of Dr. Felipe San Martin, Ing. Juan Olazabal and Antonina Cano. At CONACS we would like to thank Alfonso Martinez, Jorge Herrera, Marco Antonio Zuñiga, Marco Antonio Escobar, Alex Montufar, Roberto Bombilla and all others who assisted with sampling.

References

Barratt, E.M., J. Gurnell, G. Malarky, R. Deaville and M.W. Bruford, 1999. Genetic Structure of fragmented populations of red squirrel (*Sciurus vulgaris*) in Britain. Mol. Ecol. S12: 55-65.

Belkhir, L., 1999. GENETIX v4.0. Belkhir Biosoft, Laboratoire des Genomes et Populations, Université Montpellier II.

Benzécri, J.P., 1973. L'Analyse des Données: T. 2, l'Analyse des correspondances. Paris: Dunod.

Bruford, M.W., O. Hanotte, J.F.Y. Brookfield and T. Burke, 1998. Single and multilocus DNA fingerprinting. In: Molecular Genetic Analysis of Populations: A Practical Approach, 2nd Edition, A.R. Hoelzel (ed.), Oxford University Press, Oxford.

Kadwell, M., M. Fernandez, H.F. Stanley, J.C. Wheeler, R. Rosadio and M.W. Bruford, 2001. Genetic analysis reveals the wild ancestors of the llama and alpaca. Proc. R. Soc. Lond. B. 268: 2575-2584.

Kumar, S., K. Tamura and M. Nei, 1993. MEGA: Molecular Evolutionary Genetics Analysis, version 1.01. The Pennsylvania State University, University Park, PA 16802.

Lang, K.D.M., Y. Wang and Y. Plante, 1996. Fifteen polymorphic dinucleotide microsatellites in llamas and alpacas. Anim. Genet. 27: 293.

Nei, M., 1987. Molecular Evolutionary Genetics. Columbia University Press, New York.

O'Ryan, C., E.H. Harley, M.W. Bruford, M.A. Beaumont, R.K. Wayneand M.I. Cherry, 1998. Microsatellite analysis of genetic diversity in fragmented South African buffalo populations. Anim Cons. 1: 124-131.

Penedo, M.C.T., A.R. Caetano and K.I. Cordova, 1998. Microsatellite markers for South American camelids. Anim. Genet. 29: 411-412.

Raymond, M. and F. Rousset, 1995. GENEPOP (version 1.2): population genetics software for exact tests and ecumenicism. J. Hered. 86:248-249.

Saitou, N. and M. Nei, 1987. The neighbor-joining method: A new method for reconstructing phylogenetic trees. Mol. Biol. Evol. 4:406-425.

Wheeler, J.C., 1995. Evolution and present situation of the South American Camelidae. Biol. J. Linn. Soc. 54: 271-295.

PART III
Reproduction and pathology

Transvaginal embryo biometry in alpaca (*Lama pacos*): Preliminary report

G. Catone[1], M. Basile[1], O. Barbato[2] and C. Ayala[3]
[1]Department of Veterinary Science, University of Camerino, Via Circonvallazione 93/95, 62024 Matelica (MC), Italy; giuseppe.catone@unicam.it
[2]Department of Biopathologic Science, University of Perugia, Via S. Costanzo 06126, Perugia, Italy.
[3]Universidad Publica de El Alto, Carrera de Veterinaria, Bolivia.

Abstract

In human obstetrics the transvaginal sonography by means of high frequency transducers has been described as a repeatable procedure for the earlier and more accurate identification of embryo/fetal structures. In view of these advantages, a study was designed to measure the gestational sac diameter (GSD) and crown-rump-length (CRL) by means of transvaginal ultrasonography during the early pregnancy in alpaca. Five adult female alpacas were examined from 12 to 65 days after copulation, in the Maridiana farm in Umbertide (Perugia, central Italy). The pregnancy assessments were done by a microconvex 6,5 MHz transvaginal probe. The gestational age (GA) valued by ultrasonic measurement of the GSD and CRL were in line with the real gestation day, calculated considering day 0 as the mating day: the mean shifting resulted 0.00 ± 2.24 days by the CRL, and 0.00 ± 4.12 days by the GSD. The regression analysis that describes the relationship between GSD and GA is: $GA = 7.7335 \times GSD^{0.4214}$ ($R^2 = 0.8246$, $P<0.01$). The regression formula that describes the relationship between CRL and GA is: $GA = 10.328 \times CRL^{0.4427}$ ($R^2 = 0.9173$, $P<0.01$). These preliminary results indicate that the transvaginal sonography could be an alternative method to the transrectal ultrasonography for the early pregnancy diagnosis in alpacas. In fact it is non-invasive, well tolerated and repeatable, offering high quality image, precocity and reliability of the pregnancy diagnosis.

Resumen

En obstetricia humana, la ultrasonografia por via transvaginale, por medios transductores de alta frecuencia, ha estado descrita como un procedimiento precoz y mas seguro para el diagnostico de la gestacion, en cuyo procedimento identifica las estructuras enbrionarias/fetales. En virtud a este procedimiento se ha desarrollado un estudio en 5 alpacas hembras de la Hacienda Maridiana, Perugina – Italia. Medidas que comprenden el diametro de la membrana gestacional (GSD) y el largo del saco que comprende la cabeza-cola (CRL), a traves de una sonda ultrasonografica transvaginal, durante el primer periodo de gestacion de la alpaca. Para establecer el grado de prenez se ha utilizado una sonda microconvex transvaginale da 6,5 MHz, el diagnostico fue realizado de 12vo al 65mo dia, despues del acoplamiento. La edad de gravidez (GA) a estado calculada a traves de la medidas ultrasonograficas del GSD y CRL, los resultado son muy proximos al estado real del dia de prenez, que va desde el dia de la monta considerado el dia 0 donde: El valor medio es de 0.00 ± 2.24 dias utilizando il CRL y de 0.00 ± 4.12 dias utilizando il GSD. La formula que descrive la correlacion entre GSD y GA es la siguiente: $GA = 7.7335 \times GSD^{0.4214}$ ($R^2 = 0.8246$, $P<0.01$) y la correlacion entre el CRL y el GA es el seguiente: $GA = 10.328 \times CRL^{0.4427}$ ($R^2 = 0.9173$, $P<0.01$). Resultados que indican que la ultrasonografia por via transvaginal, puede representar una alternativa valida en comparacion a la ultrasonografia por via transrectal, para un diagnostico precoz de la prenez en la alpaca. En realidad se trata de una tecnica no invasiva, bien tolerada y con oposiones de repetibilidad, el cual ofrece una optima imagen, de los casos diagnosticados precozmente.

Keywords: alpaca, pregnancy diagnosis, ultrasonography, transvaginal probe

Introduction

In the last decade, there has been an increase in the commercial rearing of South American Camelids (SAC) in Australia, New Zealand, USA and expecially in Europe, for the companion animal industry as well as for the production of high quality fibre. Thus, this has drawn attention to investigate and improve the reproductive efficiency of this species.

Reproductive physiology in alpacas differs from other domestic animals. Alpaca is a species in which ovulation is induced by mating stimulus (San-Martin *et al.*, 1968, England *et al.*, 1969). The length of gestation lasts approximately 345 days, generally resulting in the birth of a single offspring (cria twins are rare) (San-Martin *et al.*, 1968; Novoa, 1970; Sumar, 1988). Implantation is thought to take place about 20 days after breeding (Olivera *et al.*, 2003) and maintenance of the corpus luteum (CL) is necessary throughout the entire pregnancy (Fernández-Baca *et al.*, 1970a; Sumar, 1988). They present low fertility rates in comparison to other domestic animals (Fernández-Baca, 1970b; Sumar, 1988). It is estimated that 50% of adult alpacas fail to produce young each year. The highest incidence of embryonic loss occurs during the first month of pregnancy (Fernández-Baca *et al.*, 1970a, b; Sumar, 1988). The early and accurate diagnosis of pregnancy in alpacas is essential to attain high levels of reproductive efficiency. Parraguez *et al.* (1997) described the use of transrectal ultrasonography in order to diagnose early pregnancy diagnosis and the relation between Gestational Sac Diameter (GSD) and Gestational Age (GA) in alpacas and llamas from 9 to 30 days post mating. Later Bravo *et al.* (2000) have used a 5 MHz transrectal probe to characterise the embryonic growth in alpacas from day 7 to 45 after copulation. Although, these procedures has been shown to be efficient, there are some disadvantages: *i.e.*, removing faeces from rectum, an adaptor for transrectal ultrasound transducer and two people to restrain the animal. However, to the best of our knowledge, there have been no reports on the use of Transvaginal Ultrasonography (TVU) for pregnancy diagnosis in alpacas.

In human obstetrics the TVU by means of high frequency transducers has been described as a non-invasive and repeatable procedure for the earliest and accurate identification of embryo/fetal structures. In view of these advantages, a study was designed to measure the Gestational Sac Diameter (GSD) and Crown-Rump-Length (CRL) by means of TVU from 12 to 65 days of pregnancy in alpacas.

Materials and methods

Five adult female alpacas under normal breeding condition were examined from 12 to 65 days after copulation, once a week, in the Maridiana farm in Umbertide (Perugia, central Italy).

Ultrasound scans were performed on real-time ultrasound equipment (Logiq 100 PRO, General Electric Medical Systems, USA) with a 6.5-MHz endovaginal micro-convex probe with a angle of vision of 114°Connected to a video cassette recorder (SVHS Video cassette recorder SR-S388E JVC, Japan). All the examinations were performed by authors.

Before examination, each alpaca was restrained in the standing position by one person. The operator's leg had to be between the alpaca's hindlegs to lift the abdomen in order to improve the visualization of the uterus (Figure 1). The vulva and perineal area were cleaned with tap water and 70% ethanol. Latex non-lubricated condoms, individually wrapped, were partially unrolled and filled with 3-5 ml of water-based ultrasound gel; then, they were fully unrolled

a

b

Figure 1. Restrained (a) and position of the operator to lift the abdomen during TVU (b).

onto a vaginal ultrasound probe. Undue stretching of the condom was avoided. Another 1-2 ml of gel was applied to the tip of the condom as a lubricant before vaginal insertion. The head of the probe was advanced into the vagina, positioned to the left or the right of the external os of the cervix. Different planes of scan (sagittal, transversal and oblique) were easily obtained with simple movements of the probe (Figure 2).

A small non-echoic area, circular or oblong, within the uterus was considered indicative of Gestational Sac (GS). Dorsoventral gestational sac diameter (GSD) was measured and its evolution followed until 45 days of pregnancy. In addition, after the appearance of the embryo, its development and its CRL were monitored (22-65 days). GSD and CRL measurements analysed by descriptive statistics. Power regression analysis was used to obtained the curves of GSD and CRL versus GA.

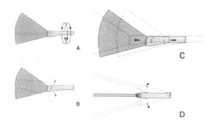

Figure 2. Schematic presentation of different scanning planes by use of the transvaginal probe.

Results

The GS was detectable in all the animals on day 12 (Figure 3). The GS rapidly increased until day 45 of gestation; after this time its measurement was not more reliable (Figure 4). According to Bravo *et al.* (2000) the GS began to spread into right uterine horn between days 19 and 24 in all pregnant alpacas. The embryo image was visible on day 22 (6 mm in length) and was viewed as a small echoic mass along the ventral border of the GS. The heartbeat was deteced between days 24 and 26 as a fluttering movement within the echogenic mass of the embryo. The embryo starts to move away from the ventral border during the next few days and by day 33 it was completely surrounded by fluid (Figure 5). After day 45 the umbilical cord, elongated from the dorsal pole, permitting the fetus to dangle at the end of the umbilical in dorsal recumbency during ultrasonigraphic examinations. Allantoic membrane was observed on approximately day 27 as a thin, slightly floating membrane. From day 36 to 45 we observed changes that transformed

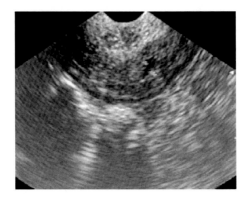

Figure 3.TVU image in alpacas at 13 days of gestation: X, dorso-ventral diameter of the GSD.

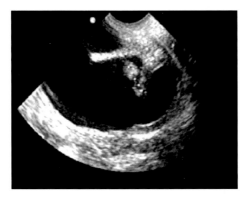

Figure 4.TVU image in alpacas at 36 days of gestation: X, crown-rump length of the embryo.

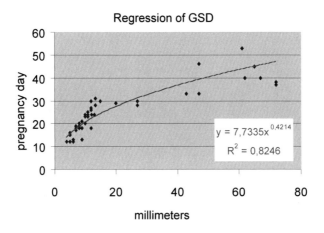

Figure 5. Growth rate of GSD in alpacas from days 12 to 45.

the undifferentiated primitive embryo into a fetus with a distinct body form. In this period the forelimbs buds become visible and after few days the hindlimbs buds and the hooves were visualized. The major change observed after day 45 was the visualization of the ribs on day 56 and the echogenicity increase of the cranial area, in particular the maxillary and mandibular regions.

The regression analysis describing the relationship between GSD and GA was:
$GA = 7.7335 \times GSD^{0.4214}$ ($R^2 = 0.8246$, $P<0.01$)

The regression analysis describing the relationship between CRL and GA was:
$GA = 10.328 \times CRL^{0.4427}$ ($R^2 = 0.9173$, $P<0.01$)

All ultrasonographic measurements showed a good correlation and GSD and CRL were clearly correlated with GA.

The GA, valued by ultrasonic measurement of the GSD and CRL, were in line with the true gestation days calculated by considering day 0 as the mating day: the mean shifting resulted 2.02 ± 2.59 days considering evaluated by the CRL, and 3.17 ± 2.58 days the GSD.

Discussion

Our study showed that the TV probe was particularly suited to the alpacas and well tolerated by all the animals. It guaranteed good results in terms of reliability and precocity in absence of complications. It also offered other advantages:
* an improved resolution;
* absence of artefacts;
* the possibility to carry out a higher numbers of scans due to the handiness of the transducer (Figure 2);
* the possibility to repeat ultrasound scans after short intervals of time;
* safety for the animals (all the females had normal gestation and parturition).

During early pregnancy (Day 12), the GS was detectable in all female alpacas, as reported before (Parraguez *et al.*, 1997; Bravo *et al.*, 2000) and in cows (Curran *et al.*, 1986) by transrectal ultrasonography. In TVU the GS appeared as a circular or elongated anechoic area about 6 mm surrounded by the homogeneous uterine tissue. The embryo was visible on day 22 (length, 6 mm) as a small echoic spot along the ventral border of the GS. The heartbeat was seen between day 24 and 26 and represent an important criterion for the assessment of embryonic viability and for exclusion of a pathological pregnancy. During this earliest phase particular attention must be paid to visualize the allantoic membrane. This thin floating membrane was been visible until day 36 and not for only 3 days as reported in alpacas (Bravo *et al.*, 2000) and in cattle (Curran *et al.*, 1986). The measurement of CRL (Figure 6) represented a valid and reliable approach to the valuation of the GA in this animal as well as in other species (Pierson & Ginther, 1984; Ginther, 1998). From day 22 to 65 the size of the CRL increased linearly (Figure 6). The embryo changed even in shape, as observed in human beings (Timor-Tritsch *et al.*, 1988) and in other species (Kastelic *et al.*, 1988; Ginther, 1998). Initially it had the appearance of a short straight line, developing into a C-shape on day 24-26. This was due to the cephalic and caudal flexures and general curvature of the back. Around day 40 the optic vesicles become visible as a small spherical anechoic areas. The amniotic vescicle was first observed on approximately day 35. It forms a thin, arched, hyperechoic line which surrounds the embryo. At this time was observed two parallel rows of echos with a central hypoechoic canal runs from the cranial to the caudal

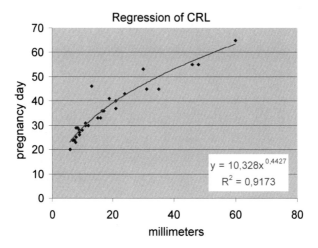

Figure 6. Growth rate of CRL in alpacas from days 22 to 65.

end of the embryo. This was presumed to represent the spinal cord as formed by closure of the neural groove.

The TVU showed to be less invasive and painful than transrectal ultrasonography; it avoided the rectal injures which were sometimes caused by the transrectal approach.

The improved resolution achieved by combining the transvaginal approach with the high-frequency transducer made possible to obtain an extremely detailed morphology and a more precise and reliable biometric evaluation of the embryo/fetus in the early pregnancy. To our knowledge, this was the first research conduced on the transvaginal pregnancy diagnosis within the field of the veterinary science. However, this method has already been used in human medicine, for the early pregnancy diagnosis (Timor-Tritsch *et al.*, 1988). In addition, the position of the operator's leg between the alpaca's hindlegs to lift the abdomen during examination, significantly improve the accuracy of the TVU for detect early pregnancy. This may be explained by the fact that the uterus tops over the pelvic brim and descends into the abdominal cavity at an early stage of pregnancy, especially in large sized pluriparous alpacas as reported in sheep (Buckrell, 1988; Breztlaff *et al.*, 1993; Karen *et al.*, 2004).

It is reassuring that no embryonic loss was observed during our study but, for validation of this method, could be important to assess the plasma concentrations of $PGF_{2\alpha}$ metabolite during the ultrasound examination, considering the statement of Aba *et al.* (1998): *"mechanical stimulation of the cervix and vagina may induce prostaglandin release in Camelidae"*.

In conclusion, these data demonstrate that TVU in early pregnancy diagnosis in alpacas could be a reliable alternative to the transrectal ultrasonography.

Acknowledgements

We thank Gianni Berna for his skilful technical assistance and Prof. Alessaandro Malfatti for the revision of the manuscript.

References

Aba, M.A., J. Sumar, H. Kindhal, M. Forsberg and L.E. Edqvist, 1998. Plasma concentrations of 15-ketodihydro-PGF$_{2\alpha}$, progesterone, oestrone sulphate, oestradiol-17β during late gestation, parturition and early post partum period in llamas and alpacas. Animal Reproduction Science 50: 111-121.

Bravo, W.P., M.M. Mayta and C.A. Ordonez, 2000. Growth of the conceptus in alpacas. American Journal Veterinary Research 61: 1508-1511.

Bretzlaff, K., K.J. Edwards, D. Forrest and L. Nuti, 1993. Ultrasonographic determination of pregnancy in small ruminants. Veterinary Medicine 88: 12-24.

Buckrell, B.C., 1988. Application of ultrasonography in reproduction in sheep and goats. Theriogenology 29: 71-84.

Curran, S., R.A. Pierson and O.J. Ginther, 1986. Ultrasonographic appearance of the bovine conceptus from days 20 through 60. Journal of the American Veterinary Medical Association 189: 1295-1302.

England, B.G., W.C. Foote, D.H. Matthews, A.G. Cardozo and S. Riera, 1969. Ovulation and corpus luteum function in llama (*Lama glama*). Journal of Endocrinology 45: 505-513.

Fernandez-Baca, S., W. Hansell and C. Novoa, 1970a. Corpus luteum function in the alpaca. Biology of Reproduction 3: 252-261.

Fernandez-Baca, S., W. Hansell and C. Novoa, 1970b. Embryonic mortality in the alpaca. Biology of Reproduction 3: 243-251.

Ginther, O.J., 1998. Ultrasonic imaging and animal reproduction: cattle. Book 3. Cross Plains, WI, USA, Equiservices.

Karen, A., K. Szabados, J. Reiczigel, J.F. Beckers and O. Szenci, 2004. Accuracy of transrectal ultrasonography for determination of pregnancy in sheep: effect of fasting and handling of the animals. Theriogenology 61: 1291-1298.

Kastelic, J.P., S. Curran, R.A. Pierson and O.J. Ginther, 1988. Ultrasonic evaluation of the bovine conceptus. Theriogenology 29: 39-54.

Novoa, C., 1970. Reproduction in Camelidae: A review. Journal Reproduction Fertility 22: 3-20.

Olivera, L.V.M., D.A. Zago and C.J.P. Jones, 2003. Developmental changes at the materno-embryonic interface in early pregnancy of the alpaca, *Lama pacos*. Anatomy Embryology 207: 317-331

Parraguez, V.H., S. Cortéz, F.J. Gazitúa, G. Ferrando, V. Macniven, L.A. Raggi, 1997. Early pregnancy diagnosis in alpaca (*Lama pacos*) and llama (*Lama glama*) by ultrasound. Animal Reproduction Science 47: 113-121.

Pierson, R.A. and O.J. Ginther, 1984. Ultrasonography for detection of pregnancy and study of embryonic development in heifers. Theriogenology 22: 225-233.

San-Martin, M., M. Copaira, J. Zuniga, R. Rodreguez, G. Bustinza and L. Acosta, 1968. Aspects of reproduction in the alpaca. Journal Reproduction Fertility 16: 395-399.

Sumar, J., 1988. Removal of the ovaries or ablation of the corpus luteum and its effect on the maintenance of gestation in the alpaca and llama. Acta Veterinaria Scandinavica 83: 133-141.

Timor-Tritsch, I.E., D. Farine and M.G. Rosen, 1988. A close look at early embryonic development with the high-frequency transvaginal transducer. American Journal of Obstetric and Gynecology 159: 676-681.

Embryo mortality and its relation with the phase of follicular development at mating in alpaca

W. Huanca[1], M.P. Cervantes[1] and T. Huanca[2]
[1]*Laboratory of Animal Reproduction, Faculty of Veterinary Medicine, San Marcos University, P.O. Box 03–5034, Salamanca, Lima, Peru; whuanca@appaperu.org*
[2]*Quimsachata Research Station, ILLPA-INIA, Puno, Peru.*

Abstract

The objective of this study was to evaluate the relation between the phase of follicular development (growth, maintenance or regression) during mating with embryo mortality to day 35-post mating. 116 alpacas ≥ 3 years of age were examined daily by transrectal ultrasonography (Aloka SSD 500) using a 7.5 MHz linear-array transducer to evaluate follicular status and then assigned to the following groups G1: Growing follicle 5 – 6 mm (n = 27); G2: Growing ≥ 7 mm.(n=30); G3 Maintenance ≥ 7 mm. (n = 30); G4 Regressing follicle ≥7 mm. (n= 29). Ultrasounds examinations were performed on days 0, 2, 9, 20, 25, 30 and 35 to determine occurance of ovulation (day 2); Corpus Luteum diameter (CL (day 9) and presence of embryonic vesicle or embryo (day 20 to 35). Male acceptance, ovulation and conception rate was compared by Chi Square test between groups. Acceptance rate was 100% in groups G2,G3 and G4 but only 81.5% of animals of G1 with follicle of 6 mm. accepted mating. Ovulation rate was 95.5; 96.7; 100.0 and 96.6% in G1; G2; G3 and G4, respectively (P > 0.05) and no differences were detected between groups in CL diameter. Conception rate on day 20 was 57.1; 68.9; 60.0 and 50.0% for G1; G2; G3 and G4 (P > 0.05). Embryonic loss rates from day 20 to 35 were not significantly different between groups. This result suggest that the status of follicular development at mating would not have influence on embryo mortality rate between day 20 and 35 of gestation in alpacas and that animals that accepted mating in all phase, including the one with follicle of 6 mm diameter present similar reproductive performance.

Resumen

El objetivo del presente estudio fue evaluar la relación entre la fase de desarrollo folicular (crecimiento, mantenimiento o regresión) al momento de la monta y mortalidad embrionaria hasta el día 35. 116 alpacas ≥ 3 años fueron evaluadas diariamente por ecografía transrectal con un ecógrafo ALOKA SSD 500 y un transductor lineal de 7.5 MHz y asignadas a uno de los grupos: G1: Folículos en crecimiento 6 mm (n = 27); G2: Crecimiento ≥ 7 mm (n = 30); G3 Mantenimiento ≥ 7 mm (n = 30) y G4 Regresión ≥ 7 mm (n = 29). Las evaluaciones fueron realizadas los días 0, 2,9, 20, 25, 30 y 35 para determinar tasa de ovulación (D 2); Diámetro del CL (D 9); presencia de vesícula embrionaria o embrión (D 20 – D 35). La tasa de aceptación al macho fue del 100% en los Grupos G2;G3 y G4 y 81.5% en las alpacas del G1. La tasa de ovulación fue del 95.5; 96.7; 100.0 y 96.6% en las alpacas de los grupos G1,G2, G3 y G4 respectivamente (P>0.05) y no se observaron diferencias en el diámetro del Cl. La tasa de concepción al día 20 fue 57.1, 68.9, 60.0 y 50.0% para los grupos G1,G2,G3 y G4 (P > 0.05). La pérdida embrionaria entre el día 20 a 35 no fueron significativamente diferentes entre grupos. Los resultados sugieren que el estadio de desarrollo folicular al momento de la monta parece no tener efecto sobre la mortalidad embrionaria entre el día 20 al 35 de gestación y que los animales que aceptan al macho en todas las fases de desarrollo folicular, incluyendo folículos de 6 mm, presentan similar comportamiento reproductivo.

Keywords: alpaca, embryo mortality, follicular status, mating

Introduction

Reproductive efficiency in alpaca is low under Peruvian condition and an important factor that affects the reproductive performance is the embryonic mortality (Novoa, 1991; Fernández-Baca et al., 1971). Alpacas as other camelids are reflex ovulators and mating is required to induce ovulation. A wave pattern of follicular development was reported in alpacas using the laparoscopic techniques (Bravo et al., 1991; Sumar, 1997) and in llamas (Adams et al., 1990) with the use of ultrasound technique. Pattern of follicular growth with a phase of growing, maintenance and regression of the largest (apparently dominant) follicle required an average of 4 days for each phase (total of 12 days) and female can accept mating in all follicular phases (Bravo et al., 1991; Sumar, 1997).

The females camels no exposed to male develop continuous follicular waves in three phases from a group of follicles that are recluted then one is selected to growing to ovulatory stage while the others regressing (Bravo et al., 1991; Fernández Baca, 1993). Alpacas can remain in state of constant receptiveness even for 40 days, attributed to the presence of one oestrogenic follicle that develops in continuous follicular waves (Novoa, 1991; Fernandez Baca, 1993).

Ovulation in camelids occurred between 26–30 hours after copulation (San Martín et al., 1968; Adams et al., 1990; Huanca et al., 2001). Ovulation in mammals involves pulsatile release of GnRH from the medio-basal nuclei of the hypothalamus into the hypophyseal portal system with subsequent release of LH from the gondotrophs of the anterior pituitary into the systemic circulation (Karsch, 1987; Karsch et al., 1997). Elevated circulating concentrations of LH elicit a cascade of events within the mature follicle culminating in follicle wall rupture and evacuation of its fluid and cellular contents (Richards et al., 2002).

Ovulatory failures post copulation might be attributed to a diminished sensibility of the follicles at LH circulating level by variations in the stage of follicular maturation (Fernández Baca, 1971; Novoa, 1989). It has been reported that alpacas with small follicles release less amount of LH and that ovulation not occurred in animals with follicles <7mm of diameter at the mating (Bravo et al., 1991), which would be related with the acquisition of LH receptors in advanced stages of follicular development (Gore-Langton & Armstrong, 1994).

The fertilization rates 3 days after mating are more than 85% but the percent of pregnant alpacas are smaller at 30 days post mating with an embryonic loss of approximately 50 percent (Fernández Baca et al., 1970b; Fernández Baca, 1971).

The progesterone secreted by the CL is necessary for the maintenance of the pregnancy in the alpaca (Novoa, 1991; Sumar, 1997) at least until the last third of pregnancy (Novoa, 1991). The establishment of CL is possible by the release of LH and a adequate release of LH can reinforces the formation of CL and an efficient secretion of progesterone (Leyva & García, 1999). If the development of a CL come from growing or regressing follicles, the pregnancy would not continue and occurred early embryonic mortality (Sumar, 1997); even more it is indicated that follicles in regression are not susceptible to ovulation (Hafez, 1996; Bravo et al., 1991) but if the CL proceed from matured follicles it would have a adequate functioning (Sumar, 1997).

The objective of this study was to evaluate the effect of the stages growing, maintenance and regression of the dominant follicle at mating on ovulation rate and embryonic survival in alpacas and would to explain some factors relations with the embryo mortality in alpacas.

Materials and methods

The study was conducted at the Quimsachata Experimental Station ILLPA-INIA, in the Department of Puno – southwest of Peru, located at 15°04' S, 70°18' W, and 4,200 m above sea level during the breeding season (January – March).

Mature non-lactating female alpacas (n = 116), ≥3 years of age, without problems at calving and resting period ≥ 15 day and weighing an average of 70 kg, were examined daily by transrectal ultrasonography (Aloka SSD 500) using a 7.5 MHz linear-array transducer by a minimal time of two weeks and then assigned randomly to 4 groups. All alpacas received the same condition of handling and were kept on natural pastures.

Completed the post-partum resting period, daily evaluation were made to determine the follicular development stage (growth, maturation or regression). The animals were assigned randomly to the following groups:
G1 (n = 27): dominant follicle in growth stage (diameter = 6 mm).
G2 (n = 30): dominant follicle in growth stage (diameter ≥ 7 ≤ 12 mm).
G3 (n = 30): dominant follicle in static stage (diameter ≥ 7 mm).
G4 (n = 29): dominant follicle in regression stage (diameter ≥ 7 mm).

The alpacas assigned to groups G1 and G2 were those animals that presented a growing dominant follicle in at least three consecutives days of evaluation until in the last day it get a diameter = 6 mm (G1) and ≥ 7 ≤ 12 mm (G2). To group G3 were assigned alpacas that presented a dominant follicle with a diameter ≥ 7 mm and that kept it without variation in diameter in at least three consecutives days. In the group G4 were included alpacas with a longer period of observation and the onset of follicular regression was defined as the first day on which the dominant follicle began a progressive decrease in diameter during three days of the observational period with a diameter ≥ 7 mm.

Alpacas were examined by transrectal ultrasonography on Day 0, 2, 9, 20, 25, 30 and 35 post mating (Day 0 = Mating day). Ovulation was determined on Day 2 and was defined as the sudden disappearance of a large follicle present on Day 0. Evaluation on Day 9 was made to detect the presence of a corpus luteum (CL). Evaluations on Day 20, 25, 30 and 35 were made to determine the presence of embryonic vesicle and/or embryo. Additionally on day 15, the sexual receptiveness test was performed, classifying alpacas how receptive and possible not pregnant to those who adopted the position of copulation; and not receptive and possible pregnant to those who rejected the male.

Statistical analysis

The information was analyzed using the statistical bundle STATA 8.0 (Statistical Analysis, 2003). The one-way analysis of variance test (ANOVA) was used to analyze the difference among groups with regard to the size of the corpus luteum at Day 9 post copulation. The relationship among the ovulation with the study groups were evaluated with CHI Squared test (X^2). The embryonic survival in the different days of observation and final was analyzed by means of Kaplan Meier survival curves. To compare the final survival among groups was used the Log-Rank test. The level of significance employed for the analysis was $p < 0.05$.

Results and discussion

Ovulation and embryonic survival rate

The ovulation rate was of 97.3% (n = 111) of the 116 alpacas that were mated at Day 2 post mating, with presence of corpus luteum on day 9. The rate of ovulation was 100% in alpacas (n=30) of group G3; in the others groups the ovulation rates were 95.45% (G1) 96.67% (G2) and 96.56% (G4) without significant differences (p<0.05) among the groups. (Figure 1). 5 alpacas from G1 non accepted mating and were not included to determine the ovulation rate.

The embryonic survival rate progressively continued to diminish until the last day of observation. No significant differences were found among the study groups (Table 1), still at Day 35 post mating a tendency to a greater rate of survival was registered for group G2 (65.52%) in comparison with groups G1 (52.38%), G3 (53.33%) and G4 (42.86%).

These results suggest that the ovulation and embryonic survival rate are not affected by the stage of follicular development at mating in alpacas. Alpacas from Group G2 present a numerical difference to embryonic survival rate in respect to the others groups at Day 35. These response can be explained by the interval from mating to ovulation of 26 to 30 hours (San Martin *et al.*, 1968; Huanca *et al.*, 2001) and would be related with the acquisition of LH receptors in advanced stages of follicular development (Gore-Langton & Armstrong, 1994) and can permit a better establishment and maintenance of the corpus luteum and improve the embryonic survival according reported by Sumar (1997). Similar explain can be made to the animals of group G3, that present a low embryonic survival to Day 35, because the dominant follicle would continue its growing and possible to the proximity of the regression stage.

Results found in the group G4 (regression) show ovulation followed by fertilization that differs from the study realized by Bravo *et al.* (1991), where regressing follicles were luteinized instead of being ovulated, this probably might suggest that the oocyte of atresic follicles in alpacas would be resistant to degeneration and with capacity to fertilize and to develop an embryo, which is similar to what was found in sheeps by Greenwald and Roy (1994). Nevertheless, depending on the degree of alteration of the quality of the dominant follicle, despite not being affected significantly in its ovulatory capacity, there might be a reduction in the subsequent pregnant rates

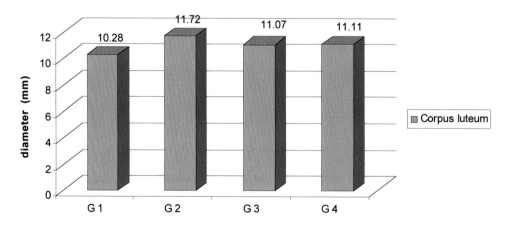

Figure 1. Size of the corpus luteum at day 9 post mating.

Table 1. Embryonic survival rate in alpacas with follicles in growth phase (G1 and G2), static phase (G3) and regression phase (G4) at moment of copulation.

Groups	G1	G2	G3	G4	Total
n	27	30	30	29	116
Mated alpacas	22	30	30	29	111
% Ovulation (n)	95.45 (21)	96.67 (29)	100 (30)	96.55 (28)	97.3 (108)
Embryonic survival rate (%)					
Day 15 post mating*	66.67	72.41	60.00	50.00	62.27
Day 20 post mating	57.14	68.97	60.00	50.00	59.03
Day 25 post mating	57.14	65.52	60.00	42.86	56.38
Day 30 post mating	57.14	65.52	56.67	42.86	55.55
Day 35 post mating	52.38	65.52	53.33	42.86	53.52

G1: growth (= 6 mm); G2: growth ($\geq 7 \leq 12$mm); G3: static (≥ 7mm); G4: regression (≥ 7mm).* Sexual receptiveness.

(Mihm *et al.*, 1999). This would be concordating with our results because we found an ovulation rage of 96.55% and a smaller embryonic survival rate (42,86%).

The occurrence of ovulation in alpacas from group G1 would be explained by the changes in the follicular structure of the dominant follicle and by the proliferative activity that occurs in the interval between the peak of LH and the ovulation (Gore-Langton & Armstrong, 1994), which is suggested at 26- 30 hours after mating (San Martín *et al.*, 1968; Huanca *et al.*, 2001), acquiring the ovulatory capacity because the increased expression of LH receptors on granulose cells of the dominant follicle (Sartori *et al.*, 2001). On the other hand, the ovulatory failure post mating would agree with results found by Bravo et al (1991) who suggested no occurrence of ovulation on follicles smaller than 7 mm of diameter at the moment of mating. It possibly due to the inadequate acquisition of LH receptors because these might be increased with the progress of the follicular stage (Gore-Langton & Armstrong, 1994) or less liberation of LH after copulation by small follicles (Bravo *et al.*, 1991).

Corpus luteum size at day 9 post mating

Average size of the corpus luteum was of 10.3 ± 1.7 mm (G1), 11.7 ± 1.7 mm (G2), 11.1 ± 1.9 mm (G3) and 11.1 ± 2.2 mm (G4) without significant difference among (Figure 1). These results suggest that the formation of corpus luteum would be independent of the stage of the follicle from which develop but not its maintenance and that might explain the differences in the embryonic survival rate among the study groups.

Sexual receptiveness test

At mating 5 alpacas from group G1, did not show sexual receptiveness to male and only 22 alpacas were mating (Table 1). The sexual receptiveness to male in the mating day was 100% in alpacas from groups G2, G3 and G4, and 92.54% in alpacas from group G1, supporting the mentioned by Bravo *et al.* (1991) and Sumar (1997) with regard to the acceptance to male by females camels in any of the follicular stages. On the other hand the no acceptance to male by alpacas from group G1 could be explained because the little ability of its follicles to secret significant quantities of oestrogen, as indicated by Gore-Langton and Armstrong (1994).

Sexual receptiveness behavior on Day 15 (95.2%) post mating present relationship with pregnancy rate with transrectal ultrasonography at Day 20 (82.6%) post mating (Figure 2). The differences can be attributed to the dominance of male on the female or possible deficiencies in the progesterone production by the quality of the corpus luteum, according Cardenas *et al.* (2001).

Conclusions

* The ovulation rate was minimal 95%, independently of the follicular development stage at the moment of mating in alpacas.
* The follicular stage does not have a significant effect on the embryonic survival rate in alpacas.
* Alpacas with growing dominant follicle $\geq$ 7 mm at mating, had a better embryonic survival rate (65,5%) and the alpacas with regressing follicles, the least embryonic survival (42,9%).

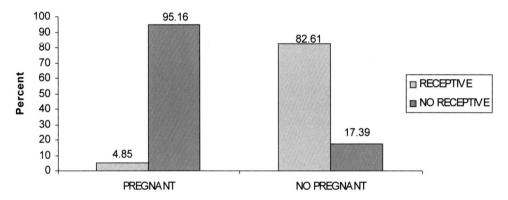

Figure 2. Sexual receptiveness percent on day 15 post mating related to the condition of pregnancy on day 20 post copula.

References

Adams, G., J. Sumar and O. Ginther, 1990. Effects of lactational and reproductive status on ovarian follicular waves in llamas (Lama glama). J. Reprod. Fertil. 90: 535-545.

Bravo, P.W., G. Stabenfeldt, B. Lasley and M. Fowler, 1991. The effect of Ovarian follicle size on pituitary and ovarian responses to copulation in domesticated South American Camelids. Biol. Reprod. 45: 553-559.

Cárdenas, O., M. Ratto, A. Cordero and W. Huanca, 2001. Determinación de la fertilidad en llamas con un servicio, mediante conducta sexual y ecografía. Rev. Inv. Vet., Lima. Supl 1, pp. 467-469.

Fernández Baca, S., 1971. La Alpaca. Reproducción y Crianza. Boletín N° 7. IVITA. Fac. Nac. Mayor de San Marcos. Lima, Peru, 43 pp.

Fernández Baca, S., 1993. Manipulation of reproductive functions in male and female New World camelids. Anim. Reprod. Sci. 33: 307-323.

Fernandez-Baca, S., W. Hansell and C. Novoa, 1970a. Corpus luteum function in the alpaca. Biology of Reproduction 3: 252-261.

Fernandez-Baca, S., W. Hansell and C. Novoa, 1970b. Embryonic mortality in the alpaca. Biology of Reproduction 3: 243-251.

Gore-Langton, R. and D. Armstrong, 1994. Follicular steroidogenesis and its control. In: The Physiology of reproduction, J. Knobil and J. Nelly (eds.). Raven Press, New York, pp. 571-628.

Greenwald, G. and S. Roy, 1994. Follicular development and its control. In: The Physiology of reproduction, J. Knobil and J. Nelly (eds.). Raven Press, New York, pp. 650.

Hafez, E., 1996. Reproducción e Inseminación Artificial en Animales. 6ta ed. Editorial Interamericana McGraw-Hill, México. 525 pp.

Huanca W., O. Cárdenas, C. Olazábal and G. Adams, 2001. Intervalo entre estimulo hormonal y no hormonal a la ovulación en llamas. Rev. Inv. Vet., Lima. Supl 1, pp. 470-473.

Karsch, F.J., 1987. Central actions of ovarian steroids in the feedback regulation of pulsatile secretion of luteinizing hormone. Ann. Rev. Physiol. 49: 365-382.

Karsch F.J., J.M. Bowen, A. Caraty, N.P. Evans and S.M. Moenter, 1997. Gonadotropin-releasing hormone requirements for ovulation. Biol. Reprod. 56: 303-309.

Leyva, V. and W. García, 1999. Efecto de la GnRH sobre la Fertilización y Sobrevivencia Embrionaria en Alpacas. In: II Congreso Mundial sobre Camélidos, Resumen, Cusco, Peru, pp. 90.

Mihm, M., N. Curran, P. Hyttel, P. Knight, M. Boly and J. Roche, 1999. Effect of dominant follicle persistence on follicular fluid oestradiol and inhibin and on oocyte maturation in heifers. J. Reprod. Fertil. 116: 293-304.

Novoa, C., 1991. Fisiología de la Reproducción de la hembra. In: Avances y Perspectivas del conocimiento de los Camélidos Sudamericanos. Cap III, S. Fernández Baca (ed.), Santiago, Chile, pp. 93-103.

Richards J., D.L. Russell, S. Ochsner and L.L. Espey, 2002. Ovulation: New dimensions and new regulators of the inflammatory-like response. Ann. Rev. Physiol. 64: 69-92.

San-Martin, M., M. Copaira, J. Zuniga, R. Rodreguez, G. Bustinza and L. Acosta, 1968. Aspects of reproduction in the alpaca. Journal Reproduction Fertility 16: 395-399.

Sartori, R., P. Fricke, J. Ferreira, O. Ginther and M. Wiltbank, 2001. Follicular Deviation and Acquisition of Ovulatory Capacity in Bovine Follicles. Biol. Reprod. 65: 1403–1409.

Sumar, J., 1997. Avances y Perspectivas en Reproducción de Camélidos. In: Memorias del I Symposium Internacional Avances en Reproducción de Rumiantes. Lima, Peru, pp. 30-44.

The influence of high ambient temperature on thermoregulation, thyroid hormone and testosteron levels in male llamas (*Lama glama*) depending on their fibre length

A. Schwalm[1], G. Erhardt[1], M. Gerken[2] and M. Gauly[2]
[1]*Institute of Animal Breeding and Genetics, University of Giessen, Ludwigstr. 21b, 35390 Giessen, Germany.*
[2]*Institute of Animal Breeding and Genetics, University of Göttingen, Albrecht-Thaer-Weg 3, 37075 Göttingen, Germany.*

Abstract

The aim of the study was to evaluate the influence of heat stress (high ambient temperature) on physiological parameters, body surface temperature, thyroid hormones (triiodothyronine=T_3 and tetraiodothyronine=T_4) and testosterone levels in llamas in relation to the body fibre. Therefore 12 fertile male llamas were housed in heated stables during the study. 5 of the animals were shorn (< 1cm), 5 animals were left unshorn and 2 animals were shorn partly (barrel cut). After a short period of acclimatization, the ambient temperature was elevated up to 30°C for a period of 4 weeks. Afterwards the temperature was decreased over one week to 20°C. The animals were kept for at least 7 weeks this temperature. Rectal temperature, heart rate and respiratory rate were measured daily. In addition the body surface temperature was measured by infrared thermography. Blood samples were taken to estimate serum thyroid hormone and serum testosterone levels once a week. All animals showed higher respiratory rates (p<0.001) and rectal temperatures in the heat-period when compared with the recovery-period. Respiratory rates were above the physiological values during the heat period. The rectal temperature stayed within the physiological range. Shorn animals were able to cope better with the high ambient temperature and showed significantly lower rectal temperatures and respiratory rates during the heat-period when compared with the other animals. The body surface temperature was significantly lower in the unshorn regions of the animals when compared with the shorn parts. The heat loss in the unshorn animals was concentrated on the ventral body regions (the thermal windows), which shows that effective thermoregulation can only take place in this part of the body. In the heat-period the thyroid hormone levels (T_3 and T_4) were both significantly lower (p<0.001, p<0.01) when compared with the recovery-period. The two thyroid hormone levels were significantly correlated. The serum testosterone level showed a decrease one week after the heat-period, with a minimum level 2 weeks later, followed by a slow increase with the levels still below the initial values until 6 weeks after the heat-period. There were no significant differences in the testosterone levels between shorn and non shorn animals. The high ambient temperature showed significant effects on the measured physiological parameters and the thyroid hormone levels in all llamas. In unshorn animals an effective thermoregulation can only take place over the thermal windows. Shorn animals seems to tolerate heat better because of the heat loss over the hole of the body surface

Keywords: llamas, heat stress, thyroid hormones, testosterone

Introduction

The thermoregulation of an animal can be divided into heat loss and heat production (Jessen, 2000). Heat production can be managed by metabolic heat production, active movement, shivering and by brown adipose tissue. Animals can loss heat by conduction, convection (free and forced), radiation and evaporation (sweating, panting). Llamas can perform heat loss by vasodilatation with concurrent increased blood flow to the skin, particulary in the perianal area, medial thigh and ventral abdomen (Pugh *et al.*, 1997) and sweating. Sweat glands are present all over the skin, showing higher densities in the ventral body region (Fowler, 1998). Evaporative cooling efficiency is determined by skin temperature and insulation. The fibre layer of llamas is an efficient insulation layer. The ventral body regions are sparsely covered with hair, providing a thermal window for dissipation of heat (Fowler, 1998). Postural changes can modify the exposed surface of the thermal windows (Schmidt-Nielson, 1999).

South American camelids (SAC) are kept in very different climatic regions. High environmental temperature can cause heat stress in SAC. SAC are probably more suspicious to heat stress when compared with other domesticated species outside South America, because their later evolutionary development took place in cool climate and breeders selected heavily fibred animals.

Shearing assists heat dissipation, resulting in non sheared alpacas being less heat tolerant than sheared animals (Navarre *et al.*, 2001). In llamas the body surface temperature is elevated up to 23% after shearing (Gerken, 1997).

The thyroid hormones triiodothyronine (T3) and tetraiodothyronine (T4) increase the basal metabolic rate, resulting in an increase of body temperature and a stimulation of the lipid- and carbohydrate- metabolism. The blood concentration of these hormones is dependent on the ambient temperature, feeding, diseases, *etc.* In llamas the plasma concentration of these hormones is negatively correlated with the ambient temperature (Gauly, 1997). As a result the two thyroid hormones can be used as indicators for the metabolism depending on temperature and thermoregulation. Testosterone is produced in the Leydi`s cells in the testis. It regulates the development of the secondary male sex characteristics, stimulates the spermatogenesis and promotes the protein biosythesis. Gauly (1997) described a positive correlation between testosterone-concentration and the ambient temperature in llamas.

The aim of the study was to estimate the effects of high ambient temperature on heat-adapting thermoregulatory reactions including various physiologic parameters (rectal temperature, respiratory rate, heart rate), the body surface and scrotal temperatures and the metabolic rate, and the hormonal levels (thyroid hormones, testosterone) in relation to the fibre length *e.g.* the type of shearing of the animals.

Material and methods

12 fertile male llamas were housed in heated stables during the study. 5 of the animals were shorn (< 1cm), 5 animals were left unshorn (last shearing ≥ 2 years) and 2 animals were shorn partly (barrel cut).

After a short acclimatization period, the temperature was elevated for one week (adaptation-period 1) up to 30°C for 4 weeks. Afterwards the temperature was lowered for one week (adaptation-period 2) down to 20°C, then the animals were allowed to recover for at least 11 weeks at 20°C. Rectal temperature, heart rate and respiratory rate were measured daily.

In addition the body surface and scrotal temperature was measured once a week by infrared-thermography (Thermovision® 900 System AGEMA Infrared Systems AB (Sweden)) (Barow, 1998). The infrared pictures of the body were analysed with a system software. The body surface temperature was evaluated with three circles of the same size for "front" (right shoulder), "middle" (right thorax) and "back" (right thigh) of the body. The scrotum was taken and measured as an area of the right and left testicle.

Blood was collected once a week and serum thyroid hormone and serum testosterone levels were evaluated by radioimmunoassay.

Results and discussion

In the heat and recovery period the shorn animals showed significantly (p<0.001) lower rectal temperatures compared to the other two groups. Llamas with the barrel cut had significantly (p=0.037) higher rectal temperatures during the heat period when compared with the unshorn ones. This difference was not visible any more in the recovery period (Figure 1c).

All animals showed an increase in the rectal temperature in the first adaptation-period with a relatively constant high level during the heat-period. The shorn animals showed a slower increase and a faster decrease in the two adaptation-periods compared to the other two groups. Rectal temperatures were within the physiological range (Fowler, 1998) during the whole study.

Respiratory rates showed an analogous development to rectal temperatures. In the heat-period, the shorn animals showed significantly (p<0.001) lower respiratory rates compared to the other two sheartypes, (p > 0.05). In the recovery-period the shorn animals (p<0.001) and the llamas with the barrel cut (p<0.001) had significantly lower respiratory rates than the unshorn animals (Figure 1b).

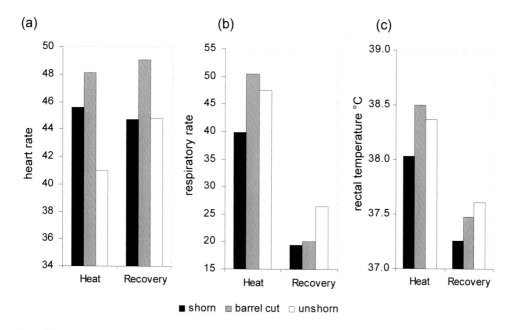

Figure 1. Lsq-means of heart rate (a), respiratory rate (b) and rectal temperature (c) of the three sheartypes (shorn, barrel cut, unshorn) in the heat- and recovery-period.

All llamas showed an increase in the respiratory rate in the first adaptation period, with the shorn animals showing the slowest increase. In the second adaptation period after the heat period the respiratory rates of the shorn animals declined fastest.

The respiration rates were above the physiological range during the heat-period but were nearly within the physiological values in the recovery-period.

In the heat period the llamas with the barrel cut showed the highest heart rate, which was significantly different from the heart rate of the shorn (p<0.05) and the unshorn llamas (p<0.001). The shorn llamas had higher heart rates during the heat period when compared with the unshorn (p<0.001). In the recovery-period the llamas with the barrel cut showed again a significantly higher heart rate as the other two sheartypes (p<0.001), which showed no differences in their heart-rates in the recovery period (Figure 1a). The heart-rates were within physiological ranges during the study.

Body surface temperatures were significantly higher (p<0.001) during the heat period when compared with the recovery period. Within the sheartypes there was only a difference between the shorn and unshorn parts of the body, so that the shorn animals were significantly warmer (p<0.001) in all of the three measured body regions compared with the unshorn. The heat dissipation in unshorn llamas was restricted on the thermal windows (Figure 2a) as earlier described by Gerken (1997). Only in these areas an effective thermoregulation seems to be possible. However in the shorn animals the heat transfer is via the entire body surface temperature (Figure 2b).

During the heat period the scrotum of the shorn animals was significantly cooler (p<0.05) when compared with the other groups, which showed no differences. During the recovery period the shorn animals and the llamas with the barrel cut showed lower (p<0.01) scrotal surface temperatures than the unshorn llamas. A significant correlation was estimated between the body surface and scrotal and rectal temperatures.

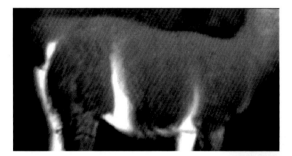

a

b

Figure 2. Thermography of an unshorn (a) and a shorn (b) llama. Darker regions show the cooler surface temperatures.

Testosterone level was not significantly different between the heat and the recovery period and the three groups.

T_3 (p<0.001) and T_4 (p<0.01) were both significantly lower in the heat-period when compared with the recovery-period. T_3 levels were not significantly related to the sheartypes, whereas the levels of T_4 were dependent on the type of shearing. The shorn llamas had significantly (p=0.038) lower T_4 levels in the heat period compared to the unshorn llamas. The differences in the recovery period in the T_4 levels between the shorn llamas and the other two sheartypes was significant (p<0.01). The llamas with the barrel cut and the unshorn animals showed no significantly different T_4 levels neither during the heat nor during the recovery period. The two thyroid hormones were positively correlated (r=0.66, p<0.001).

All animals had elevated rectal temperatures, respiratory rates, body and scrotal surface temperatures during the heat period. However all llamas were able to maintain rectal temperature within normal levels probably by increasing their respiratory rate, which is a thermoregulatory mechanism of llamas to use the evaporative cooling of the respiration tract (Pugh et al., 1997). All animals showed lower T_3 and T_4 levels during the heat period as expected.

There were no significant differences in testosterone-levels in the heat and the recovery period. An effective thermoregulation can probably only take place via the thermal windows. Shorn animals showed lower rectal temperatures, respiratory rates and scrotal surface temperatures but higher body surface temperatures during this study compared to the unshorn animals. There were no differences between the sheartypes in the T_3 and testosterone levels. However completely shorn animals had lower T_4 levels compared to the other two sheartypes.

Conclusions

Healthy llamas can cope with ambient temperatures up to 30°C. They are able to maintain their rectal temperature within normal levels independent of the sheartype, by increasing their respiratory rate.

An effective thermoregulation in unshorn animals can only take place via the thermal windows. At 30°C the llamas with a barrel cut have only marginal thermoregulatory benefit compared to the unshorn animals. At 30°C only complete shearing assists thermoregulation enough to reduce rectal temperature and respiration rate.

References

Barow, U., 1998. Methodische Untersuchungen zur Erfassung der Thermoregulation bei Mutterkühen unter Feldbedingungen.Dissertation. Institute of Animal Breeding and Genetics, University of Göttingen.

Fowler, M.E., 1998. Medicine and Surgery of South American Camelids. Second Edition, Iowa State University Press.

Gauly, M., 1997. Saisonale Veränderungen spermatologischer Parameter und der Serumkonzentration von Testosteron, Oestradiol 17β, Thyroxin sowie Trijodthyronin männlicher Neuweltkameliden (Lama glama) in Mitteleuropa. Inaugural Dissertation, Institute of Animal Breeding and Genetics, University of Giessen.

Gerken, M., 1997. Application of infra-red thermography to evaluate the influence of the fibre on body surface temperature in llamas. European Fine Fibre Network, Occasional Publication No. 6.

Jessen, C., 2000. Wärmebilanz und Themperaturregulation. In: Physiologie der Haustiere, W. Engelhardt and G. Breves (eds.). Enke im Hippokrates Verlag GmbH, Stuttgart.

Navarre, C.B., A.M. Heath, J. Wenzel, A. Simpkins, E. Blair, E. Belknap and D.G. Pugh, 2001. A comparison of physical examination and clinicopathologic parameters between sheared and nonsheared alpacas (Lama pacos). Small Rumin. Res. 39: 11-17.

Pugh, D.G., C.N. Evans, J. Hudson and A. Kennel, 1997. Heat stress in llamas. International Llama Association Educational Brochure # 11.

Schmidt-Nielsen, K., 1999. Physiologie der Tiere. Spektrum Akademischer Verlag Heidelberg, Berlin.

Skin lesions in UK alpacas (*Lama pacos*): Prevalence, aetiology and treatment

G.L. D'Alterio

Farm Animal Practice & Hospital, Department of Clinical Veterinary Science, University of Bristol, Langford, Bristol, United Kingdom; gianlorenzod@yahoo.it

Abstract

The members of the two British camelid breeders' associations were surveyed by means of a postal questionnaire. Given the lack of up-to-date data, the survey aimed to characterise the UK South American camelid population, and the prevalence and features of skin disease. A total of 3,520 camelids were counted, of which 77.2 per cent were alpacas and 20.6 per cent llamas (*L. glama*). Occurrence of skin disease was reported by 51 per cent of respondents. Zinc deficiency and ectoparasitism, presumptively diagnosed, were listed as the cause of disease by 31.9 per cent and 26.4 per cent of respondents, respectively. Based on the results from the previous study and the recurrent isolation of *Chorioptes spp.* mite in conjunction with skin lesions from case material referred to the author, a study on the prevalence of mite infestation in alpacas was initiated. A total of 209 alpaca in nine units in the south-west of England were included in the study. Forty-seven (47/209; 22.5%) showed signs of skin disease. In the sampled population, 33 alpaca (33/83; 39.8%) were positive for *Chorioptes spp.* mite. The efficacy of eprinomectin *vs.* ivermectin, and the field efficacy of eprinomectin for the treatment of chorioptic infestation in naturally infested alpacas were assessed in two studies. No localised or systemic side effects were observed in either trial. The eprinomectin protocol employed in Study B proved highly effective at controlling parasitic burden in alpacas naturally affected by chorioptic mange mites.

Keywords: camelid, United Kingdom, skin disease, *Chorioptes spp.*, avermectins

Introduction

South American camelids, especially alpacas, are increasingly popular in the United Kingdom. Reports received from breeders and veterinarians suggested that skin disorders in alpacas were common in the UK population, and treatment often unrewarding.

The distribution and type of skin lesions reported were similar to those observed by the author on case material referred. Animals were referred for a variety of medical and surgical conditions, were examined at the clinic or on farm, and detection of skin lesions was often an incidental finding. In cases that were referred specifically for a dermatological complaint, no abnormalities other than skin lesions were found at clinical examination. Pruritus did not appear to be a predominant feature of the skin condition.

A full dermatological diagnostic work up was instigated in all these cases, in collaboration with the University of Bristol Veterinary Dermatology Unit.

Ancillary tests, including sampling for bacterial and fungal culture and standard haematological and serum blood biochemistry panels were often unrewarding. *Staphylococcus aureus* was occasionally isolated from epidermal swabs. An increase in total white cell count, with associated neutrophilia and eosinophilia, was observed in some animals. In contrast, histopathological examination of skin biopsies from affected regions consistently yielded orthokeratotic hyperkeratosis, a variable severity of perivascular or diffuse "eosinophilic" dermatitis, with varying degrees of lymphocyte and plasma cell infiltrate and, less frequently, intra-corneal pustule formation (Cecchi *et al.*, 2003).

The most specific result obtained in conjunction with the detection of skin lesions was the isolation of *Chorioptes spp.* mite from skin scrapings. Given the significance of *Chorioptes spp.* mite in camelids (Rosychuk, 1989; Fowler, 1998; Petrikowski, 1998; Wernery & Kaaden, 2002), this finding warranted further investigation.

A series of studies were undertaken in order to establish the prevalence of the skin condition/s, define a plausible aetiology, and consequently develop a protocol of treatment. This paper aims at summarise the results from three separate studies, which were instigated in a sequence based on the conclusions drawn from each previous study.

Study 1 provides the most up to date and comprehensive information on South American camelids farming in the UK available so far. These data were obtained through a postal questionnaire aimed primarily at characterisation of the extent and type of skin disease in the camelid population. Additionally, the study achieved the goal to fill in the gap in knowledge on population census, husbandry, and preventative medicine procedures adopted by camelid breeders in the UK. For brevity, only the data referring to the skin condition are presented. For the full body of information please see D'Alterio *et al.* (2006).

Study 2 presents a study on the prevalence of *Chorioptes spp.* mite infestation in alpacas in the south-west of England. Through this study, the dermatological complaint was better defined in epidemiological and clinical terms, and an aetiological association established with the presence of the mite. Findings from this study suggest that the recurrent dermatological complaint affecting alpacas in the UK could be termed as chorioptic mange.

Study 3 bridges two treatment trials aimed at establishing an effective, safe, easy to administer and cost effective treatment against *Chorioptes spp.* mite infestation in alpacas. The first study (Study A: a single-centre, randomised, treatment-controlled, blinded field trial) aimed at comparing the efficacy of topical eprinomectin against that of systemic ivermectin for the treatment of natural *Chorioptes spp.* infestation in a group of alpacas. Based on the results from Study A, the second study (Study B: a single-centre, open, un-controlled field trial) aimed at verifying the efficacy of a refined protocol based on topical eprinomectin for the treatment of the same ectoparasite in the same naturally infested host species, under field conditions.

Material and methods

Study 1: Postal questionnaire

Members of the British Llama and Alpaca Association (BLAA; currently known as British Camelid Ltd) and the British Alpaca Society (BAS) were contacted by means of a three page, double-sided questionnaire, cover letter and a return-addressed, stamped envelope.

The questionnaires were posted in December 2000, to be returned by January 31st 2001, with 316 copies sent directly to BLAA members and 380 with the society's periodical to BAS members, to give a total of 696 postal questionnaires posted. The questionnaire was divided in three sections: population statistics and husbandry procedures, preventative health measures and occurrence of disease, and skin conditions, respectively.

In the section on skin disease 13 questions were asked, ranging from the presence of skin disease within the herd, to species, number, and sex and age class of the affected animals. Divided into two questions, six different clinical signs and 13 anatomical regions were listed, allowing multiple answers on type and distribution of skin lesions. Owners were asked whether a veterinary surgeon had investigated skin disease, and if so what diagnosis was formulated, what treatment if any was instigated and the response to treatment.

The data from the survey questionnaires were entered into a Microsoft Excel spreadsheet and used to calculate summary statistics and cross-tabulations.

Study 2: Prevalence of Chorioptes spp. mite infestation

The study was carried out from November 2001 to January 2002 on nine different alpaca units in the south-west of England. The units were recruited through a local breeders association (South West Alpaca Group) with all but one unit reporting the presence of skin lesions in some of their animals. Detailed herd histories were taken from each unit, including feeding, vaccination, endo- and ectoparasite control regimes. All the alpacas from each unit were clinically examined for the presence of skin lesions. All alpacas presenting signs of skin disease, as well as approximately one in five clinically healthy, randomly selected, in-contact alpaca were included in the sampled population. For each sampled animal, an individual history was taken, and age (categorised as < 24 month-old; 24 to 48 month-old; > 48 month-old), sex, colour of the fleece and body condition recorded. Body condition was scored from one to five, with one being poor and five being obese, according to the technique described by Fowler (1998). Body regions where the skin lesions were detected were also noted and recorded.

Superficial skin scrapings were taken from each animal included in the sampled population from six different sites: bridge of the nose, ear pinna, axilla, medial thigh, perineum and forefoot (just proximal to the interdigital cleft). The first five sampling sites were selected on the basis of occurrence of lesions, as previously observed on referred case material, while the forefoot was selected because it appears to be a predilection site for *Chorioptes spp.* in several host species (Sweatman, 1957). A dry swab was also taken from the ear canal. Samples were maintained at room temperature and examined for the presence of ectoparasites within 24 hours of collection by light microscopy using an x10 eyepiece and x4 objective.

All data were entered into a database (Microsoft Excel). Where appropriate, pairs of variables (*e.g.* presence of mite *vs.* presence of lesions) were tested for association using the Chi-square test in the MINITAB statistical software package. Potential confounding factors, such as age, sex, colour of the fleece and body condition were investigated in a similar manner by applying the Chi-square test to the relevant contingency tables. The adopted significance level was $p < 0.05$ for all statistical analysis.

Study 3: Treatment trials

For Study A, thirty entire male alpacas were selected on the farm of origin, a large alpaca commercial unit in West Sussex, UK, experiencing considerable problems with skin disease in the herd. The animals were selected on the basis of being: a) positive for the presence of *Chorioptes spp.* mite in skin scrapings taken on the farm from multiple body sites, b) negative for other ectoparasites, c) concomitantly affected by skin lesions consistent with chorioptic mange, as described in the literature (Cremers, 1985; Petrikowski, 1998) and as observed by the authors during the course of clinical and research work (D'Alterio *et al.*, 2005, 2006). Additionally, no clinical evidence of systemic or non-parasitic skin disease was detected by a general clinical examination. All animals had received no treatment against endo- or ectoparasites in the three months preceding the study. Following the assessment on farm, they were isolated from the parent herd, travelled as one group and were re-located to a farm near the School of Clinical Veterinary Science at Langford, Bristol. Following the trial procedures on day 0 (see below), the alpacas were moved to a field, where they were maintained in isolation from other camelids and any other livestock species for the entire duration of the trial. On day 0 of the trial (8th August 2002), all animals were clinically examined and weighed for accurate therapeutic dosing. One skin scraping was collected from each animal from the interdigital space of one of the front feet. This body site was chosen based on the results of previous studies that showed there is a greater chance of recovering mites than any other body site (D'Alterio et al., 2005) and the scraping repeated on days 7, 14, 28, 42 and 63. Samples were maintained at room temperature and examined for the presence of ectoparasites within 24 hours of collection by light microscopy using an x10 eyepiece and x4 objective. Mites were counted up to 100 individuals per slide. For data analysis, individual animals that had greater than the maximum number of mites counted (100) were assumed to have the maximum number of mites.

On days 0, 14, 28, 42 and 63 alpacas were given a clinical score. Clinical scoring consisted of assessing the severity (absent = 0; mild = 1; moderate = 2; severe = 3) of three types of skin lesions (alopecia, scaling and thickening) at seven different body sites (face, ear, ventral abdomen, antebrachium, caudal aspect of the hindlimb, foot and perineum). The maximum possible score was 3 x 3 x 7 = 63. The skin scraping and lesion scoring was carried out by an investigator blinded to the treatment the animals had received. Animals were partially assessed on day 7 with skin scrapings only. While being handled for clinical scoring and skin scrapings, animals were also checked for localised and systemic signs of any adverse reactions to the treatment. For the duration of the trial, all animals were checked twice daily from a close distance for evidence of adverse reactions to treatment and any other sign of ill health.

On day 0, alpacas were randomly allocated to two treatment groups identified by ear tag numbers. The 15 alpacas in group I received a single, topical application of 0.5% eprinomectin formulation (Eprinex pour-on; Merial Animal Health, UK) at the dose rate of 500 mcg/kg bodyweight. By opening the fleece, the product was applied directly onto the skin of the dorsum at multiple sites.

The 15 alpacas in group II were treated with a subcutaneous injection of 1% ivermectin formulation (Panomec injection; Merial Animal Health, UK), in the skin fold caudal to the elbow, at the dose rate of 400 mcg/kg. Treatment in group II was repeated on days 14 and 28.

Mite counts from each animal were summed for each time point and a mean value obtained. Clinical scores for the different types of lesion affecting the seven different body sites under examination were summed and a total clinical score obtained for each animal at each time point. Within each group, mean mite counts and total clinical score at each time point were compared using non-parametric repeated-measures analysis of variance (Friedman test) using statistical software (SPSS v12.0, SPSS Inc.). The same analysis was repeated for differences across time in clinical scoring, irrespective of treatment. A non-parametric Mann-Whitney's test analysis of variance was used to test for differences in clinical score between the two groups, at each time point.

For Study B, nineteen alpacas were selected. They were maintained as a closed herd at one location in North Somerset, England. They were selected on the basis of five out of nineteen animals showing skin lesions consistent with chorioptic mange, the detection of *Chorioptes spp.* mites in the apparently unaffected herd members, and failure to detect other ectoparasites during the course of clinical work previously carried out on farm by the author. None of the animals had received treatment against endo- or ectoparasites for at least two months preceding the study. On day – 3 of the trial (10th February 2003), all alpacas were identified by individual name, and clinically examined. All abnormalities, dermatological and non-, were recorded. One skin scraping was collected from each animal from the interdigital space of one of the front feet, and repeated on days 14, 28, 42, 60 and 117. Skin scrapings were collected and samples examined for mites, as described for Study A. Mites were counted up to 100 individuals per slide. For data analysis, counts above 100 mites were considered as in Study A. All alpacas were checked for localised and systemic signs of adverse reaction to treatment whilst they were handled for skin scrapings and administration of treatment. For the duration of the trial, animals were checked twice daily by the owner or a helper, who had been instructed to report immediately any abnormality to the authors. All 19 alpacas were treated on day 0 by topical application of 0.5% eprinomectin formulation (Eprinex pour-on; Merial Animal Health, UK) at the dose rate of 500 mcg/kg, which was repeated on days 7, 14 and 21. Frequency of administration was adopted following preliminary analysis of data from Study A. A non-parametric repeated-measures analysis of variance (Friedman test) was performed to compare the mean value of the mite counts at each time point within the group.

The adopted level of significance was $P < 0.05$ for all statistical analysis in both studies.

Results

Study 1: Postal questionnaire

Of 696 questionnaires sent, 225 were returned by the 31st of January 2001, giving a response rate of 32.3%. Five respondents did not keep camelids, one replied from Tasmania and one from Switzerland. These seven responses were not included in the study, resulting in a usable response rate of 31.3% (218/696).

The presence of skin disease was reported by 51.1% (111/217; one non-respondent) of respondents. A total of 317 animals were reported to be affected (three non-respondents), ranging from one to 43 affected animals per unit, representing 9.0% (317/3520) of the total population surveyed. Of the 317 affected camelids, 81.4% (258/317) were alpacas, 17.3% (55/317) llamas and 1.3% (4/317) guanacos. The affected alpacas, llamas and guanacos represented 9.5% (258/2719), 7.6% (55/726) and 6.4% (4/62) of their respective surveyed species population. Females and adults (over 18 months of age) represented 79.5% (252/317) and 89.9% (285/317) of the affected population.

Camelids exclusively bred in the UK, imported into the UK or both, and affected by skin disease were found on 44.3% (47/106; five non-respondents), 35.8% (38/106) and 19.8% (21/106) of herds, respectively.

The percentage of respondents who thought the skin condition to be more prevalent in spring, summer, autumn, winter or all year round was 9.0% (10/111), 40.5% (45/111), 12.6% (14/111), 15.3% (17/111) and 34.2% (38/111), respectively. When asked to report whether the affected camelids were predominantly solid white, all other colours or both, 63.3% (69/109; two non-respondents) opted for all other colours, 18.3% (20/109) for solid white, and 18.3% (20/109) for both.

Table 1 shows the type of skin lesions reported and their proportions, and Table 2 summarises the occurrence of body regions reportedly affected.

Veterinary advice was sought by 68.2% (75/111; one non-respondents) of respondents in order to address the dermatological condition. Of the owners reporting skin disease in their camelids, 64% (71/111) reported a presumptive diagnosis being formulated by their veterinarian.

Table 1. Type of skin lesions and percentage of reporting respondents.

Type of lesion	Percentage of respondents
Hair loss/alopecia	77.4
Crusty/scabby	55.8
Scaly/thickened	46.8
Itchy/pruritic	33.3
Reddened	18.9
Bleeding/oozing	13.5

Table 2. Distribution of skin lesions and percentage of reporting respondents.

Anatomical region	Percentage of respondents
Nose	51.3
Ears	50.5
Periorbital	38.7
Inner thigh	37.8
Axilla	34.2
Dorsum	25.2
Abdomen	21.6
Tail	14.4
Flank	12.6
Neck	5.4
Outer thigh	4.5
Feet	3.6

A total of 16 different aetiological diagnoses were counted, and no diagnosis was made in some cases. In the majority of cases skin lesions were thought to be the result of zinc deficiency or ectoparasite infestation, reported by 31.9% (23/72) and 26.4% (19/72) of respondents. All other 14 diagnoses were far less common.

Irrespective of whether veterinary advice was sought, and/or a diagnosis was reached, 73.9% (82/111) of respondents who reported skin disease initiated treatment.

Of those who treated their animals, 71.9% (59/82) reported an improvement following treatment, 13.4% (11/82) did not, and 14.6% (12/82) felt it was too early to judge the response. Of those respondents who observed an improvement in skin condition following treatment, 45.8% (27/59) thought the improvement to have been permanent, 37.3% (22/59) thought that the improvement was only temporary, and the remaining 16.9% (10/59) felt that it was premature to report whether the improvement was permanent or temporary. Of those observing an improvement, 67.8% (40/59) considered the improvement to have involved all treated animals, 25.4% (15/59) thought only some of the treated animals improved, and 6.8% (4/59) were not yet able to say.

Study 2: Prevalence of Chorioptes spp. mite infestation

A total of 209 alpacas from nine units were included in the study, ranging from seven to 76 animals per unit.

All the alpacas were kept on pasture all year round and received supplementary hay for part of the year (mostly in winter). Varying regimes and formulation of supplementary concentrate feeding were adopted in all units. Currently, alpaca in all the units were vaccinated by means of different multi-valent anticlostridial vaccines, either once or twice a year. Macrocyclic lactone compounds were used once or twice a year for endoparasite control. One unit alternated the use of macrocyclic lactones with a benzimidazole compound.

Of the 209 alpacas clinically examined, 47 (22.5%) showed signs of skin disease, ranging from mild alopecia, thickening, crusting and scaling of the skin of the pinnae, to severe and diffuse similar lesions affecting most commonly the ears, axilla, face and dorsum. The affected animals did not appear overtly pruritic during the periods they were observed for clinical examination and sample collection. In addition, 36 unaffected, in-contact alpacas were examined for ecto-parasite.

Thirteen out of the 47 alpacas affected by skin lesions (27.7%) were concurrently positive for *Chorioptes spp.* mite, whilst 20/36 (55%) unaffected sampled alpacas were also positive for this mite. Four out of 36 (11%) unaffected, in contact alpacas were positive for *Psoroptes spp.* mite, and an unidentified species of louse was detected from 4 animals in the same category (11%). When detected, *Chorioptes spp.* mite count ranged from 1 to 22 mites per glass slide examined, and the mite was more commonly isolated from the fore foot (22/33; 66.7%) and the axilla (13/33; 39.4%). Cumulatively, in 29 out of 33 positive cases (87.9%) *Chorioptes spp.* mites were detected in scrapings taken from the forefoot and/or the axilla.

A Chi-squared test showed the association between *Chorioptes spp.* mite and skin lesions to be statistically significant ($p = 0.01$), in that affected animals tended not to be positive for the mite whilst un-affected animals tended to be positive for the mite. Sex, colour and body condition were not identified as confounding factors. The contingency table for the age category < 24 month-old showed that the most common finding for this group was "affected by mites but not lesions" (12/27; 44.4%). This finding was in contrast to that observed in the overall sampled population, where the most common finding was "affected by lesions but not mites" (34/83; 41.0%). The tendency observed was therefore for an increase in the presence of skin lesions and a decrease in the presence of mites with increasing age. A Chi-square test for lesions *vs.* age and mite *vs.* age yielded a highly significant association between these factors ($p = 0.007$ and $p = 0.01$ respectively).

Study 3: Treatment trials

No local and general side effects, or other signs of ill health, were observed in any of the treated animals in either study. Some light coloured animals treated with eprinomectin showed darkening of the fleece over the areas of topical administration. This appeared to be linked to rolling in dirt, a normal behaviour in alpacas, and consequent adhesion of soil particles to the fleece presumably coated by the drug formulation. No ectoparasites, other than *Chorioptes spp.* mites, were detected from either study.

On day 0 of Study A, all alpacas in both groups were positive for *Chorioptes spp.* mites. Seven days after treatment, five alpacas in group I and six in group II were found negative for mites, whereas all but one alpaca were positive for mites on day 14. The trial ended on day 63 with all but one animal positive for mites.

There was a statistically significant difference (P < 0.001) in mite counts across time within treatment group I and treatment group II, with mite counts decreasing significantly on day 7 if compared to day 0 in both groups. Mite counts increased again on day 14 and remained high for the duration of the trial in both treatment groups (Figure 1).

No statistically significant differences in mite counts were found within either group I and group II from day 14 onward. Comparison of mite counts at each point in time between the two treatment groups showed no statistically significant differences.

Although skin lesions affected all alpacas, clinical signs were relatively mild. No significant differences across time in either group were found for clinical scoring of alopecia, scaling and thickening, nor significant differences in clinical scoring between groups at any time point. Testing for overall differences across time in clinical scoring, irrespective of treatment, showed a significant reduction in alopecia (P = 0.022) and scaling (P = 0.005). No statistically significant changes across time were observed in thickening of the skin.

On day –3 of Study B, 10 out of 19 alpacas were found positive for *Chorioptes spp.* mite. On day 14, four out of nineteen animals were still positive, but the reduction in mite count was highly significant if compared to day –3 (P = 0.008; Figure 2). One week after the last treatment, only two animals were still positive for mites, but by day 117 it had increased to four animals. From day 14 onward mite counts remained very low, and were not significantly different throughout the remainder of the study.

Discussion

Occurrence of skin disease in the herd was reported by just over half of the respondents to the postal questionnaire (51.1 per cent; 11/217), resulting in 9.0 per cent (317/3520) of the surveyed population being affected. Alopecia (77.4 per cent; 86/111) of the head, namely bridge of the

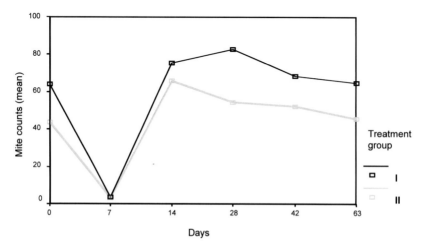

Figure 1. Study A. Changes in mite counts (mean) for the duration of Study A. On day 7 there was a statistically significant reduction in both treatment group when compared to any other point in time (P < 0.001).

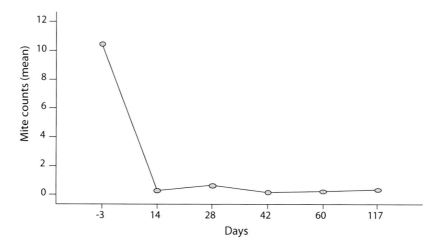

Figure 2. Changes in mite counts (mean) for the duration of Study B. A significant drop in mite counts was observed on day 14 compared to day -3 (P = 0.008). Mite counts remained low throughout the duration of the study.

nose (51.3 per cent; 57/111) and ears (50.5 per cent; 56/111), were the most commonly described type of skin lesion and area of distribution, respectively. The high number of reported different diagnoses (16) was probably a reflection of uncertainties in the cause of the skin disease. Zinc deficiency (31.9 per cent; 23/72) and ectoparasite infestation (15.3 per cent; 17/72) were the two most commonly reported, presumptive diagnoses in this survey.

Evidence of different types of ectoparasitic infestation in camelids abounds in the literature, as reviewed by Rosychuk (1994), Fowler (1998), Wernery and Kaaden (2002), and for UK camelids, by Bates *et al.*, (2001). In contrast, evidence in the literature confirming zinc deficiency in llamas is lacking (Rosychuk, 1994).

Considering the type and distribution of skin lesions described by respondents, and the relatively poor response to treatment, the latter a factor which had to be interpreted carefully because of the lack of definitive diagnoses, it seemed prudent to establish the significance of both ectoparasite infestation and zinc deficiency (zinc-responsive dermatosis) before committing to further aetiological investigations. Also on the basis of clinical findings from referred cases, we as a research group decided to focus on a possible link between the described skin condition and the presence of *Chorioptes spp.* mite.

As in other livestock species, in terms of prevalence and economic losses, parasitic skin disease caused by mange mites is the most important dermatological condition affecting camelids in South America (Leguia, 1991). *Chorioptes bovis* is an obligatory, non-burrowing mite that feeds on epidermal debris (Sweatman, 1957). *C. texanus* is the only other species known to occur in the genus (Sweatman, 1957; Essig *et al.*, 1999). In the present study, speciation of the mites isolated from alpacas was not performed beyond genus level, and therefore the term *Chorioptes spp.* was adopted throughout the manuscript. Chorioptic mites had been observed in the llama (Sweatman, 1957; Cremers, 1985; Rosychuk, 1989; Fowler, 1998) and the alpaca (Young, 1966; Cremers, 1985; Rosychuk, 1989; Petrikowski, 1998; Bates *et al.*, 2001). Interestingly, we failed to find reports of chorioptic mange in camelids in South America. However, Rosychuk (1989) and Fowler (1998) consider chorioptic mange a rare condition in both host species, whilst Cremers (1985) suggests that the mite is common in camelids, but the mange lesions are not.

In our survey, lesions and mites were both highly prevalent among the sample population, with 22.5% of the alpacas examined affected by skin lesions, and 39.8% of the population sampled for ectoparasite positive for *Chorioptes spp.* Furthermore, we observed that approximately one in four alpacas affected by skin lesions and approximately one in two healthy, in-contact individuals were positive for the mite. Indeed, the prevalence of skin lesions that we observed was more than double that previously estimated from responses to the postal survey. In addition, the study reported here had enough statistical power to suggest a negative association between *Chorioptes spp.* mite and the presence of the above described skin lesions. Collection of superficial skin scrapings from the feet and the axilla of in-contact, healthy individuals (preferably juveniles) gives almost a 90% chance of successfully recovering *Chorioptes spp.* mite, if present.

The results from Study A provide evidence that a single administration of topical, 0.5% eprinomectin formulation at the dose rate of 500 mcg/kg body weight produces only a transitory reduction in parasitic load but is ineffective at controlling a natural infestation of *Chorioptes spp.* mite in alpacas. Three parenteral administrations of 1% ivermectin formulation, at the dose rate of 400 mcg/kg body weight every 14 days proved equally ineffective. The dramatic decrease in mite burdens observed on day 7 of Study A for both treatments, and the subsequent sharp increase detected on day 14, suggested that multiple, weekly administrations of an avermectin compound should be tested. Jarvinen *et al.*, (2002) found that in llamas the administration of subcutaneous 1% ivermectin at the dose rate of 200 mcg/kg, and the administration of a 0.5% topical formulation of ivermectin at the dose rate of 500 mcg/kg resulted in a similar pharmacological behaviour. Both protocols resulted in peak serum concentration above 2 ng/ml occurring approximately one week after treatment.

The administration of a 0.5% formulation of eprinomectin at the dose rate of 500 mcg/kg body weight, once every 7 days for four weeks, proved highly effective at reducing the *Chorioptes spp.* mite burden in naturally infested alpacas. Since no control animals were available, these findings must be interpreted with caution. Movement of animals to clean pasture following treatment, and cleaning and disinfecting of fomites was not implemented in the present study. It can be speculated that if stricter biosecurity measures had been implemented, eradication of the mite from the group might have been successful. The above studies indicate that, despite high dosage or frequency of treatment, both avermectins tested appear safe for use in camelids. Findings from the two studies also reiterate the inadequacies of extrapolating drug dosages from one ruminant species to another.

To the author`s knowledge, this is the first report on a large scale monitoring the response to treatment in a camelid naturally affected by chorioptic mange. This is also the first report on the use of eprinomectin in a camelid species

Acknowledgements

Aiden Foster, Toby Knowles, Anna Jackson, Erin Eknaes, Ida Loevland, Clare Callaghan, Chris Just and Alex Manner-Smith are kindly acknowledged for co-researching these topics. I have been the recipient of a Millenium Production Animal Scholarship from the RCVS Trust.

References

Bates, P., P. Duff, R. Windsor, J. Devoy, A. Otter and M. Sharp, 2001. Mange mite species affecting camelids in the UK. Veterinary Record 149: 463-464.

Cecchi, R., G.L. D' Alterio, G.R. Pearson and A.P. Foster, 2003. Retrospective histopathological study of some skin disorders of alpacas (*Lama pacos*). In: Proceedings of the 41st International Symposium on Zoo and Wildlife Diseases, J. Wisser, H. Heribert and K. Frolich (eds.), Rome, Italy, May 28 to June 01, 2003, pp. 179-183.

Cremers, H.J.W.M., 1985. *Chorioptes bovis* (Acarina: Psoroptidae) in some camelids from Dutch zoos. Veterinary Quarterly 7: 198-199.

D'Alterio, G.L., C. Callaghan, C. Just, A. Manner-Smith, A.P. Foster and T.G. Knowles, 2005. Prevalence of *Chorioptes sp.* mite infestation in alpaca (*Lama pacos*) in the south-west of England: implications for skin health. Journal. of Small Ruminant Research 57: 221-228.

D'Alterio, G.L., T.G. Knowles, E.I. Ekneas, I.E. Loevland and A.P. Foster, 2006. Postal survey of the population of South American camelids in the United Kingdom in 2000/01. Veterinary Record 158: 86-90.

Essig, A., H. Rinder, R. Gothe and M. Zahler, 1999. Genetic differentiation of mites of the genus Chorioptes (Acari: Psoroptidae). Experimental and Applied Acarology 23: 309-318.

Fowler, M.E., 1998. Medicine and surgery of South American camelids: llama, alpaca, vicuña and guanaco. (2nd ed.) Ames, Iowa State University Press.

Leguia, G., 1991. The epidemiology and economic impact of llama parasites. Parasitology Today 7: 54-55.

Jarvinen, J.A., J.A. Miller and D.D. Oehler, 2002. Pharmacokinetics of ivermectin in llamas (*Lama glama*). Veterinary Record 150: 344-346.

Petrikowski, M., 1998. Chorioptic mange in an alpaca herd. In: K.W. Kwochka, T. Willemse, and C. von Tscharner (eds.), Advances in Veterinary Dermatology, vol 3, Butterworth Heinemann, pp.450-451.

Rosychuk, R.A.W., 1989. Lama dermatology. Veterinary Clinics of North America-Food Animal Practice 5: 203-215.

Rosychuk, R.A.W., 1994. Lama dermatology. Veterinary Clinics of North America-Food Animal Practice 10: 228-239.

Sweatman, G.K., 1957. Life history, non-specificity, and revision of the genus chorioptes, a parasitic mite of herbivores. Canadian Journal of Zoology 35: 641-689.

Wernery, U. and O-R. Kaaden, 2002. Infectious diseases in camelids. Blackwell Wiss.-Verl., Berlin and Vienna.

Young, E., 1966. Chorioptic mange in the alpaca, *Lama pacos*. Journal of the South African Veterinary Medical Association 37: 474-475.

Evaluation of the immune response to vaccination against *C. Pseudotubercolosis* in an alpaca herd in Italy: Preliminary results

D. Beghelli[1], G.L. D'Alterio[2], G. Severi[3], L. Moscati[3], G. Pezzotti[3], A. Foglini[3], L. Battistacci[3], M. Cagiola[3] and C. Ayala Vargas[4]

[1]Facoltà di Medicina Veterinaria, Via Circonvallazione 93/95 62024 Matelica (MC), Italy; daniela.beghelli@unicam.it.

[2]Libero Professionista, Studio Veterinario associato "Vannucci, Casarosa & D' Alterio", Manciano, Grosseto, Italy.

[3]Istituto Zooprofilattico Sperimentale dell'Umbria e delle Marche, Via Salvemini 1, 06129 Perugia, Italy.

[4]Università Publica de El Alto, La Paz, Bolivia.

Abstract

The applicability of some assays to evaluate the effectiveness of the immune function, as previously used in other animal species (pigs, cows and sheep), was tested in an alpaca (*Lama pacos*) herd affected by Caseous Lymphadenitis *(Corynebacterium pseudotubercolosis)* and subject to a vaccination programme against the causative agent. The vaccination programme was initiated following failure of controlling the outbreak by isolation and treatment of the initially affected animals. An homotypic, autogenous inactivated and adjuvated (aluminium hydroxide) vaccine, administered subcutaneously at day 0 (A) and after 21 days (B) was used. Blood samples were collected by venipuncture of the jugular vein on days 0, 21 and 42 (C). Samples in EDTA, collected on day A and C, were analysed for standard haematological parameters. Serum obtained from whole blood with no anticoagulant on days A, B and C was tested for: Lysozyme serum titration (expressed in µg/m), serum Bactericidal assay (expressed in %), semiquantitative complement titration (expressed in CH50) and immunoglobuline (by electrophoresis) and IgGs concentrations (by Radial Immunodiffusion tests).

Resumen

La aplicación de algunos ensayos para evaluar la eficiencia inmunológica funcional, como previamente fue aplicada en otras especies animales (cerdos, vacunos y ovinos), estas fueron ensayadas un rebaño de Alpacas (*Lama Pacos*), por la enfermedad denominada Lifoadenitis Caseosa (*Corynebacterium pseudotuberculosis*). Sujeto a un programa de vacunación contra el agente causal, se inicio el programa de vacunación, pese a las fallas de control durante el brote de la enfermedad, en el aislamiento y tratamiento de los animales inicialmente afectados. Una autogénesis hemotípica inactiva y diluida en (hidróxido de aluminio) fue la vacuna, administrada por vía subcutánea el día 0 (A) y después el día 21 (B). Muestras de sangre fueron colectadas por venoclisis de la vena yugular, los días 0, 21 y 42 (C). Las muestras colectadas en EDTA en los días A y C, se analizaron bajo parámetros de los estándares hematológicos, y con el suero obtenido de la sangre con anticoagulante, durante los días A, B y C, fueron examinadas con: Suero con lizosima titratium (expresado en µg/m), ensayo de suero bactericida (expresado en %), complemento semicuantitativo de titration (expresado en CH50) y inmunoglobulina (por electroforesis), y la concentración de inmunoglobulinas gama (por el test de radioinmuno difusión).

Keywords: alpaca, *C. pseudotubercolosis*, vaccination, immune response

Introduction

Caseous lymphadenitis (*Corynebacterium pseudotubercolosis*) is one of the most common causes of contagious abscess in sheep and goats and represents an important cause of economic loss in small ruminants (Stoops *et al.*, 1984; Gezon *et al.*, 1991.). This disease has also been diagnosed in South American Camelids (Wernery & Kaaden, 2002), although this appears to be a sporadic cause of contagious abscess in these species (Anderson, 2003). An outbreak of CLA first started in April 2003, following the introduction of 54 alpacas (25 males and 29 females) imported from Germany, into an established herd of 28 in Central Italy. While in quarantine, two imported alpacas were found affected by a sub-mandibular and mandibular abscess, respectively. Despite treatment, these two animals died after 32 days since the lesions were first noticed, followed by a third animal two months later. At post-mortem examination, several abscesses were found affecting multiple internal organs, from which *C. pseudotuberculosis* was isolated. Failure in controlling the spread of the disease by isolation and treatment of the affected animals resulted in additional cases being detected among the imported animals, and even more worryingly, among the original nucleus. At today, a total of 24 alpacas have shown signs of CLA, of which 18 died.

An inactivated, autogenous vaccine was developed as part of the disease control and eradication strategy.

The aims of this study were to assess the innate immune response in alpacas naturally affected by CLA, and following vaccination.

Material and methods

A strain of *C. pseudotuberculosis* was isolated from the abscesses of different animals that died of CLA in this herd. The sub cultured micro-organisms were used to prepare a herd specific auto vaccine (Kutschke *et al.*, 2000; Cameron & Bester, 1984). The micro-organism was concentrated for 24 h at 37°C ± 2°C in Tryptone soya agar tubes (OXOID, CM 0131) and then was grown for 24 h at 37°C ± 2°C in Biotone agar plates (Biolife, 401145). After the second incubation, the cultures were collected by centrifugation at 800 g for 30 minutes at room temperature (22°C ± 5°C), and diluted in physiological solution to the final concentration of $1,5x10^9$/ml (Nolte *et al.*, 2001). After being inactivated in the presence of 0.05% (vol/vol) formaldehyde (Panreac, Quimica) for 48 h at 37°C ± 2°C, 20% of adjuvant (aluminium hydroxide) was added. The vaccine was assessed for sterility under both aerobic and anaerobic culture conditions for 7 days. The auto vaccine was administered to 64 animals (24 males and 40 females) subcutaneously in the thoracic skin fold caudal to the elbow following the schedule given in the Table 1.

Table 1. Vaccination schedule.

Vaccine administration	Volume
Day0 (A)	2 ml
Day21 (B)	2 ml
Day150 (D)	2 ml*

* Not yet administered at the time of writing

A third dose of vaccine will be administered in October 2004. Blood samples (n.64; affected group) were collected by venipuncture of the jugular vein on days 0 (A), 21 (B) and 42 (C). The progression of disease and any signs of adverse reactions were monitored on each vaccination and/or blood collection day (Paton *et al.*, 1995; Eggleton *et al.*, 1991). Samples in EDTA, collected on day A and C, were analysed for standard haematological parameters. Serum obtained from whole blood with no anticoagulant on days A, B and C was tested for innate immune response, in this study defined as lysozyme serum titration (expressed in μg/m), serum bactericidal assay (expressed in %) and semiquantitative complement titration (expressed in CH50). Immunoglobuline (by electrophoresis) and IgGs concentrations (by Radial Immunodiffusion tests) were also tested on serum. Given the lack of normal reference range, serum samples obtained from healthy animals (16 samples from one herd at Bristol, U.K.; 8 samples from another flock from central Italy) were also analysed in order to determine control values for the innate immune response (control group).

Serum lysozyme was measured by the lyso-plate assay (Osserman & Lowlor, 1966), carried out at 37°C for 18 h, in a humidified incubator. The method is based on the lysis of *Micrococcus lysodeikticus* in 1% agarose. The diameter of the lysed zones was determined using a measuring viewer and it was compared with the lysed zones of a standard lysozyme preparation (Sigma®. Milan, Italy, M 3770). Its concentration value is expressed in μg/ml.

The serum bactericidial assay was performed by culturing a strain of *E. coli* with the serum under test, and assessing the variation in optical density by photometry (Amadori *et al.*, 1997). Briefly, a dilution (at 10^{-2}) of *E. coli* suspension (giving an optical density of 250 to 400 lambda at λ 590) are delivered into a 96 wells plate and incubated with 50 μl of serum under test, 100 μl of Brain Heart Infusion and 50 μl of Veronal Buffer (VB). The reading is performed after 18 to 20 h of incubation, at 37°C in a humid chamber, at λ 690. The value is expressed as the percentage of serum bactericidial activity.

The haemolytic complement assay (Barta & Barta, 1993) was carried out in microtitre plates; a volume of 25 μl of 6% suspension of rabbit red blood cells (RaRBC) was added to pig serum (1:4) serially diluted in 150 μl VB. The plate was incubated at 37°C for 30 min and centrifuged at 2,000 rpm for 2 min. The extent of the haemolysis was estimated by measuring the optical density of the supernatant at 550 nm. The total haemolysis value was given by the optical reading of supernatant from 25 μl of RaRBC added to 150 μl of distilled water. The complement titre (ACP 50 units/ml) was the reciprocal of the plasma dilution causing 50% lysis of RBC. This test (Seyfarth, 1976) measures the alternative pathway of complement activation (ACP). Its concentration is expressed as CH 50 and it is the result of a mathematic formula in which are considered only the values of optical density of those samples ranging between 100% and 10% of standard haemolysis.

The electrophoretic separation of serum proteins was performed using a fully automated instrument, housing both the electrophoretic apparatus, and the densitometer (MICROTECH 648 ISO, InterLab, Italy); whereas the blood serum total proteins were determinated by reagents and automated analyser (Hitachi 704).

Radial Immunodiffusion Test for quantitation of camelid IgG in serum was performed only on ten animals on days A, B, and C (Triple J Farms, Bellingham, WA). This test is based on the diffusion of antigen from a circular well radial into a homogeneous gel containing specific antiserum for each particular antigen. The precipitation rings depend on the reaction of each protein with its specific antibody and their diameter is a function of antigen concentration. The unknown IgGs serum concentrations of these animals were measured by reference to the standard curve.

Data were analysed using ANOVA (SPSS) for repeated measures. Pearson's coefficient was applied for correlation analysis.

Results

No systemic adverse reactions were observed following vaccination. In two animals, a soft and circumscribed swelling at the site of injection was detected.

Standard haematological parameters at days A and C were within normal range for the species (Fowler, 1998). Comparing values at day A with day C, there was a significant decrease in haemoglobin and white blood cell, with a relative decrease in neutrophils and eosinophils, whereas basophils were relatively increased (Table 2).

In measuring the innate immune response, values obtained from the affected group for lysozime and serum bactericidial activity were lower than those from the control group. More specifically, lysozime was significantly lower on day C, whereas serum bactericidial activity was significantly lower throughout the sampling period. No significant variations were recorded for the semiquantitative complement titration between the affected group and the control group. Table 3 summarise these results.

Serum electrophoresis findings are reported in Table 4. Over the sampling period, values for total protein, $\alpha 2$ and β globulins were significantly higher on day B than on days C; whereas significant higher values on day B *vs.* day A were also registered for total protein and γ globulins. No significant differences were noticed for the fractions expressed as percentages. Specific alpaca IgG assessment showed a gradual increase in values throughout the sampling period, albeit not statistically significant (Table 4).

Table 2. Hematological parameters of the affected group.

Parameters	Units	Day A	Day C	P<
RBC	$\times 10^6\ \mu l$	12.6 ± 2.4	12.9 ± 1.9	n.s.*
Hb	g/dl	10.7 ± 1.8	10.1 ± 1.3	0.05
Hct	μl %	29.4 ± 5.3	29.8 ± 4.0	n.s.
WBC	$\times 10^3\ \mu l$	18.2 ± 6.6	15.1 ± 4.4	0.05
Lymphocytes	$\times 10^3\ \mu l$	3.6 ± 1.4	3.1 ± 1.2	n.s.
Monocytes	$\times 10^3\ \mu l$	1.0 ± 0.7	1.0 ± 0.6	n.s.
Neutrophils	$\times 10^3\ \mu l$	10.3 ± 3.9	8.2 ± 2.7	0.001
Basophils	$\times 10^3\ \mu l$	0 ± 0	0.14 ± 0.02	0.001
Eosinophils	$\times 10^3\ \mu l$	3.4 ± 1.9	2.6 ± 1.4	0.05
Lymphocytes	%	19.7 ± 4.6	20.9 ± 4.9	n.s.
Monocytes	%	5.5 ± 3.0	6.5 ± 3.7	n.s.
Neutrophils	%	56.1 ± 8.5	54.6 ± 9.6	n.s.
Basophils	%	0 ± 0	0.9 ± 1.4	0.001
Eosinophils	%	18.7 ± 7.9	17.0 ± 5.9	n.s.

*n.s. = not significant

Table 3.Values of the innate immune response in affected and control group. Different letters (a, b) mean significant differences (P<0.05).

	Complement titration CH50	lysozime µg/m	% bactericidial activity
Day A	20 ± 10.7[a]	1.32 ± 0.7[b]	69.2 ±39[a]
Day B	20.8 ± 6.9[a]	1.29 ± 0.6[b]	83.9 ± 25[a]
Day C	22.0 ± 10.5[a]	0.99 ± 0.5[a]	75.6 ± 29[a]
Control	19.8 ± 5.1[a]	1.58 ± 0.6[b]	98.6 ±1.9[b]
P<	n.s.	0.05	0.05

Table 4. Results of serum electrophoresis in the affected group during the trial period. Different letters (a, b) mean significant differences (P<0.05).

	Total protein mg/dl	albumin mg/dl	α globulins mg/dl	α1 globulins mg/dl	α2 globulins mg/dl
Day A	7.1 ± 0.7[a]	4.1 ± 0.4	0.4 ± 0.1	0.1 ± 0.03	0.33 ± 0.1[a]
Day B	7.4 ± 0.5[b]	4.1 ± 0.4	0.4 ± 0.1	0.1 ± 0.02	0.33 ± 0.09[a]
Day C	7.2 ± 0.7[a]	4.1 ± 0.4	0.4 ± 0.1	0.1 ± 0.06	0.28 ± 0.1[b]
P<	0.05	n.s.	n.s.	n.s.	0.05

	β globulins mg/dl	β1 globulins mg/dl	β2 globulins mg/dl	γ globulins mg/dl	Alpaca IgG mg/dl
Day A	1.4 ± 0.3[b]	0.8 ± 0.3	0.6 ± 0.2	1.1 ± 0.3[a]	3,477 ± 501
Day B	1.5 ± 0.3[b]	0.9 ± 0.3	0.6 ± 0.2	1.3 ± 0.4[b]	3,513 ± 566
Day C	1.3 ± 0.3[a]	0.8 ± 0.3	0.5 ± 0.2	1.2 ± 0.4[ab]	3,900 ± 1,024
P<	0.05	n.s.	n.s.	0.05	n.s.

Discussion

The outbreak of CLA described in the present paper appear more similar in epidemiology, clinical signs and pathological aspects to those described in sheep (Gilmour, 1991) than to the few described in alpacas (Anderson, 2003). In sheep, the most effective way for controlling and attempt the eradication of the disease from an infected flock includes culling of the affected animals. However, given the high value of the individual alpaca outside of South America, this practice appears unsuitable under the present circumstances.

Vaccination against *C. pseudotuberculosis* is a main component of any protocol aimed at stopping the spread of the disease among an infected flock (Paton *et al.*, 1995). While some vaccines are commercially available for sheep, none are currently specifically formulated for alpacas, thus the need for an autogenous vaccine. The lack of data regarding the safety and efficacy of such vaccine in alpacas affected by CLA has prompted the present study.

The reduction in WBC, neutrophils, eosinophils and α2 globulins, as observed on day C of the study, can be interpreted as a change in the inflammatory status affecting the group. Since a number of alpacas were concomitantly affected by chronic chorioptic mange, and treatment for the mange was given in the interval between the first and the second vaccine administration, amelioration in the skin condition might have contributed to the above changes in haematological parameters.

Instead, the significant changes in total proteins (day B vs. days A and C), β (day B vs. C) and γ globulins (day A vs. B), together with the variation in specific IgG, could be explained by the vaccine`s effect. In fact, even if on the one hand the reduction in antigenic stimulus given by the chorioptic mite might have caused a relative reduction in IgE levels (part of the γ globulins fraction), on the other hand the increase in total proteins and β and γ globulins recorded on day B could be linked with and increased production of IgM (part of the β and γ globulins) induced by the vaccine. No significant increase on day C was recorded for γ globulins by electrophoresis, while the alpaca IgG radial immunodiffusion has shown values progressively higher throughout the study. The shift of immunoglobulins toward the mammary gland associated with the late stage of pregnancy, also considering the predominance of females over males in the affected group, could explain this otherwise contrasting result.

Lysozyme is a strong antibacterial enzyme carrying out a synergic action with immunitary humoral response and factors of the complements. Its determination can provide information concerning the intensity of non-specific immune reaction (phagocytosis) against infection (Degorski & Lechowski, 1982). The serum bactericidal assay in micromethod too is a major parameter for evaluation of the non- specific immunitary system's activity. The capacity of the serum to inhibit bacteria growth is determined by the presence of complement factors and moderate concentration of natural antibodies directed towards more diffused enviromental bacterial agents, mainly enterobacteriacee.

In the present study, lysozime and serum bactericidal values in the affected group have been consistently lower than in the control group. Phagocytes reduction, and even more a prolonged "consumption" of these factors could explain these findings.

The haemolytic complement assay gives indications on the defense mechanisms of the animals that involve the activation of the complement system. A test that highlights possible complement deficiencies is of great help in assessing the risk of infection or the severity of pathologies already taking place (subacute/chronic inflammatory disease).

No variations in complement activity in the affected group have been recorded at this stage of the study, with the values observed similar to those in the control group. In this respect alpacas appear to behave similarly to other ruminants species, where the haemolytic complement parameter is the last parameter to change following an antigenic stimulus.

Conclusion

Three and a half months after the beginning of the vaccination protocol no more deaths have been recorded in the affected herd. However, four more alpacas have shown signs of CLA, with lymphoadenopathy and abscessation of superficial lymph nodes, particularly the retro-mandibular. One additional vaccine administration, and monitoring of the immune response has to be carried out in order to complete the study and fully assess the immune and clinical response to the autogenous vaccine.

In this type of farming enterprise (high value of the individual animal), and under the circumstances described in this paper, vaccination and early detection of infected animals are fundamental to stop the spreading of the disease within the herd. Efforts have been put into the development of an in-house ELISA test for early diagnosis of CLA in alpacas.

The described outbreak of CLA in alpacas appears not only characterised by high morbidity, but by high mortality too. Considering that the incubation period for CLA can be longer than the standard quarantine period, serological screening for CLA should be considered for all new in-coming animals.

Acknowledgements

The authors gratefully acknowledge Prof. A. Valbonesi and Dott. M. Antonini, and the technical support of Ms M. Timi and Ms E. Serri.

References

Amadori, M., I.L. Archetti, M. Frassinelli, M. Bagni, E. Olzi, G. Caronna and M. Lanterni, 1997. An Immunological Approach to the Evaluation of Welfare in Holstein Frisian Cattle. Journal of Veterinary Medicine B 44: 321-327.

Anderson, D.E., 2003. Periapical tooth roots infection in South American camelids. In: Proceedings of tenth meeting of the British Veterinary Camelid Society, Wallingford, Oxfordshire, 24-26 October, pp.1-10.

Barta, V. and O. Barta, 1993. Testing of Hemolitic Complement and its components. In: O. Barta (ed.) Veterinary Clinical Immunology and Laboratory, Bar-Lab, Blacksburg, USA.

Cameron, C.M. and F.J., Bester, 1984 An improved *Corynebacterium pseudotuberculosis* vaccine for sheep. Onderstepoort Journal Veterinary Research 51: 263-267.

Degorski, A. and R., Lechowski, 1982. Relationship of serum lysozime activity to peripheral leucocyte count in calves. Journal of the Veterinary Medicine B 29: 320-323.

Eggleton, D.G., H.D., Middleton, C.V., Doidge and D.W., Minty, 1991. Immunisation against ovine caseous lymphadenitis: comparison of Corynebacterium pseudotuberculosis vaccines with and without bacterial cells. Australian Veterinary Journal, Oct; 68 (10): 317-9.

Fowler, M.E., 1998. Medicine and surgery of South American camelids: llama, alpaca, vicuña and guanaco. (2nd ed.) Ames, Iowa State University Press.

Gezon, H.M., H.D. Bither, L.A. Hanson and J.K., Thompson, 1991. Epizootic of external and internal abscess in a large goat herd over a 16 year period. J. Am. Vet Med Assoc. 198: 257-263.

Gilmour, N.J.L., 1991. Caseous lymphadenitis. In: Disease of sheep. W.B. Martin and I.D. Aitken (eds.), Blackwell Scientific Publications, Oxford, pp. 58-65.

Kutschke, L., M. Ganter and J. Kaba, 2000. Efficacy of a flock-specific pseudotuberculosis vaccine in goats. Dtsch Tierarztl Wochenschr. 107: 495-500.

Nolte, O., J. Morscher, H.E. Wiss and H.G. Sonntag, 2001. Autovaccination of dairy cows to treat post partum metritis caused by *Actinomyces pyogenes*. Vaccine 19: 3146-3153.

Osserman, E.F. and D.P. Lawlor, 1966. Serum and urinary lysozyme (muramidase) in monocytic and monomyelocyticleukemia. Journal Exp. Med. 124: 921- 952.

Paton, M.W., S.S. Sutherland, I.R. Rose, R.A. Hart, A.R. Mercy and T.M., Ellis, 1995. The spread of *Corynebacterium pseudotuberculosis* infection to unvaccinated and vaccinated sheep. Aust Vet J. 72: 266-269.

Stoops, S.G., H.W., Renshaw, and J.P., Thilsted, 1984. Ovine caseous lymphadenitis: disease prevalence, lesion distribution, and thoracic manifestations in a population of mature culled sheep from western United State. Amer J. Vet Res. 45: 557-561.

Seyfarth, M., 1976. Virologische Arbeits-Methoden. Fisher, Jena, pp. 145 148.

Wernery, U. and O.R., Kaaden, 2002. Infectious diseases in camelids. Blackwell Wiss.-Verl., Berlin and Vienna, pp. 134-138.

Course of gastro-intestinal parasite and lungworm infections in South American camelids on a farm in central Germany

Simone Rohbeck[1], M. Gauly[2] and C. Bauer[1]
[1]*Institute of Parasitology, University of Giessen, Germany.*
[2]*Institute of Animal Breeding and Genetics, University of Goettingen, Albrecht-Thaer Weg 3, 37075 Goettingen, Germany; mgauly@gwdg.de*

Abstract

The *Eimeria* fauna was in agreement with other studies. Even if no clinical signs of coccidiosis have been shown during the study the high prevalences indicating the potential importance of this parasites. However prevalence and oocyst counts are influenced by the age of the animals and probably the season. *Trichostrongylus/Ostertagia spp.* and *H. contortus* were the dominant species. The prevalences were high, but intensities were low. It has to be taken into account, that no informations are available for SAC between the correlation of FEC and worm burden. A low prevalence of *Moniezia spp.* and no liverflucks were found, whereas *Trichuris* and *D. viviparus* infections were present in the herd. This infections may cause serious problems as can be seen in other species.

Keywords: gastro-intestinal parasites; lung worm infection

Introduction

The domesticated South American camelids have become increasingly popular outside South America were llamas and alpacas are mainly kept as pets or packers. In Germany between 10,000 to 15,000 (*Lama glama, L. pacos*) animals are kept mostly in small herds.

Not much is known about the importance and prevalence of parasite infections under European conditions. Main problems caused by parasite infections and treatments are economical losses (reduced productivity, drug costs, animal losses), development of drug resistance and animal welfare problems because animals may suffer from pain and may even die.

The aim of the study was the evaluate the incidence and prevalence of helminth and protozoan infections in a herd of llamas (*L. glamas*) and alpacas (*L. pacos*) in central Germany. Therefore a longitudinal study was performed to obtain data on the saisonal and age-depending course of infections with gastro-intestinal parasites and lungworms in llamas and alpacas.

Material and methods

120 llamas and alpacas were included in the study. All animals were kept outdoor on pasture during the study on one single farm. Faecal samples were taken individually from all animals. Crias were sampled until an age of six weeks once per week and adults once per month, respectively. One anthelmintic treatment was given during the study.

Individual faecal egg counts (FEC) and oocyst counts (opg) were carried out using the modified McMaster method with saturated sugar solution ($\delta = 1.3$) as the flotation fluid. Faecal samples were pooled for culture to produce larvae which were then identified.

The *Eimeria* species were differentiated by morphological signs. Baermann technique was used for separating *Dictyocaulus ssp.* larvae. Faecal slides with Carbolfuchsin stain were prepared to identify *Cryptosporidum parvum*. SAF (sodium acetate formaldehyde) concentration technique was used for the detection of *Giardia ssp.*

SPSS 10.0 (Statistical Package for the Social Sciences) for Windows Release 6.3.1. was used for data analyses.

Results

Protozoal infections

The following protozoan parasites were detected during the study: *Giardia spp.* and *Eimeria punoensis, E. alpacae, E. lamae, E. macusaniensis* and *E. ivitaensis.* The prevalences of *Eimeria* oocysts in crias and adults are shown in Figure 1 and Figure 2.

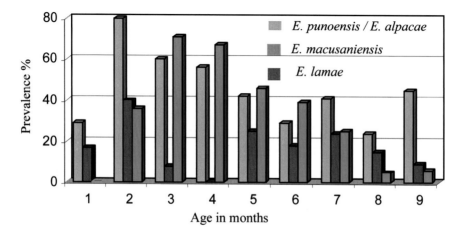

Figure 1. The prevalence of Eimeria *oocysts in crias.*

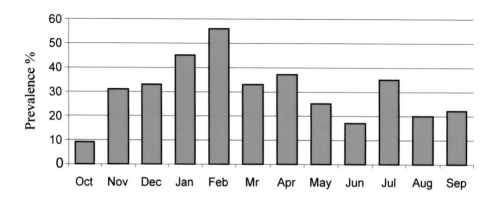

Figure 2. The prevalence of Eimeria *oocysts in adults.*

Both the highest prevalence and intensity of total *Eimeria* oocyst shedding were observed, independent from the season, during the first months of life and were significantly higher in crias (mean maximum: 16,000 oocysts per gram faeces, OPG) than in older animals (600 OPG). First, oocysts of *E. alpacae*, *E. punoensis* and *E. lamae* were detected in the faeces; shedding of *E. macusaniensis* oocysts started in the 2nd month of life. *E. ivitaensis* oocysts were sporadically found. The intensity of oocyst shedding is shown in Figure 3.

Nematode infection in the flock with *Moniezia spp.*, *Strongyloides* and the gastrointestinal helminths *Nematodirus battus, N. filicollis, Capillaria spp., Trichuris spp.* and *Dictyocaulus viviparus* were detected. The highest prevalence of trichostrongyle egg shedding (*Haemonchus, Ostertagia, Trichostrongylus, Cooperia*) was observed in late summer. Crias and yearlings showed higher trichostrongyle egg counts (maximum 450 eggs per gram faeces, EPG), on average, than adult animals (maximum 150 EPG). Eggs of *Nematodirus spp.* (*N. battus* and others), *Strongyloides, Trichuris, Capillaria* and *Moniezia* were repeatedly detected, mainly in the faeces of crias and yearlings (Figure 4). *Trichuris* egg counts were sporadically very high (maximum 2000 EPG) in some animals suggesting a possible risk for clinical disease (Figure 5).

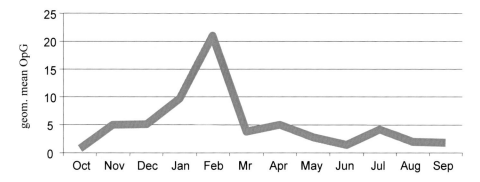

Figure 3. Intensity of oocyst shedding (OpG) in adults.

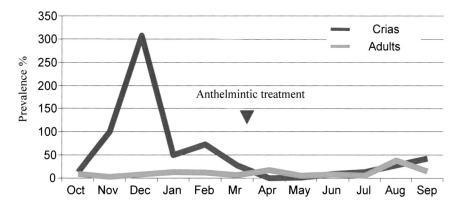

Figure 4. Intensity of egg shedding (EpG) from gastrointestinal helminths (except Nematodirus spp.)

Part III

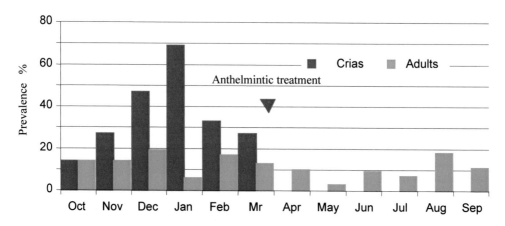

Figure 5. Prevalence of Trichuris.

Patent infections with the bovine lungworm (*Dictyocaulus viviparus*) were found in dams and yearlings, and a few animals showed slight signs of respiratory disease. Maximum shedding of lungworm larvae (14% in dams, 47% in yearlings) was observed during the winter months. There were hints that several animals imported from the Netherland had been the source of the lungworm infection on this farm. After an anthemintic treatment the prevalence decreased to zero. In adults the prevalence was between 5 and 18% from November to August. No treatment was given.

Trichostrongylus/Ostertagia spp. were the dominant species followed by *H. contortus* and *Cooperia* (Figure 6).

Acknowledgements

The authors want to thank the breeders Lydia and Konrad Kraft, Walter Egen and Ursula Brinkmann and the German Breeders association (Verein der Züchter, Halter und Freunde von Neuweltkameliden e.V.) for technical and financial support.

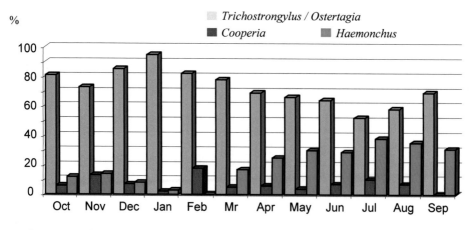

Figure 6. Gastrointestinal helminths: relation of different species.

Evaluation of laboratory results of camelids, made for import, export or participation in shows

I. Gunsser[1], A. Aigner[2] and C. Kiesling[1]
[1]*Römerstr. 23, 80801 München, Germany; ilona.gunsser@t-online.de*
[2]*Am Hagen 6b, 83339 Chieming, Germany.*

Abstract

Although camelids belong to the separate suborder *Tylopoda* within the order *Artiodactyla*, they are quite frequently misclassified as ruminants. There are many differences between camelids and ruminants. One is, for example, that they are quite resistant to many ruminant diseases, regulated by governmental laws. No special tests exist to diagnose ruminant diseases in camelids. For this reason tests are used which are usually applied for other species. In this study, altogether 9,391 tests of 2,546 llamas and alpacas were evaluated. The laboratories which provided these results were in Europe (Switzerland, Germany, France, Italy, Sweden, and Finland), the United States, Chile and Peru. This investigation concentrates on different laboratory tests for Brucellosis (2,535), Leucosis (1,499), BHV1 (1,619) and Tuberculosis (914). In addition, results on tests for other ruminant diseases in smaller amounts of animals are presented. The results of this study were negative in the majority of the diseases tested. Some of the tests were false-positive, for example 0.9% of the Brucellosis tests. Among the diseases which were tested in a smaller number of animals, some were false positive (*e.g.* BVD), and some positive tests for Leptospirosis were found.

Keywords: camelids, blood tests, government-regulated ruminant diseases

Introduction

Over the last ten years llamas and alpacas have become more and more popular in Europe. For this reason camelids are being imported in large numbers from countries outside of Europe, or are relocated within the EU-region. Although camelids belong to the separate suborder *Tylopoda* ("padded foot") within the order *Artiodactyla* (Zeuner, 1963; Wheeler. 1995), they are quite frequently misclassified as ruminants. Until today there is a protocol to test for ruminant diseases in llamas and alpacas before they are allowed to pass the EU-borders.

Camelid evolution began in North America in the early Eocene epoch (Webb, 1974) some 40-50 million years ago. Separation of the *Tylopoda* and *Ruminatia* occurred early in the evolutionary process when the progenitors of both groups were still small animals with simple stomachs. Foregut fermentation, complex multi-compartmentalized stomachs, food regurgitation and re-chewing are not limited to ruminants but are found also in other species as diverse as kangaroos and non-human primates (Hume, 1982). Foregut fermentation and multi-compartment stomachs are also seen in many species like hippopotamus, colobus monkey and peccary (Feldhammer, 1999).

There are many differences between camelids and ruminants, especially considering the blood, foot, digestive system, reproduction system and urinary system (Wernery *et al.*, 1999). Here are some more detailed examples: Camelids have no gallbladder, and the anatomy and physiology of the female reproduction system shows a placenta diffusa and an induced ovulation. They copulate in prone position. The red blood cells are elliptical and small, and they are quite resistant to many government-regulated ruminant diseases (Fowler, 2004). No special tests exist to diagnose ruminant diseases in camelids. For that reason tests are used which are usually applied for other species.

Material and methods

For this evaluation, owners of llamas and alpacas were asked to send in blood test results of their camelids which were made between 1994 and 2004. The reason for the laboratory tests were imports and exports or participation at shows. The laboratory tests which were provided were made in Austria, Chile, Germany, Finland, France, Italy, Peru, Sweden, Switzerland and USA. Altogether 9,391 test results were collected based on blood probes taken from 2,546 llamas and alpacas.

Results

Distribution of tests among different countries

The evaluated tests were made in 10 different countries. Most blood tests were made in Switzerland (1,391 tests). The reason for this is that, due to import restrictions into other countries, most of the imported animals from South America and North America went to a quarantine station in Switzerland. The tests made in the other countries were made to export animals to different countries in Europe or for participation at shows. The numbers of tests made in the different countries are shown in Table 1.

Bovine diseases tested in camelids

Tests were made for 15 different ruminant diseases. Most blood tests (> 1,500) were made for Brucellosis, BHV1 and Leucosis, followed by Tuberculosis, FMD, and Blue tongue (> 500). Further test were made of BVD, Border disease, Jones disease (paratuberculosis), Salmonellosis, Vesicular stomatitis, Leptospirosis, *Trypanosomum vivax*, *Trypanosomum evansi* and *C. Burnetii*. The tested ruminant diseases are shown in Table 2.

Test techniques performed

The test used most frequently was the ELISA test. For Brucellosis, also the RBT test was applied. Table 3 shows all the tests, as declared on the evaluated laboratory sheets, which were applied.

Results of tested bovine diseases

All test results were negative, with 3 exceptions: In Brucellosis, 23 tests were false-positive (0.9%), and in the results of the BVD tests there were 7 false-positive (9.3%). Only among the Leptopiroses tests which were made in 1996-1998 there were 13 tests positive (5.2%). None of the Leptospirosis tests were repeated. Table 4 shows the statistics on the diseases which led to non-negative test results.

Table 1. Numbers of tests made in different countries. The question mark indicates that the type of South American camelid was not marked and is therefore unknown.

Country	D	A	CH	F	I	S	Fin	Peru	Chile	USA	Sum
Llamas	70	51	0	77	0	0	0	0	0	0	198
Alpacas	133	5	1,032	30	0	14	65	150	161	196	1,786
A/L	0	0	359	26	0	0	4	0	0	0	389
?	0	0	0	38	135	0	0	0	0	0	173
Sum	203	56	1,391	171	135	14	69	150	161	196	2,546

Table 2. Ruminant diseases tested in camelids. The question mark indicates that the type of South American camelid was not marked and is therefore unknown.

	Brucel.	Leuko.	BHVI	BVD	Border	M. para.	TB	FMD	Salmo.	Blue t.	Ves. St.	Lepto.	T. viv.	T. ev.	C. bur.
Llamas	241	46	115	0	2	0	94	3	0	0	0	5	0	0	0
Alpacas	1,956	1,239	1,270	21	18	42	812	555	52	507	311	210	311	150	368
A/L	230	183	187	0	0	0	8	180	0	0	0	0	0	0	0
?	108	31	47	54	0	0	0	0	0	0	0	35	0	0	0
Sum	2,535	1,499	1,619	75	20	42	914	738	52	507	311	250	311	150	368

Brucel. = Brucellosis; Leuko. = Leukosis; Border = Border disease; M. para. = Mycobacterium paratuberculosis; Salmo = Salmonellosis; Blue t. = Bluetongue; Ves. St. = Vesicular Stomatitis; Lepto. = Leptospirosis; T. viv. = Trypanosoma vivax; T. ev. = Trypanosoma evansi; C. bur. = Coxiella burnetii.

Table 3. Tests used for diagnosing the various ruminant diseases. The question mark indicates that the test used was not marked and is thus unknown.

Brucellosis		KBR	Elisa	RBT	SLA	?
	Llamas	89	0	101	3	48
	Alpacas	151	711	523	40	531
	A/L	15	185	26	0	4
	?	18	12	78	0	0
	Sum	273	908	728	43	583

Leukosis		ID	Elisa	?		
	Llamas	25	11	10		
	Alpacas	435	188	616		
	A/L	0	183	0		
	?	0	31	0		
	Sum	460	413	626		

BHVI		Elisa	gBElisa	SNT	full V	?
	Llamas	32	24	3	0	56
	Alpacas	271	26	199	3	771
	A/L	180	0	0	0	7
	?	47	0	0	0	0
	Sum	530	50	202	3	834

BVD		BVD Antigen	Elisa	PCR	Cell cult	?
	Llamas	0	0	0	0	0
	Alpacas	3	3	0	1	14
	A/L	0	0	0	0	0
	?	0	47	7	0	0
	Sum	3	50	7	1	14

Table 3. Continued.

Border.		Antibody	NPLA	
	Llamas	2	0	
	Alpacas	15	3	
	A/L	0	0	
	?	0	0	
	Sum	17	3	
M. para.		Elisa	Ziehl	?
	Llamas	0	0	0
	Alpacas	3	1	38
	A/L	0	0	0
	?	0	0	0
	Sum	3	1	38
TBC		M. avium	M. bovis	?
	Llamas	2	25	67
	Alpacas	50	50	712
	A/L	4	4	0
	?	0	0	0
	Sum	56	79	779
FMD		SNT	?	
	Llamas	3	0	
	Alpacas	5	550	
	A/L	0	180	
	?	0	0	
	Sum	8	730	
Salmo.		?		
	Llamas	0		
	Alpacas	52		
	A/L	0		
	?	0		
	Sum	52		
Blue T.		C-Elisa	?	
	Llamas	0	0	
	Alpacas	196	311	
	A/L	0	0	
	?	0	0	
	Sum	196	311	

Table 3. Continued.

Lepto.		ALT	
	Llamas	0	
	Alpacas	100	
	A/L	0	
	?	0	
	Sum	100	
T. viv.		Elisa	?
	Llamas	0	0
	Alpacas	150	161
	A/L	0	0
	?	0	0
	Sum	150	161
T. ev.		Elisa	
	Llamas	0	
	Alpacas	150	
	A/L	0	
	?	0	
	Sum	150	
Ves. St.		?	
	Llamas	0	
	Alpacas	311	
	A/L	0	
	?	0	
	Sum	311	
C. bur.		Elisa	KBR
	Llamas	0	0
	Alpacas	367	1
	A/L	0	0
	?	0	0
	Sum	367	1

Discussion

In the evaluated material almost all tests yielded negative results. Only the tests for Leptospirosis, which were made in Switzerland in 1996 and 1998 using the test ALT, showed some positive results. These results, however, were not verified by a second test. False-positive results were found in the tests for Brucellosis with the Elisa test in a total of 23 from 2,535 blood samples. Further false-positive results from tests for BVD with Elisa were found in 7 of 75 blood samples.

Table 4. False-positive (Brucellosis, BVD) and positive (Leptospirosis) tests.

	Brucel	BVD	Lepto
Llamas	1	0	0
Alpacas	12	0	13
A/L	10	0	0
?	0	7	0
Sum	23	7	13
percent	0.9	9.3	5.2

In these cases the tests were repeated (Brucellosis: Elisa, BVD: PCR), but with a negative result. One has to keep in mind that no specific tests exist with validation for camelids.

Conclusion

Blood tests for ruminant diseases in camelids do not exist. The blood tests which were carried out with tests, validated for other species, were all negative. There was one exception, namely Leptospirosis. The corresponding tests were not repeated. This investigation (2,535 blood tests for Brucellosis, 1,499 blood tests for Leucosis and 1,619 blood tests for BHV1) yields a very small probability to find an infection of these ruminant diseases in camelids, amounting to 0.04% for Brucellose, 0.07% for Leukosis and 0.06% for BHV1.

References

Feldhammer, G.A., L.C. Dickamer, S.H.Vessey and J.F. Merritt, 1999. Mammalogy, Adaption, Diversity and Ecology. Boston, McGraw-Hill.

Fowler, M.E., 2004. Regulated Ruminant Diseases – Where do camelids fit in? Ohio State University camelid conference.

Hume, I.D., 1982. Digestive Physiology and Nutrition of Marsupials. Cambridge, Cambridge University Press, pp. 112–119.

Webb, S.D., 1974. Pleistocene Mammals of Florida. Gainesville, University of Florida Press.

Wernery, U., M.E. Fowler and R. Wernery, 1999. Color Atlas of Camelid Haematology. Berlin, Blackwell Science.

Wheeler, J.C., 1995. Evolution and present situation of the South American Camelidae, Biological J. Of the Linnean Society 54:271-295.

Zeuner, F.E., 1963. New world species. In: A History of Domestic Animals. New York, Harper and Row Publishers, Chapter 21.

Case reports in South American camelids in Germany

R. Kobera[1] and D. Pöhle[2]
[1]Tierarztpraxis in Possendorf, Kreischaer Str. 2a, 01728 Bannewitz, Germany; ralph.kobera@ freenet.de
[2]Landesuntersuchungsanstalt Sachsen, Jägerstr.10, 01099 Dresden, Germany.

Three different kinds of case studies in alpacas are presented.

Firstly the aetiopathology of an acute kidney failure in a 6 months old female alpaca, caused by a systematic yeast infection is described. The animal had a major disorder to urinate. The urine analysis showed high contents of *Candida albicans*. Although the animal could urinate again during the treatment, it died after two weeks. As the pathological and histological analysis revealed, there was a high inflamation, caused by a endomycosis, especially in the kidney, pancreas, spleen and myocardium tissue as well as in the cerebric.

The second case describes the application of a herd specific vaccine against pseudotuberculosis in an alpaca herd of 8 animals. *Corynebacterium pseudotuberculosis* was isolated during the bacteriological examination of the lymph knots. From this strain a herd specific vaccine was made. During the vaccination period the animals with increased lymph knots showed distinct visual signs of improvement. After 2 animals died at the beginning of the vaccination period, no further deaths occurred.

The third case describes the clinical aetiopathology of a 3 year old male alpaca infected by the borna virus. While the animal showed distinct signs of lameness on the hind limbs, the blood parameters had almost normal values. X-rays suggested hypoplastic cristas of the knees. Since during the course of the disease the animal showed increasing signs of ataxy it had to be put down. The pathohistological examination of the brain revealed the existence of a highly, non-sanious Encephalitis, which was characterised by distinct lymphocytic and perivascular infiltrates. PCR and immune histology confirmed the occurrence of the Borna virus. The examination of the three remaining camelids of the herd for Borna virus antibodies produced negative results.

Part IV
Nutrition

DECAMA-project: determination of the lactation curve and evaluation of the main chemical components of the milk of llamas (*Lama glama*)

C. Pacheco and A. Soza
Centro de Estudios y Promoción del Desarrollo (DESCO), Malaga Grenet 678, Umacollo, Arequipa, Peru; arequipa@descosur.org.pe

Abstract

The objectives of the present work, with respect to llamas (*Lama glama*), were: to characterise the lactation curve, to determine values for the main components of milk and to evaluate the growth rate of their cria offspring during the lactation period measured. Six mature llamas entering their second lactation and their progeny were maintained under conditions of traditional high Andean pasture management. The volume of milk produced and the percentages of total solids, protein, fat, ash and whey protein, in addition to the growth rate of cria were evaluated. Milk samples were taken on days 1, 7, 30, 60, 90, and 120 post partum. The taking of the samples was carried out between 12 am and 2 pm after fixing a muzzle to the cria in order to prevent suckling. Milking was carried out manually by placing the animal in a lateral position. The weights of the cria were recorded between 6 and 7 hours after the extraction of the milk. The results obtained up to day 120 post partum show that on the average, milk volume reached its peak by day 60 (207.33 ml) and then steadily declined. The percentages of total solids, protein, ash and whey protein decreased from day 1 to day 7 and showed limited variation thereafter. In contrast, the percentage of fat increased up to day 7 and decreased by day 60 to values which were then essentially maintained. The average body weights of the cria increased from 10.75 kg at birth to 35 kg at 120 days post partum.

Resumen

Los objetivos del presente trabajo fueron analizar la curva de lactación, conocer las características de los principales componentes de la leche de llama, (lama glama) y evaluar el incremento de peso de las crías durante el periodo de lactancia. El volumen de leche, los porcentajes de sólidos totales, proteínas totales, grasa, cenizas y proteínas del lacto suero, así como el incremento de peso en crías fueron evaluadas en 6 llamas adultas, de segundo parto, segunda lactación, y sus crías correspondientes mantenidas bajo condiciones de pastoreo tradicional alto andino. Los días 1, 7, 30, 60, 90 y 120 post parto se tomaron muestras de leche. La toma de muestras se realizó entre 12 y 14 horas después de colocar a la cría un bozal para evitar la succión (ingesta de leche). El ordeño fue realizado manualmente con el animal de cubito lateral. El peso de las crías fue registrado entre las 6 y 7 horas antes de que estas ingieran leche. Los resultados obtenidos hasta el día 120 post parto muestran que en promedio, el volumen de leche logra su pico mas alto el día 60 (207.33 ml), disminuyendo en forma sostenida hasta el día 120. Los porcentajes de sólidos totales, proteínas, cenizas, y proteínas del lacto suero disminuyen al transcurrir los días de lactación, observándose que la variación de estos a partir del día 7 es mínima; lo que no ocurre con el porcentaje de grasa el cual se incrementa hasta el día 7 post parto. El peso de las crías se incrementa en forma sostenida, llegando el día 120 a un 225% del peso inicial.

Keywords: llama, milk, lactation curve, chemical composition of milk

Introduction

The most substantial advances in knowledge of lactational physiology and chemical composition of milk have been made in cattle, goats and laboratory rodents. The more limited scientific information describing lactation in camelids has largely been derived from old world camels. The capacity to produce an adequate milk yield with the required quality of its components. are both factors of vital importance for optimal post-natal development in mammals. The aims of the present study were to characterise the lactation curve and to evaluate the concentration of the main constituents in milk during lactation in llamas.

Materials and methods

Samples of milk were drawn from animals belonging to the Centre of Alpaca Development of Toccra (Centro de Desarrollo Alpaquero of Toccra, CEDAT), located at an average altitude of 4,400 metres in Toccra, District of Yanque, Province of Caylloma, Department of Arequipa, Peru. This farm has areas of pampa and eroded slopes of low productivity, in addition to *bofedales* (wetland), pasture of A and B class quality and *tolares*. The habitat corresponds to the Dry Puna, with an annual rainfall of less than 400 mm.

Six second lactation female llamas maintained under traditional high Andean husbandry conditions were used in the study. Milking was carried out manually on days 1, 7, 30, 60, 90 and 120 post partum between 12 am and 2 pm following a previously established system. The total milk volume of 12 h was collected. The cria were fitted, prior to milking, with a muzzle made from llama fibre which prevented them from suckling. The body weights of the cria were recorded between 6 and 7 hours after the extraction of the milk

For the milk extraction, the animals were placed in a lateral position on a blanket with the posterior limbs extended. While one operator held the neck of the animal on the floor, another fastened the posterior limbs, so that two teats on one side were accessible for manual milking. Subsequently, the animal was turned onto the other side, and a similar procedure applied to permit access to the other two teats. Manual milking was made by gently massaging the gland from the base to the tip and pressing the teats with the fingers.

The milk samples were refrigerated and transferred to the laboratory within 12 hours for further analysis for the following components:
* Fat percentage by Gerber's method.
* Protein percentage by Macro Kjeldahl method.
* Whey proteins in the serum fraction (following removal of caseins by precipitation with glacial acetic acid), by Macro Kjeldahl method.
* Percentages of ash and total solids by gravimetric methods.

Results

Changes over the lactation period, in the chemical components of milk, the volume collected and the increases in body weight of the cria are summarized in Tables 1 and 2 respectively.

Analysis of the lactation curve showed that the highest milk production was observed on 60 day (average of 207.33 ml), with progressive reductions thereafter to give the lowest value on the last day of recording at 120 days (90.50 ml). There was a steady increase in weight gain of the crias throughout the entire lactation period, reaching a body weight of 35 kg by day 120 post partum.

Table 1. Llama milk constituents (means ± SEM) across the lactation period (n=6).

| | Days pp | | | | | |
Traits	1	7	30	60	90	120
Total solids (%)	20.41 ± 2.71	17.74 ± 2.34	15.53 ± 1.72	14.48 ± 1.13	14.44 ± 1.24	14.34 ± 1.66
Ash (%)	1.18 ± 0.42	0.83 ± 0.05	0.78 ± 0.05	0.76 ± 0.03	0.77 ± 0.05	0.76 ± 0.05
Fat (%)	2.26 ± 1.70	4.88 ± 1.50	4.06 ± 1.38	3.47 ± 0.64	3.65 ± 0.62	3.43 ± 0.72
Total Protein (%)	11.21 ± 3.16	5.93 ± 0.49	4.68 ± 0.69	4.67 ± 1.11	4.20 ± 0.27	4.25 ± 0.6.3
Whey protein (%)	1.97 ± 0.46	1.14 ± 0.28	0.95 ± 0.35	0.97 ± 0.21	1.28 ± 0.47	0.57 ± 0.29

Table 2. Body weight of crias and volume of milk collected (means ± SEM) (n=6).

| | Days pp | | | | | |
Traits	1	7	30	60	90	120
Body weight (kg)	10.75 ± 1.29	13.58 ± 1.28	18.00 ± 2.34	24.83 ± 3.06	29.00 ± 2.83	35.00 ± 1.83
Milk volume (ml)	101.67 ± 37.24	167.83 ± 41.64	190.33 ± 111.19	207.33 ± 115.42	126.00 ± 105.41	90.50 ± 61.02

With regard to chemical components, the highest percentages of total solids, total proteins, whey protein and ash, and the smallest percentage of fat, which are typically expected of colostrum (Ramirez et al., 1981; Bustinza, 2001), were found on day 1 of sampling.

Following changes in composition to normal milk production by day 7 post partum, the percentages of total solids, fat, total protein, whey protein and ash tended to decrease until day 30 or 60. The values then remained relatively stable with minimal variations in the following days of sampling.

Discussion

In considering the morphology of the mammary gland, our observations are in agreement with the descriptions provided by Arpi (1994) and Bustinza (2001). These authors describe the mammary gland in the alpaca and the llama as being relatively small and heart-shaped in its location in the inguinal zone. The anterior part is slightly wider that the posterior part. The skin of the udder is covered with small fine hairs and it has a pigmentation which changes between pink to the grey. The nipples are hairless with a conical cylindrical form, and colouration which changes from cream to grey.

One of the main problems found at the beginning of this study was to identify the best methodology for milk extraction. The method chosen differs significantly from that applied by Jimenez (1970) and Leyva et al. (1983) who used separation of the cria, hormonal stimulation, blockade of the nipples and measurement of the milk volume produced in 4 hours, for extrapolation to 24 hours.

The milk volume collected after 12 hours was found to be highly variable (ranging from 50 to 450 ml) between the individual animals studied. These results are in accordance with the report of Jimenez (1970) who studied the chemical characteristics of milk in 200 lactating alpacas (75

Huacayas and 125 Suris) and observed milk volumes between 40 to 1,200 ml after 12 to 13 hours. Such variability between animals raises the possibility of selecting animals for milk yield.

The highest milk volume in the present study was found in the measurements taken at day 60 post partum. These results contrast with those of Leyva *et al.* (1983) which indicated that the highest milk production in llamas occurs in the third week of lactation. However, it is important to consider that these authors worked with llamas aged 6 to 10 years and extracted the milk by applying stimulation by oxytocin (10 UI/ml).

With regard to milk composition, Bustinza (2001) describes the percentage of colostrum protein at parturition and on day 20 as 23.10% and 4.71% respectively. These results compare with the present study where the concentration of total proteins between days 1 and 30 post partum reduce from 11.21 to 4.25%, respectively.

In addition, Ramírez *et al.* (1981), kept 12 llamas aged between 3 and 14 years on native pastures at 4,200 m of altitude and extracted milk daily from the beginning of lactation for a further 9 weeks. They observed that the milk of the first 96 hours had the characteristics of colostrum due to its high content of protein (11.9 to 6.5%), fat (8 to 4.2%), total solids (22.7 to 16%) and Dornic acidity (47.7 to 18.6°). After that period, protein and fat varied between 5.9 to 6.9%, and 3.7 to 4.3% respectively and the Dornic acidity ranged between 16.7 to 21.5°.

These results may be compared with results from the present study for the first day post partum which gave average values for protein, fat and total solids of 11.21%, 2.26% and 20.41%, respectively. With the exception of percentage of fat, which increased until day 7 post partum and then decreased, values for all other major constituents markedly decreased with subsequent samplings throughout the study.

Conclusions

1. Milk production steadily increased until day 60 post partum and decreased during the following days of lactation.
2. There is a high variability in the milk production among the individual animals studied. This result offers the possibility for animal selection for this trait.
3. Compared with values at day 7, values for samples taken on the first day after parturition for percentages of total solids, total protein, whey protein and ash were lower, while the percentage of fat was markedly increased.
4. Following day 7 post partum, changes in values for milk components were relatively small.
5. Average body weight of the crias increased from 10.8kg at birth to 35kg at 120 days post partum.

Acknowledgements

The investigation team of the present work acknowledges, with thanks, the following: DESCO (Center of Studies and Promotion of Development), Arequipa, Peru, which made the execution of the present study possible; EU Project DECAMA which provided economic support; MVZ Víctor Vélez UCSM - Peru, for collaboration in the statistical analysis of the data; MVZ Ana Orihuela UCSM Peru, for assistance; the Cayllahua family for their invaluable cooperation in milk extraction. Our special gratitude is also given to Professor Dr. Martina Gerken, University of Goettingen, Germany, who made it possible to present this work during the 4th European Symposium on South American Camelids.

References

Arpi, C.F., 1994. Tamaño de la glándula mamaria y composición química de la secreción láctea antes y después del parto en alpacas. Tesis F. M. V. Z. Universidad Nacional del Altiplano, Puno, Peru.

Bustinza, V., 2001. La Alpaca, conocimiento del gran potencial andino. Primera edición. Tomo I y II. Oficina de recursos del aprendizaje, Sección Publicaciones, Universidad Nacional del Altiplano, Puno.

Jiménez, O.R., 1970. Estudio de algunos aspectos sobre la leche de alpacas (*Lama pacos*). Bol ext IVITA (Peru) 4: 60-70.

Leyva, V, E. Franco and N. Condorena, 1983. Determinación de la curva lactacional en alpacas y llamas en condiciones de pastura natural. Resum. 6ta. Reunion Cient. Anu. Asoc. Peruana Prod. Anim. Peru, 1983: MR-3.

Ramírez, A., A. Zegarra, A. Ogi, J. Sumar and R. Valdivia, 1981. Características Fisicoquímicas de la Leche de Llama. *Lama glama*. Reum. 4ta. Conv. Int. Sobre Camélidos Sudamericanos. Chile: Punta Arenas.

Meta-analysis of glucose tolerance in llamas and alpacas

C.K. Cebra[1] and S.J. Tornquist[2]
[1]*Department of Clinical Sciences, Oregon State University College of Veterinary Medicine, 105 Magruder Hall, Corvallis, OREGON 97371, USA; christopher.cebra@oregonstate.edu*
[2]*Department of Biomedical Sciences, Oregon State University College of Veterinary Medicine, 105 Magruder Hall, Corvallis, OREGON 97371, USA.*

Abstract

To investigate whether observed differences in glucose tolerance testing in New World camelids could relate to the species tested or the gender of the subjects, data were compared from different trials using similar protocols and methods of analysis. In total, 5 adult female llamas, 9 adult llama geldings, and 22 adult gelded alpacas were administered 0.5 g/kg glucose as an intravenous bolus after an overnight fast. Blood was withdrawn for glucose determination before the bolus and 15, 120, and 240 minutes afterward from all camelids, and also 30 and 60 minutes afterward from all but 8 alpacas. Additionally, theoretical volume of distribution for glucose at the 15 and 30 minutes time points and the fractional turnover rate of glucose over each sampling interval were calculated for all camelids. Concentrations, the differences between the measured concentration and baseline concentration, and the fractional turnover rate were each compared for differences between the three groups based on species and gender. Theoretical volumes of distribution were compared between the 3 groups for the 15 and 30 minute time points. Gelded alpacas had significantly lower absolute glucose concentrations and changes in glucose concentration than female llamas from 15 through 120 minutes after glucose administration, and than gelded llamas from 15 minutes through the conclusion of the trial. There were no differences in absolute glucose concentration or change in glucose concentration between the 2 genders of llamas. Alpacas had a significantly greater estimated volume of distribution for glucose at both 15 and 30 minutes than either gender of llama. The fractional turnover rate of glucose was not different between groups. Estimated volume of distribution was not different between genders of llama at either time point. In the absence of data suggesting faster glucose clearance in alpacas, these findings suggest that alpacas may have a greater volume of distribution for glucose than llamas, and hence potentially require different dosing regimens for medications that distribute throughout the extracellular fluid compartment. They also suggest that the lower insulin response to hyperglycemia identified earlier in alpacas may be the result of lower peak glucose concentrations, not a greater degree of pancreatic insufficiency.

Keywords: glucose, extracellular fluid compartment, energy metabolism

Introduction

Recent investigations of glucose tolerance in New World camelids have revealed many unique features of camelids, including that they have slower glucose clearance, a weaker insulin response to intravenous glucose challenge, and partial insulin resistance compared to sheep, cattle, horses, and other domestic species. (Arraya *et al.*, 2000; Cebra *et al.*, 2001a, b; Cebra *et al.*, 2004; Ommaya *et al.*, 1995) These experiments have also suggested certain other characteristics of New World camelids, including that alpacas have an inferior insulin response to llamas and that there may be differences in volume of distribution for glucose between llamas and alpacas or potentially between male and female camelids.(Cebra *et al.*, 2001a)

The poor insulin response in camelids has potentially been linked to pathologic syndromes, including hyperosmolar syndrome in crias (Cebra, 2000), stress hyperglycemia (Garry, 1989), diabetes mellitus-like conditions (Stehman *et al.*, 1997), hepatic lipidosis, (Anderson *et al.*, 1994; Tornquist *et al.*, 1999; Van Saun, 2000) and hyperlipemia. Although there are numerous clinical reports on these disorders, none of them suggest that alpacas are at greater risk than llamas.

The difference in volume of distribution may be of greater importance. Because glucose distributes predominately through the extracellular fluid compartment, a difference in volume of distribution would suggest that medication regimens devised for one species or gender of New World camelid might not be applicable to other species or the other gender, particularly when the medication involved also distributes freely through the extracellular fluid compartment. Certain antibiotic trials appear to have confirmed the findings from our glucose tolerance trial.

We have completed several recent glucose tolerance trials in both llamas and alpacas in recent years. Because the model was relatively consistent between trials, we believe these data may be compared and may be useful in answering questions about gender and species differences in glucose tolerance in New World camelids. The hypotheses of this trial were that alpacas had a greater volume of distribution for glucose than llamas, and that male camelids have a greater volume of distribution than females.

Materials and methods

Data were collected from four separate trials.(Cebra *et al.*, 2001a, b, 2004; Ueda *et al.*, 2004) Altogether, intravenous glucose tolerance tests were performed on 9 adult llama geldings, 5 adult female llamas, and 22 adult alpaca geldings. All camelids were judged to be healthy by physical examination and basic clinicopathologic evaluation. A 16-gauge double-lumen intravenous catheter was placed into the right jugular vein of each camelid at least 2 days before the trial. Camelids were housed in groups to minimize stress.

All studies were conducted with approval of the Institutional Animal Care and Use Committee of Oregon State University. After an 8 hour overnight fast, camelids received 0.5 gram of glucose in a 50% solution per kilogram of body weight by rapid injection (< 10 seconds) through one lumen of the catheter. All subsequent blood samples were withdrawn through the other lumen after discarding the first 5 ml of withdrawn fluid, and collected into either clot tubes (all llamas and 5 alpacas) or tubes containing anticoagulant (17 alpacas). The timing of sample collection varied in the different trials, but samples were drawn from all camelids immediately before glucose injection and at 15, 120, and 240 minutes after injection. Samples were also drawn from all but one group of 8 male alpacas at 30 and 60 minutes.

Anticoagulant samples were placed on ice immediately, and plasma was separated within 20 minutes. Clot samples were stored at room temperature, and serum was removed within 20 minutes. All plasma and serum samples were analyzed for glucose content using an automated chemistry analyzer; plasma and serum results were considered comparable and pooled.(Stockman & Scott, 2002) Additionally, change in glucose from baseline was calculated for each sample. The fractional turnover rate (k) of glucose was calculated for each interval using the following formula (Kaneko, 1997):

$k(\%/min) = (\ln[\text{glucose}]_1 - \ln[\text{glucose}]_2)/ \text{interval}_{min} \times 100$

Theoretical volume of distribution (V_d) per kilogram of body weight was calculated for 15 and 30 minutes by dividing the administered dosage of glucose (0.5 gm/kg) by the measured increase in glucose concentration at those time points. This was divided by the body weight to provide an estimate of the percentage of body weight that was extracellular fluid.

Absolute glucose concentration and change in glucose concentration from the three alpaca trials were compared using two-way ANOVA for repeated measures. On finding no differences between the trials, the alpacas were pooled for subsequent comparison to the llamas of each gender. Absolute glucose concentrations, changes in glucose concentrations, and fractional turnover of glucose were compared among gelded alpacas, gelded llamas, and female llamas using two-way ANOVA for repeated measures. Differences between mean values were detected by use of the Tukey test. Estimated volumes of distribution were compared among alpacas and the two genders of llama at the 15 and 30 minute time points using one-way ANOVA. Differences between mean values were detected by use of the Tukey test.

Results

All comparisons were considered significant at $P < 0.05$. All data are expressed as mean $\pm$ standard deviation. Both glucose concentration (Figure 1) and change in glucose concentration (Figure 2) were significantly different between alpacas and both gelded and female llamas at 15 minutes ($P < 0.001$ for all) and 120 minutes ($P < 0.001$ for change in glucose concentration and $P = 0.001$ to 0.003 for glucose concentration) after glucose administration. At 240 minutes, both values were different between gelded alpacas and gelded llamas ($P < 0.013$), but were no longer different between gelded alpacas and llama females. There were no significant differences between the two genders of llama at any time point.

Expansion of the analysis to include the 30 and 60 minute time points revealed significant differences between alpacas and llamas of both genders at those time points, and failed to reveal a difference between the two genders of llama.

There were no differences between species or genders in fractional turnover rate of glucose. From 15 to 120 minutes, turnover in alpacas was $0.49 \pm 0.04\%$/min, in gelded llamas was $0.52 \pm 0.06\%$/min, and in female llamas was $0.43 \pm 0.08\%$/min.

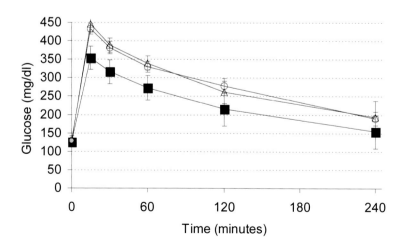

Figure 1. Blood glucose concentrations in gelded alpacas (squares), gelded llamas (circles), and female llamas (triangles) after intravenous administration of 0.5 g/kg glucose in a 50% solution.

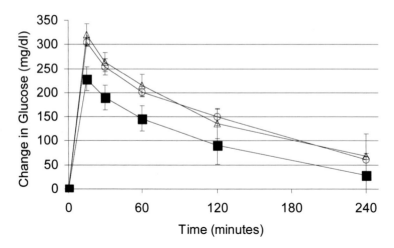

Figure 2. Changes in blood glucose concentrations from baseline in gelded alpacas (squares), gelded llamas (circles), and female llamas (triangles) after intravenous administration of 0.5 g/kg glucose in a 50% solution.

Alpacas had a significantly greater estimated volume of distribution for glucose at 15 minutes (22.1 ± 2.6% of body weight) than either male llamas (15.7 ± 1.1%; P < 0.001) or female llamas (16.4 ± 0.5%; P < 0.001). Alpacas also had a significantly greater estimated volume of distribution for glucose at 30 minutes (26.9 ± 3.7% of body weight) than either male llamas (19.1 ± 1.5%; P < 0.001) or female llamas (19.8 ± 1.4%; P < 0.001). Estimated volume of distribution was not different between genders of llama at either time point.

Discussion

Gelded alpacas reached lower glucose peaks, had a smaller change in blood glucose concentration from baseline, and had a greater estimated volume of distribution than either gender of llama. These findings confirmed those of an earlier study (and part of this one) (Cebra *et al.*, 2001a), that gelded alpacas had lower glucose peaks after administration than female llamas, and expanded those findings to demonstrate that gelded alpacas are different than gelded llamas.

These findings could relate to faster glucose clearance in alpacas or a real difference in volume of distribution. Analysis of organ weights in llamas and alpacas has revealed that vascular soft tissue organs make up a greater proportion of body weight than gastric structures and contents, providing a potential explanation for a greater extracellular fluid compartment in alpacas (Cebra *et al.*, 2006). There is currently no evidence that alpacas have faster glucose clearance than llamas, and the current trial demonstrated similar fractional turnover between gender and species. Alpacas have been shown to have less of an insulin response to glucose challenge than alpacas (Cebra *et al.*, 2001a), suggesting possibly less active mechanisms promoting clearance, but evidence for the greater volume of distribution in alpacas also brings into question whether the weaker insulin response to glucose challenge in that species is actually an effect of the lower peak concentrations. Insulin is produced in the pancreas and released based on glucose concentration in the extracellular fluid, and hence alpacas with their greater volume of distribution for glucose would be expected to have less pancreatic stimulation.

The identification of a difference in volume of distribution for glucose between llamas and alpacas could have wide-reaching effects. Glucose is distributed throughout the extracellular fluid compartment, usually within 30 minutes. In most species, glucose is not the ideal marker to measure volume of distribution because it is cleared from the blood by a variety of mechanisms. However, it may be of greater value in species with poor glucose clearance, such as camelids. If the difference in peak glucose concentrations between llamas and alpacas accurately reflects an approximately 37.5% larger extracellular fluid compartment in alpacas, then pharmacokinetic data derived from one species would be inaccurate in the other species. Other exogenous substances that distribute predominantly throughout the extracellular fluid compartment would be expected to reach lower peak values in alpacas than in llamas administered the same dosage regimen, and hence might not reach therapeutic concentrations. Such substances include many common medications, including antibiotics and sedatives.

References

Anderson, D.E., P.D. Constable, K.E. Yvorchuk, N.V. Anderson, G. St-Jean and L. Rock, 1994. Hyperlipemia and ketonuria in an alpaca and a llama. J Vet Internal Med 8: 207-211.

Arraya, A.V., I. Atwater, M.A. Navia *et al.*, 2000. Evaluation of insulin resistance in two kinds of South American camelids: llamas and alpacas. Comp Med 50: 490-494.

Cebra, C.K., 2000. Hyperglycemic, hypernatremic hyperosmolar disorder in neonatal llamas and alpacas. J Am Vet Med Assoc 217: 1701-1704.

Cebra, C.K., S.J. Tornquist, R.J. Van Saun and B.B. Smith, 2001a. Intravenous glucose tolerance testing in llamas and alpacas. Am J Vet Res 62: 682-686.

Cebra, C.K., S.A. McKane and S.J. Tornquist, 2001b. Effects of exogenous insulin on glucose tolerance in alpacas. Am J Vet Res 62: 1544-1547.

Cebra, C.K., S.J. Tornquist, R.M. Jester and C. Stelletta, 2004. The effects of feed restriction and amino acid supplementation on glucose tolerance in llamas. Am J Vet Res 65: 996-1001.

Cebra, C.K., R.J. Bildfell and C.V. Lohr, 2006. Determination of internal organ weights in llamas and alpacas. In: South American Camelids Reasearch, Vol. 1, M. Gerken and C. Renieri (eds.), Wageningen Academic Publishers, Wageningen, The Netherlands. (this book)

Garry, F., 1989. Clinical pathology of llamas. Vet Clin North Am Food An Pract 5: 55-70.

Kaneko, J.J., 1997. Carbohydrate metabolism and its diseases. In: J.J. Kaneko, J.W. Harvey and M.L. Bruss (eds.). Clinical Biochemistry of Domestic Animals, San Diego, Academic Press, pp. 45-81.

Ommaya, A.K., I. Atwater, A. Yañez, M. Szpak-Glasman, J. Bacher, C. Arriaza, L. Baer, V. Parraguez, A. Navia, C. Oberti *et al.*, 1995. Lama glama (the South American camelid, llama): a unique model for evaluation of xenogenic islet transplants in a cerebral spinal fluid driven artificial organ. Transplant Proc 27: 3304-3307.

Stehman, S.M., L.I. Morris, L. Weisensel *et al.*, 1997. Case report: picornavirus infection associated with abortion and adult onset diabetes mellitus in a herd of llamas. In: Proceedings. 40th Annu Meet Am Assoc Vet Lab Diagn, pp. 43.

Stockham, S.L. and M.A. Scott, 2002. Glucose and related regulatory hormones. In: Fundamentals of veterinary clinical pathology. Iowa State Press, Ames, IA, pp. 487-506.

Tornquist, S.J., R.J. Van Saun, B.B. Smith, C.K. Cebra and S.P. Snyder, 1999. Histologically-confirmed hepatic lipidosis in llamas and alpacas: 31 cases (1991 - 1997). J Am Vet Med Assoc 214: 1368-1372.

Ueda, J., C.K. Cebra and S.J. Tornquist, 2004. Effects of exogenous long-acting insulin on glucose tolerance in alpacas. Am J Vet Res 65: 1688-1691.

Van Saun, R.J., 2000. Nutritional support for treatment of hepatic lipidosis in a llama. J Am Vet Med Assoc 217: 1531-1535.

Part V
Meat and fibre production

DECAMA-project: Analysis of farm income from South American camelids meat production in Latin American countries: Preliminary results of a comparison between case studies

F. Ansaloni[1], F. Pyszny[1], A. Claros L.[2], R. Marquina[3], J. Zapana Pineda[4], A. Claros J.[2] and J.L. Quispe Huanca[2]
[1]*Department of Veterinary Sciences, University of Camerino, Via Fidanza 15, 62024 Matelica (MC), Italy; francesco.ansaloni@unicam.it*
[2]*UPEA Universidad Publica de El Alto, La Paz, Bolivia, PRORECA, La Paz, Bolivia.*
[3]*DESCO, Centro de Estudios y Promocion del Desarollo, Arequipa, Perù.*
[4]*ACCRA, Asociacion de Cooperacion Rural en Africa y América Latina, Calle Sbtte, Aramayo n.1008 Esq. Jaimes Freyre, 10424 La Paz, Bolivia.*

Abstract

The scarce wealth of the Andean rural areas imposes the necessity to support the economic development, improve the income of the population, the hygien status and the quality of the products of animal origin. The main interventions in agricultural policy consist in the provision of professional skills and farm economic training services and infrastructures for the supply of the products on the market. The purpose of this paper consists in illustrating the preliminary results of a comparative analysis on the formation of the income and the calculation of the production cost of camelids meat representative of the more diffused Andean realities in Peru and Bolivia. The data are technical and economical and they refer to the year 2003. The source of the data is represented by some farm groups, homogeneous from the point of view of the productive system and the endowment of the production factors. The collection of the data has been carried out by technicians, with the use of an homogeneous questionnaire for the different countries, through business visits and direct interviews with the agricultural entrepreneurs. The results analysis consist in the elaboration of balance sheet (calculation of the income and the unitary production cost of alpaca and llama meat) in three Peruvian farm groups and five Bolivian ones.

Resumen

La riqueza escasa de los territorios rurales andinos impone la necesidad para favorecer el desarrollo económico para mejorar el ingreso de la población, el estado higiénico y la calidad de los productos de origen animal. Las intervenciones principales de política agrícola consisten en la oferta de 1 - la técnica profesional y granja los servicios de entrenamiento económicos; 2 - las infraestructuras para la oferta de los productos en el mercado. El propósito de este papel consiste en ilustrar los primeros resultados de un análisis comparativo en la formación del ingreso y el cálculo del precio de producción de carne de los camelidos en algunos casos de estudio, los grupos de 5-7 granjas con engendrar de llama y representante de alpaca de las realidades andinas más difundidas en los territorios de Peru y Bolivia. Los datos son técnicos y económico y ellos se refieren un año 2003. La fuente de los datos se representa por algunos grupos de las granjas, homogéneo del punto de vista del sistema productivo y la dotación de los factores de la producción. La colección de los datos se ha servido como técnicos, con el uso de una encuesta homogénea para los países diferentes, a través de las visitas de negocio y las entrevistas directas a los empresarios agrícolas. Los resultados del análisis consisten en la elaboración de hoja de balance (el cálculo del ingreso y el precio de producción unitario de alpaca y carne de llama) en tres grupos de las granjas de Peru y cinco grupos de las granjas de Bolivia.

Keywords: economic budget, llama and alpaca breedings, meat production, case studies.

Introduction

The aim of the DECAMA Research (Sustainable Development of Market Oriented Camelid Products and Services in Andean Region) is to encourage domestic South America camelids' (DSAC) meat chain development in Argentina, Bolivia and Peru. This research is being carried out on two parallel paths: 1) an evaluation of the opportunities provided by improved technology for animal rearing and meat processing activities; 2) economic and environmental sustainability analysis of the production systems.

The objective of the market research group it consists in the improvement of the conditions of welfare of the Andean populations through the exploitation of the direct and indirect products of the breeding of llama and alpaca.

The research plan has been divided into four interrelated sections that have been cross-referenced at the various meat chain stages (Abbadessa *et al.*, 2004). The analysis of the data is carried out with the writing up of: 1) specific reports for the different aspects of every analysis unit; 2) economic balances for the case studies.

This paper illustrates some of the first results obteined by the analysis of the economic balances of the breedings llama and alpaca farms case studies of the Andean territories devoted to the meat production and represents a initial footstep to start a check of their economic sustainability.

Analysis units

The first part is a supply and demand analysis. Demand is defined at the level of the final consumer while the supply factor has been traced back to the live animal stage. In particular, data to be collected includes farm resources, the identification of a sample and representative group of breeding farms, production, supply, market relationship with food processors or business community.

The second analysis unit illustrates the meat processing stage and market distribution. Here too, a representative sample of processing firms and commercial distribution channels will be identified.

Data collected from the preceding analysis units will allow the research team to design the main sales circuits and to determine the final commercial value of the products.

The last analysis unit will examine the main public interventions needed to encourage development of the agro-industrial sector, with special emphasis on the animal-rearing field.

In order to identify the strong and weak points of the DSAC llama meat chain, the choice of the data sources represents a crucial research element. Each data source demonstrates advantages and different costs. As far as DECAMA is concerned, the research illustrates the limited availability of resources and difficulties involved in collecting meaningful data in the territory. The choice to use only the national statistics from the various countries does not seem to satisfy research requirements. For this reason, the decision to experiment with the technical and economic comparison of more diffused Business Models in the various meat chain phases in Bolivia, Peru and Argentina may be more suitable sources of valid information.

Farm Business Model concept

The Business Model should illustrate the technical production system and income results in the environmental, professional and institutional context of the various countries. In general, it should reflect the behaviour of the majority of business people in a significant number of businesses specialised in the production of goods located in a determined territory.

The technical and economic Business Model characteristics are expressed as average values gathered from a sample group of 5-7 farms, which collaborated in the research project (Abbadessa *et al.,* 2004; Deblitz, 2002).

The main elements are: production cost analysis, description of the technical production system and a profit and loss statement. Other aspects are represented by the identification of the business community's objectives, strong points, weak points, development opportunity, problems and/or tie, private objective, private and public interventions.

In order to construct a Business Model, the first step is the identification of the specialised territory and/or the area dedicated to DSAC meat product. As far as the rearing stage is concerned, the geographic area is defined as the largest animal rearing area for the country under study and/or the one with the highest animal population density.

Business Models studied were those that reflect in greater measure the real business situation in the country under study. The definition of the appropriate dimension of productive resources of the Business Model can be drawn from statistical analysis and/or empirically.

The empirical investigation more common Business Models consist in an information survey from experienced witnesses in the productive region. The advantage that derives to empirically operate consists of eliminating the case studies of scarce importance because of their limited offer of product on the market. The risk is that devoting greater attention to the more visible Business Models on the market but with little statistical incidence in the productive context.

In the initial phase a pre-panel group consisting of two farms, one scientist and one advisor was set up. The obvious advantage here is time. It takes less time to organise a small than a full-panel because fewer persons are involved. The advisor's role is to filter out possible biases from the participating farmers. Once the Business Model has been standardised, the sample can be enlarged the to 5-7 farms necessary for a full-panel. The farms that are part of the Business Model group should be as similar to one another as possible.

The following is a list of criteria, which the farms should meet (Ansaloni, 1999): 1) available resources: in equal quantities as the Business Model; 2) highly specialised production: ratio of the amount of income derived from the product under study to total farm income; 3) production technique: same as Business Model; 4) profitability: business farms model that can stay on the market with quality products; 5) business person: market oriented and if possible relatively young, open to innovation and willing to collaborate in the research project.

Business Model data are average values of the farms in the group. None of the farmers or business persons was obliged to reveal his individual farm or company data. All figures are being determined by consensus that is by discussing the model farm not about individual farms. Feedback of data to the panel members is fundamental.

The data analysis is done by accounting classification, determination of production cost and description of production system. Pre-panels are sufficient for status quo production cost analysis of one year if there are no changes in the physical figures / production farm structure. With the full-panel analysis the research can be widened to include a description of productive strategies and adjustments to extra-business changes. A profit and loss statement as well as a summary of the most important farm/ business economic indicators are returned to the panel members for checking. This process is repeated until the panel agrees on the results achieved.

The farm data have been collected through repeated visits and interviews to the single farm producers of every group - case studies - from the authors or technicians. To this purpose a homogeneous questionnaire has been built for all farms.

The elaborate values of some business models variable, for example the heads of livestock, can introduce some decimal values because they represent the average value of the individual values of the different farms that belong to every group of business model. For some aspects, technical and financial data result esteemed. For example when, as in the case of Peru, for a variable it is present only a aggregate value that reflects more case studies. In this case the aggregate value of the variable has been divided in proportion to the number of the heads in breeding in the different various case studies.

Materials and methods

Case studies

The total number of the analyzed business models reaches 8 cases: 3 in Peru and 5 in Bolivia. The number of farms for every group changes from a minimum of 3 to a maximum of 28.

The Peruvian cases are represented by family farms and they are differentiated in relation to the dimension of the herd (number of llama and alpaca heads) and the surface of available land: small, medium and large (Table 1).

The Bolivian cases belong to two categories: the first is represented by family farms (2 cases) while the second is rural community (3 cases). In the family farms the number of llama and alpaca heads is very similar (average value of 226) while the avialable bofedale land is notably different (Table 1).

The criterions of choice of the enterprises in order to establish the farm groups for the case studies depended on their ability to represent the most common realities found in the field. In the case of Peru, the agricultural unities farming alpaca and llama breeding reach the total number of 71,614 (INEI). In the province of Caylloma, where the farms that collaborate to the investigation for the group's analysis are located, the total enterprises are 3,041 (INEI, 1994). In Peru the distribution of the properties depends on the dimension of the herd, that depends in turn from the volume of the income and from the level of the investments. In this country the distribution of the number of alpaca breeding units by size of alpaca herds is as follows: with less than 80 heads, 20-33%; 81 to 300 heads, 65-70%; with over 300 heads, 2-10%. The selected criterions to identify the farms of the groups of the case studies satisfy the necessity to reflect in the most greater possible measure the main characteristics of this reality (Marquina, pers. comm.).

The main elements of this study consist in the analysis of the availability of the resources and the farm economic results in the different case studies. We begin with the illustration of the farm resources exploited in the different breeding realities.

Table 1. List of farm breeding business models - case studies (Source: A.L. Claros, pers. comm.; R. Marquina, pers. comm.; J. Zapana Pineda, pers. comm.).

Case studies	Country, Department, Region	Enterprise types *	Farms of the group of the case studies	Land		Livestock		Labour unit (LU)	
				Bofedal	Total	Llama - alpaca	Sheep	Family	Extra-family
			n.	ha	ha	n. heads	n. heads	n. LU	n. LU
PE Small	Peru, Arequipa, Arequipa	P	9	28.3	175.99	70	53.2	3.45	0.75
PE Medium	Peru, Arequipa, Arequipa	P	28	47.2	232.30	191	57	3.45	0.77
PE Large	Peru, Arequipa, Arequipa	P	6	119.7	427.23	583	106	3.45	0.79
BOL Curahuara	Bolivia, Oruro, Curahuara de Carangas	P	5	164	522	238	28	2.63	0.33
BOL Oruro Turco	Bolivia, Oruro, Turco	P	5	25	789.40	214	25	2.62	0.45
BOL Quetena	Bolivia, Potosì, Sud Lipez	C	3	50	3,055	313	17	5.00	0
BOL Coroma	Bolivia, Potosì, Sud Oeste	C	3	49	910	191	28	2.33	0
BOL Pozo Cavado	Bolivia, Potosì, Altopiano sur	C	3	0	3,834	190	79	4.00	0

* P private family farms, C rural community

Farms resources

The first production resource considered is represented by the amount of bofedale land available per head (llama and alpaca). In general, the smaller the quantity of land per head the higher is the quantity of forage produced by the land; this represents the productive synthesis that depends on climatic and natural factors (water, rain, temperature, fertility, *etc.*). Among the considered cases a noteworthy variability is observed: BOL Pozo Cavado shows the overall absence of bofedale land and the availability of land of inferior quality (Table 2). The other cases vary from a minimum of 0.10 (BOL Oruro) to 0.62 land per head (BOL Curahuara).

The farm capital depends in an exclusive way on the herd patrimony. The capital represented by equipment and machineries for the Peruvian cases and one of the Bolivian rural communities (BOL Quetena), reaches only around 3% of the total value (Table 3). It is worth noticing that in all cases there is non re-investments of commodities produced on farm, such as forage, seed, *etc.*

Table 2. Bofedal land per head (n.ha/head).

Case studies	Land (ha)		Heads			Bofedal land (ha) per head
	Bofedale a	Total b	Llama - alpaca c	Sheeps d	Total e	f=a/e
PE Small	28.30	175.99	70	53	123	0.23
PE Medium	47.20	232.30	191	57	248	0.19
PE Large	119.70	427.23	583	106	689	0.17
BOL Curahuara	164.00	522.00	238	28	266	0.62
BOL Oruro Turco	25.00	789.40	214	25	239	0.10
BOL Quetena	50.00	3,055.00	313	17	330	0.15
BOL Coroma	50.00	910.00	191	28	219	0.23
BOL Pozo Cavado*	0	3,834.00	190	79	269	0

* In this farm group no bofedal is available for grazing

Table 3. Percentage composition of the farm capital value (US$).

	Livestock		Equipment machineries		Stock products		Total	
	Amount	%	Amount	%	Amount	%	Amount	%
PE Small	5,794.30	97.82	128.86	2.18	0.00	0.00	5,923.16	100
PE Medium	13,228.44	97.37	356.71	2.63	0.00	0.00	13,585.15	100
PE Large	37,961.38	97.22	1,085.89	2.78	0.00	0.00	39,047.27	100
BOL Curahuara	11,157.01	100	0.00	0.00	0.00	0.00	11,157.01	100
BOL Oruro Turco	12,062.38	99.49	62.42	0.51	0.00	0.00	12,124.80	100
BOL Quetena	16,083.57	96.94	507.64	3.06	0.00	0.00	16,591.21	100
BOL Coroma	4,491.00	99.58	19.10	0.42	0.00	0.00	4,510.10	100
BOL Pozo Cavado	7,482.80	99.68	24.20	0.32	0.00	0.00	7,507.00	100

The small number of total labor unit (LU) that contributes to the economic activity of the case studies shows the "family" character enterprise. This seems confirmed by the fact that a value min of 2.96 and max of 4.22 LUs it is observed and the quantity of heads for LU is extremely variable: from 17 heads around to over 137 (Table 4). It is unlikely that the differences in breeding techniques can explain this variability of heads per LU. Instead, the flexibility of labor type family seems the factor that better explains this situation. In other words, in these farms the labor could be an abundant resource that not always, mostly when the patrimony livestock is limited, is completely exploited.

Method of analysis

The method of analysis homogeneously adopted for the elaboration of the average data of all the groups of farms – case studies - consists in the calculation of the economic balance for the year 2003. The product considered of greater importance in the formation of the farm's income is represented by the meat. Now, after the examination of the availability of the resources of the farms, we examine the results of the analysis of the economic activity: earnings cash, costs and level of income. One of the limits of this study consist partly in the fact that to improve the reliability of the elaborate raw data a phase of further collection and control is needed.

Results

Earnings cash

Wide part of the earnings cash by the farms derives, directly or indirectly, by the livestock patrimony. The sale of live animals (alpaca and llama) and other products of animal origin earn cash from a 70% minimum (BOL Pozo Cavado) to a maximum of 95% (PE Large) of the total of the earnings cash (Figure 1). Only a model business show an economic activity very diversified between breeding and extra-agricultural services (BOL Coroma).

Apart from the case of BOL Coroma, from 75% to 95% of the total earning in cash derives in almost equal parts by 1) sale of live animals and 2) other products of animal origin (meat sheep, carcasses, transformed meat - charqui, chalona -, fiber, leather).

Among the case studies of Peru and Bolivia, the ratio among number of heads of llama and alpaca is very different. The alpaca are the most common in the Peruvian business models. In

Table 4. Number of heads per labor unit.

Case studies	Heads			Total Labor Unit (LU n.)	Llama - alpaca (n.heads/LU)	Total livestock n.heads/LU
	Llama - alpaca	Sheeps	Total			
PE Small	70	53.2	123	4.20	16.67	29.33
PE Medium	191	57	248	4.22	45.37	58.88
PE Large	583	106	689	4.24	137.45	162.45
BOL Curahuara	238	28	266	2.96	80.41	89.86
BOL Oruro Turco	214	25	239	3.07	69.71	77.85
BOL Quetena	313	17	330	2.96	105.74	111.49
BOL Coroma	191	28	219	3.07	62.21	71.34
BOL Pozo Cavado	190	79	269	4.00	47.50	67.25

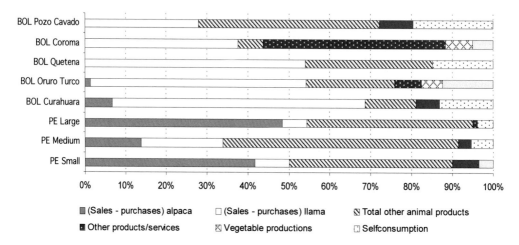

Figure 1. Earning cash per head (%).

Bolivia, instead, the alpaca are in smaller quantity while there is an increase in llama heads. Finally, in the Bolivian rural communities are present only llama heads.

The extra-agricultural services have contributed to 2 to 7% of the total volume of the earnings cash, excluded the case of BOL Coroma that, as we have already written, reaches over 40%. The types of services consist of commerce, handicraft activity (BOL Coroma), rural tourism (BOL Curahuara) and house building (BOL Oruro).

Only two cases show vegetable productive activity (BOL Oruro Turco and Coroma).

Finally, the selfconsumption of final products consumed on the farms by the owner and his family is present in all farm models and varies from a 3% minimum (PE Small and Large) to a maximum of around 20% of the total earnings cash (BOL Pozo Cavado).

In absolute value, the greater quantity of earnings cash is 9,640.65 US$ observed for PE Large while the lowest value is noticed for BOL Coroma (1,747.38 US$). In general there seems to be a positive correlation between dimension of the herds and level of earnings cash.

The greater quantity of earnings cash per head is reached by PE Small (19.69 US$) while the lower is observed for BOL Quetena (6.33 US$). An hypothesis that could explain the elevated level of earnings cash of the first quoted case could consist in the small dimension of the herd and therefore, in the intensive use of labor factor for this family enterprise (Tables 5, 6 and 7).

Production cost

Among the case studies examined the greater production cost per head per year reaches 9,41 US$ (PE Medium) while the lower is 5.21 US$ (BOL Pozo Cavado).

Particularly, the variable costs (part-time labor, transport, other) show the minimum value of 0.17 US$ (BOL Coroma) and the maximum of 4.43 US$ (PE Medium). Bolivian case studies clearly show inferior variable costs then Peru (Tables 8, 9 and 10). Bolivian rural communities in

Table 5. Earnings cash per head per year (US$) Peru[*].

Cash payment	PE Small		PE Medium		PE Large	
	Total	Per head	Total	Per head	Total	Per head
Vegetable productions	0.00	0.00	0.00	0.00	0.00	0.00
(Sales - purchases) alpaca	1,009.19	8.20	499.67	2.01	4,662.06	6.77
(Sales - purchases) llama	202.00	1.64	721.86	2.91	570.85	0.83
Sale heads sheep	216.15	1.76	308.65	1.24	253.97	0.37
Carcasses	413.68	3.36	1,187.30	4.79	2,610.30	3.79
Transformed meat (charqui, chalona)	107.79	0.88	73.30	0.30	91.98	0.13
Fiber	190.37	1.55	406.52	1.64	888.19	1.29
Other animal products (leather)	40.17	0.33	94.57	0.38	84.72	0.12
Total animal products	968.15	7.87	2,070.34	8.35	3,929.16	5.70
Other products/services (commerce and handicraft activity)	155.84	1.27	117.86	0.48	102.04	0.15
Selfconsumption	86.23	0.70	191.13	0.77	376.54	0.55
Total	2,421.41	19.69	3,600.86	14.52	9,640.65	13.99

[*] Average money rate 2003: 1 US$ = 3.479 NS

Table 6. Earnings cash per head per year (US$) Bolivia[*] private farms.

Cash payment	BOL Curahuara		BOL Oruro Turco	
	Total	Per head	Total	Per head
Vegetable productions	0.00	0.00	168.02	0.70
(Sales - purchases) alpaca	157.05	0.59	40.29	0.17
(Sales - purchases) llama	1,470.39	5.53	1,707.55	7.14
Sale heads sheep	34.52	0.13	0.00	0.00
Carcasses	127.90	0.48	628.80	2.63
Transformed meat (charqui, chalona)	0.00	0.00	0.00	0.00
Fiber	115.14	0.43	37.59	0.16
Other animal products (leather)	20.83	0.08	30.51	0.13
Total animal products	298.39	1.12	696.89	2.92
Other products/services (rural tourisme, house building and handicraft activity)	135.03	0.51	210.19	0.88
Selfconsumption	311.36	1.17	399.94	1.67
Total	2,372.22	8.92	3,222.88	13.48

[*] Average money rate 2003: 1 US $= 7.85 BS

Table 7. Earnings cash per head per year (US$) Bolivia* rural community.

Cash payment	BOL Quetena		BOL Coroma		BOL PozoCavado	
	Total	Per head	Total	Per head	Total	Per head
Vegetable productions	0.00	0.00	117.20	0.47	0.00	0.00
(Sales - purchases) alpaca	0.00	0.00	0.00	0.00	0.00	0.00
(Sales - purchases) llama	1,127.39	3.42	657.32	2.65	522.29	1.94
Sale heads sheep	0.00	0.00	74.90	0.30	14.52	0.05
Carcasses	611.46	1.85	0.00	0.00	764.33	2.84
Transformed meat (charqui, chalona)	0.00	0.00	0.00	0.00	0.00	0.00
Fiber	40.76	0.12	30.57	0.12	40.76	0.15
Other animal products (leather)	0.00	0.00	0.00	0.00	0.00	0.00
Total animal products	652.22	1.98	105.47	0.43	819.62	3.05
Other products/services (rural tourisme, house building and handicraft activity)	0.00	0.00	779.62	3.14	152.87	0.57
Selfconsumption	310.83	0.94	87.77	0.35	366.88	1.36
Total	2,090.44	6.33	1,747.38	7.05	1,861.66	6.92

*Average money rate 2003: 1 US $= 7.85 BS

Table 8. Costs per year (US$)*.

Costs	PE Small		PE Medium		PE Large	
	Total	Per head	Total	Per head	Total	Per head
Variable costs (VC)	538.99	4.38	1,099.33	4.43	1,511.24	2.19
Part-time labour	381.40	3.10	397.70	1.60	425.35	0.62
Transport	128.85	1.05	356.71	1.44	1,085.89	1.58
Other	28.74	0.23	344.92	1.39	0	0.00
Fixed costs (FC)	240.25	1.95	476.46	1.92	523.37	0.76
Tax	31.56	0.26	185.91	0.75	224.20	0.33
Maintenance. depreciation. insurance	208.69	1.70	290.55	1.17	299.17	0.43
Total (VC + FC)	779.24	6.34	1,575.79	6.35	2,034.61	2.95
Land cost	52.81	0.43	72.47	0.29	128.94	0.19
Farm capital interest (stock and advance capital)	302.94	2.46	685.27	2.76	1,856.90	2.70
Total	1,134.99	9.23	2,333.53	9.41	4,020.45	5.84

*Average money rate 2003: 1 US $ = 3.479 NS

Table 9. Costs per year (US$)*.

Costs	BOL Curahuara		BOL Oruro Turco	
	Total	Per head	Total	Per head
Variable costs (VC)	641.81	2.41	338.08	1.41
Part-time labour	36.31	0.14	42.03	0.18
Technical support	560.00	2.11	31.85	0.13
Transport	6.40	0.02	116.40	0.49
Breeding (feeding, medicinal)	39.10	0.15	145.10	0.61
Vegetable productions (seeds)	0.00	0.00	2.70	0.01
Fixed costs (FC)	109.36	0.41	239.70	1.00
Tax	107.81	0.41	1.26	0.01
Maintenance, depreciation, insurance	1.55	0.01	238.44	1.00
Total (VC + FC)	751.17	2.82	577.78	2.42
Land cost	0.00	0.00	0.00	0.00
Farm capital interest (stock and advance capital)	952.30	3.58	1,020.10	4.27
Total	1,703.47	6.40	1,597.88	6.69

* Average money rate 2003: 1 US $= 7.85 PB

Table 10. Costs per year (US $)*.

Costs	BOL Quetena		BOL Coroma		BOL Pozo Cavado	
	Total	Per head	Total	Per head	Total	Per head
Variable costs (VC)	295.54	0.90	42.54	0.17	326.11	1.21
Part-time labour	0.00	0.00	0.00	0.00	0.00	0.00
Technical support	0.00	0.00	0.00	0.00	0.00	0.00
Transport	276.43	0.84	18.34	0.07	256.05	0.95
Breeding (feeding, medicinal)	19.11	0.06	24.20	0.10	70.06	0.26
Vegetable productions (seeds)	0.00	0.00	0.00	0.00	0.00	0.00
Fixed costs (FC)	960.44	2.91	195.56	0.79	363.04	1.35
Tax	0.00	0.00	0.00	0.00	0.00	0.00
Maintenance, depreciation, insurance	960.44	2.91	195.56	0.79	363.15	1.35
Total (VC + FC)	1,255.98	3.81	238.10	0.96	689.15	2.56
Land cost	0.00	0.00	0.00	0.00	0.00	0.00
Farm capital interest (stock and advance capital)	1,550.40	4.70	412.18	1.66	711.43	2.64
Total	2,806.38	8.50	650.28	2.62	1,400.58	5.21

* Average money rate 2003: 1 US $= 7.85 PB

fact exploit pasture, don't purchase production items and do not pay paid salaries to extra-farm workers (Distaso & Ciervo, 2004).

With regards to the calculated fixed costs (tax and building, equipment and machineries maintenance, depreciation, and insurance) the lowest value amounts to 0.41 US$ (BOL Curahuara) and the maximum to 2.91 US$ (BOL Quetena).

The cost of the land factor is considered only for the case studies of Peru. In this country the average cost of the lease of the land is known, from which the amount of the remuneration of the factor has been esteemed. The lowest value amounts to 0.19 US$ (PE Large) and that the maximum to 0.43 US$ (PE Small). Besides, in the case of the Bolivian communities there is no market exchange of the land (Distaso & Ciervo, 2004).

The amount of the interest on the total farm capital is proportionate to the remuneration of the investments of the stock capital (livestock, equipment and machineries, and stock product) and advance capital. The lowest value amounts to 1.66 US$ (BOL Coroma) and the maximum to 4.70 US$ (BOL Quetena). The elevated value of the interest mainly depends on three elements: 1) elevated value of the patrimony livestock; 2) considered period for the anticipation of the farm expenses; 3) money inflation rate. For example, in the case of BOL Oruro the value of the livestock reaches 99.5% of the total farm capital (heads n. 240).

Farm net income

The analysis of the total farm net income shows that all case studies succeed in earnings a quantity of money greater than fixed and variable costs. The greatest level of total farm income reaches 7,605.82 US$ (PE Large), while the lower amounts to 834.40 US$ (BOL Quetena). In Peru, where the number of working unity doesn't sensitively change among case studies, it seems that the total farm income is proportional to the number of the animal heads (Table 11).

The higher income of BOL Oruro in comparison to that of BOL Curahaura seems justified, to parity of working unity, from an almost double sale of "other products of animal origin".

The greatest level of total farm income among the cases of the rural communities of Bolivia is reached by BOL Coroma and it seems justified by the presence of the smaller number of unity working and low costs per animal head.

From the point of view of the net income for LU, the maximum level is 1,793.83 US$ (PE Large) and the minimum 166.88 US$ (BOL Quetena).

EARNINGS CASH							
Variable cost	Fixed cost			FARM NET INCOME			
Feeding Medicinal Part-time labor Transports Technical support	Wages Tax Bank interest	Maintenance Depreciation Insurance	Rent land	Farm capital interest	Farm labor		Profit
					LABOR FARM INCOME		

Figure 2. Calculation scheme of the farm net income.

Table 11. Net Income per farm and per labor unit (LU*) per year (US$**).

Case studies	LU (n.)	Total farm net income (US$)	Net Income per LU (US$)
PE Small	4.20	1,642.15	390.99
PE Medium	4.22	2,025.10	479.88
PE Large	4.24	7,605.82	1,793.83
BOL Curahuara	2.96	1,618.80	546.89
BOL Oruro Turco	3.07	2,595.12	845.32
BOL Quetena	5.00	834.40	166.88
BOL Coroma	2.33	1,449.02	621.90
BOL Pozo Cavado	4.00	1,116.45	279.11

* Total number of labor unit (family and extra-family) that works in farm.
** Average money rate 2003 Peru: 1 US$ = 3.479 NS; 2003 Bolivia: 1 US$= 7.85 PB.

Conclusion

The breeding and agricultural production observed in the case studies of the Andean plateau illustrate how there is a lack of investment toward an increase in productive levels, and a resulting low specialisation in meat production. In fact, the most important investments are the pasture and the livestock.

The absence of specialization for the production of meat is shown by the following facts:
1. The numerousness of animals farmed, besides the llamas, is large (alpaca, sheep).
2. An extremly variable age of the animals at slaughter while the suggestions for optimal age at slaughter is 16 to 19 months for the alpaca and 19 to 21 months for llamas. Animals are in reality slaughtered at over 30 months of age. Animals are only slaughtered at a yonger age if there is a need for cash and/or area caracterised by a low fiber quality.
3. The range of the raw materials and the transformed products is notable (live heads, fresh meat, meat transformed of llama and alpaca - charqui - and ovine - chaloma).

In synthesis, from the point of view of the breeders, it seems that the animal is not seen yet how a factor that, if specialised meat production was adopted, can potentially improve the income.

The hypotheses that could be tested for improving the welfare conditions of the Andean populations are two. The first, at producers agricultural level, is that to verify the environmental and financial possibilities of realization of a greater investment, even modest, and the adoption of innovations in the productive process. The desirable technical and organizational interventions for this hypothesis concern the labor (professional formation to improve the knowledge on the techniques of breeding and the transformation of the products of animal origin), the techniques of the pastures fodder productions, of the livestock breeding and the raw materials transformation. Beside it is necessary to verify the possibility to increase the business activity of the direct commodities, and transformed, particularly through the development of Associations of producers.

The second hypothesis to be verified aims to the public institutions and consists in the choices of agricultural and social politics to help the rural development. Particularly, it includes the decision to improve the investments in infrastructures for the supply of electricity, roads and drinkable water, which in turn represent the tie that abridges the activities of transformation of the raw materials.

The collection and the control of the data of the case of studies was rather laborious because of the necessity to establish constant relationships with numerous farmers located in distant places. Our hopes consist of seeing the application of the knowledge that has been gained in order to favor the permanent adoption of the accounting tools and income analysis, and to stimulate the maintenance, during the time, of the case studies as pilot centers for the observation of the evolution of the income and demonstrative cases of breeding techniques. The investigation DECAM is still in progress. For the future, we hope to succeed in extending this analysis of the income a years 2002 and 2004.

Acknowledgement

Grateful thanks for the suggestions on the result commented by Gianlorenzo D'Alterio.

References

Abbadessa, V., F. Ansaloni, M. Antonini, S. Canese and S. Misiti, 2004. Economic and sustainability analysis of the camelids meat chain: the case of 'Decama' research methodology. Universidad Catolica de Cordoba, 4th Seminario Internacional de Camélidos Sudamericanos, Cordoba, Argentina.

Ansaloni, F., 1999. Regione Emilia-Romagna: studio dei fattori di successo delle imprese Agrozootecniche. In: Ires Materiali, Learning Region Una strategia per lo sviluppo dei sistemi locali meridionali, G. Altieri and F. Belussi (eds.), Ediesse, Roma gennaio 1997.

Deblitz, C. *et al.*, 2002. IFCN Beef report 2002. IFCN/FAL, Braunschweig, International Farm Comparison Network IFCN Type-CAL farm models; http://www.ifcnnetwork.org - FAL Germany.

Distaso M., Ciervo M., 2004. Economia rurale come economia di reciprocità. Il caso degli ayllu delle Ande boliviane. Convegno di Studi della Società Italiana di Economia Agraria (SIDEA), Roma.

INEI, 1994. Resultados definitivos del III Censo Nacional Agropecuario. Perù.

DECAMA-project: Evaluation and classification of carcass quality of alpaca

G. Condori, C. Ayala, N. Cochi and T. Rodriguez
Sustainable Development of Camelid Products and Market Oriented Services in the Andean Region, (DECAMA) Street Sanchez Lima 2340, City La Paz, Bolivia; gencond@hotmail.com.

Abstract

The object of the work was to develop a system for determining the quality of the carcass of domestic South American camelids and to improve methods for their evaluation in the commercial market. Carcasses of twenty five 16-months old alpacas were classified by a subjective method of visual evaluation of the conformation of the leg muscles and rump. A range of five classes was used. These were: 1) poor; 2) normal; 3) good; 4) very good; and 5) excellent. The results demonstrated that the majority of the animals (84 to 96%) were within the classes 2 and 3 (normal). The conformation of the internal profile of the leg in alpacas presented the form of a closed "V" in class 1, which differed from those of class 3 in which the profile tended to form a "Y", with slightly convex muscles. The classification for fat covering considered five regions each with five grades. The carcasses presented generally limited development of subcutaneous fat and there were none in classes 3, 4 or 5. The indices of compactness calculated by objective measurement of the ratio between carcass weight and length of carcass varied between 0.23 and 0.37. The ratio between rump width and length of leg varied between 0.40 and 0.49. The latter results indicated that the alpaca has a more lengthened than compact morphology. The relationships among these indices for compactness and the subjective classifications reveal that the most appropriate procedure in qualification is for the rump conformation, which was highly correlated (r = 0.61 and 0.57). Multiple regression models were developed in which the width of the tarsus (measurement F) and width of the rump were shown to influence the weights of leg and loin. The rump width had more influence than the length of the leg on the carcass weight and influenced positively the area of the muscle *Longissimus dorsi*. In contrast, the length of the carcass has a negative effect on the same muscle (P <0.05).

Resumen

A objeto de buscar un sistema que determine la calidad de las carcasas de camélidos sudamericanos y mejorar su valoración en el mercado, fueron clasificadas 25 carcasas de alpacas machos de 16 meses de edad, por el método subjetivo de apreciación visual de la conformación de los músculos de pierna y grupa. Los resultados subjetivos demostraron que la mayoría de los animales (84 a 96%) se encuentran entre la clase 2 a 3, la conformación del perfil interno de la pierna en alpacas clase 1 denominado "normal" presentan la forma de una "V" cerrada, y los de clase 3 "bueno" presentan el perfil en forma de "Y", con músculos ligeramente convexos. La clasificación por cobertura de grasa demostró que la mayoría de las carcasas de alpaca se encuentran en las Clases 1 a 3 con presencia de un escaso desarrollo de grasa subcutánea. Los índices de compacidad de peso de carcasa / longitud de carcasa, varia de 0.23 a 0.37, entre el ancho de grupa / longitud de pierna varia de 0.40 a 0.49, estos últimos le dan a la alpaca una morfología más alargada que compacta. Las correlaciones entre estos índices *vs.* clasificación subjetiva revela que el procedimiento más adecuado de calificación es por la conformación de grupa, altamente correlacionada (r = 0.61 y 0.57). Por regresión múltiple se obtuvo modelos en las cuales el ancho del tarso (medida F) y ancho de grupa influyen en el peso de pierna y lomo. El ancho de grupa tiene más influencia que la longitud de pierna en el peso de carcasa, la variable espesor del *Longissimus dorsi* tiene un efecto mayor en el peso de lomo y pierna.

Keywords: carcass, meat, alpaca, quality

Introduction

In the subject of meat production, the parameters most frequently considered of major importance are growth rate and carcass yield of the animal. These characteristics are regulated by the differential growth of anatomical components and factors such as species, age, sex and genotype. These variables determine the order and extent of development of each anatomical part and tissue. They also contribute to the differences in conformation and the chemical and anatomical composition of animals differing in live-weight and breed (Ruiz *et al.*, 2000).

The conformation of an animal is determined by the structure of bone and muscle and the shape they develop during the time period of interest. The dimensions of muscle and adipose cover in relation to the size of the skeleton are the principal indicators of carcass quality. The degree of roundness and compactness among the anatomical regions of the carcass are considered to provide a good measure of quality by De Boer *et al.* (cited by Ruiz *et al.*, 2000).

Currently, carcass weight is the most important consideration during transactions among wholesalers in the camelid meat trade. Nevertheless, retailers also value the quality of the carcass in terms of morphology of muscles and conformation as opposed to the quality of the live animal.

At the national level, the Bolivian Norm 794-97 (Bolivian Institute of Norms and Quality), exists to classify carcasses of South American camelids for slaughter according to age, sex, conformation, finish and level of infestation by sarcocystiosis (IBNORCA, 1997). On this basis the carcasses are generally classified as extra, first, second and industrial. However, there is no specific classification procedure based on conformation or fat coverage for carcasses of animals of the same age or sex. Therefore, it is important to provide a subjective classification of alpaca carcasses based on conformation and fat coverage, in addition to objective comparison of the morphology of anatomical regions.

Materials and methods

Twenty-five male alpacas originating from Turco (Oruro, Bolivia) and the same age and breed (Wacaya) were used. The animals were transported, at 10 months of age, to the Experimental station in Choquenaira, where they were kept under an extensive system on native pasture (mostly *Stipa ichu* and *Calamagrostis ssp*) in the ecoregion corresponding to the Northern *Altiplano*.

The animals were slaughtered at the age of 16 months in the slaughterhouse in Palcoco, 45 km distant, from the city of El Alto (La Paz), applying the same method used for cattle.

The classification of carcasses according to conformation was performed by applying the subjective visual appraisal technique and observing the internal profile of the leg and the conformation of the rump. The classification pattern used was based on the system developed for ovine livestock in accordance with EU rules CEE N° 461/93 and was properly modified for its use in camelids (Colomer *et al.* cited by Ruiz *et al.*, 2000). The scaling system comprised 5 classes with the following characteristics:
- Class 1 "Poor" conformation: with long and thin legs and "V"-shaped internal profile; the muscular mass of the leg is bigger than that of the buttock. The rump is only slightly bulky and stretched.

- Class 2 "Normal" conformation: with stretched but proportionate legs and "V"-shaped internal profile and outward-bending branches; the leg and the buttock portions are similar. The rump length is slightly greater than its width.
- Class 3 "Good" conformation: with rounded, fairly long and fairly thick legs and "Y"-shaped internal profile. The rump is prominent and tends to be square.
- Class 4 "Very good" conformation: with slightly short legs and with "Y"-shaped internal profile and outward-bending branches. Of these, the horizontal branch, located at the same height as the perineum, separates into two similar muscular masses. The rump is prominent and its width is greater than its length.
- Class 5 "Excellent" conformation: with relatively short legs which have atrophic muscles in comparison to those of the buttock. The internal profile is "Y" shaped with considerably outward-bending branches. The rump is bulky, wide and short.

The other parameter used was the ischio-tarsal profile which is recommended for cattle by the Scientific Association for Animal Production (Associazione Scientifica di Produzione Animale – ASPA, 1991). In this case the classification is based on a pattern created for cattle. It considers, in carcasses hanging from a gambrel (see Figure 1), the profile of the superficial muscle of the buttock and the *Semitendinosus* muscle of the leg, in relation to a straight line extending from the tuberosity of the ischium to the front face of the tarsus

The classification based on fat coverage was performed through visual appraisal and applied to the following five anatomical regions, as established and recommended by ASPA (1991) for cattle: leg (A), loin (B), ribs (C), shoulder (D) and neck (E) (see Figure 2). Each region was evaluated according to a system of five classes: Class 1, no fat coverage; Class 2, some infiltrations of a thin layer of fat; Class 3, thicker layers of fat in some parts; Class 4, mostly covered by a thick layer of fat; Class 5, completely covered by a thick layer of fat. The five regions were firstly identified in order to classify the carcasses. The average value was then calculated and expressed by one of the nine scores (1, 1.5, 2, 2.5...5) assigned to each of the five regions.

Objective measures

External measurements were made on the cold carcass hanging from the gambrels which were placed at an average distance of 9.3 cm from one another, ensuring that the tibia remained parallel. The measurements recommended by ASPA (1991) and Ruiz *et al.* (2000) were the following: rump perimeter (A), rump width (B), body length (C) and thorax width (D) (see Figure 3). Likewise, two measurements were made on the tarsus width. These were (E) for the

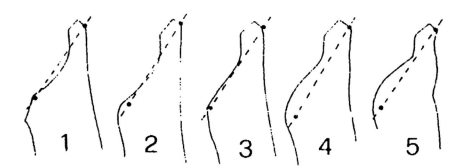

Figure 1. Classification of the carcass for the ischio-tarsal profile.

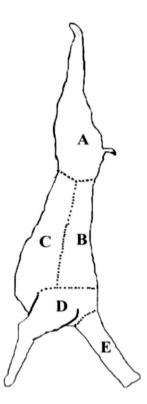

Figure 2. Regions of the evaluation of the fat coverage.

distance between the lower border of the left face of the calcaneus and the lower border of the astragal and (F) for the distance between the internal malleolus of the tibia and the malleolus located at the base of the calcaneus.

The following internal measurements were made on the half carcass hanging from the gambrel: leg length (A – B), carcass length (C – D) and thorax depth (E – F) (see Figure 4). Subsequently, the half carcasses were jointed and five prime cuts of leg, shoulder, loin, ribs and neck prepared. In the case of the dorsal (back) aspect, transversal cuts were made at the level of the 9th thoracic vertebra at the same height as the muscle, and similarly at the 12th vertebra. The larger and smaller diameters of *Longissimus dorsi* muscle were then measured. The contour of the transversal surface of the muscle was traced on transparent material and its area was calculated using image analysis methodology.

A multiple regression analysis was carried out using the weight of important commercial cuts as the dependent variable and the objective measures of the carcass as the independent variables, in accordance with the method of Backward variable elimination.

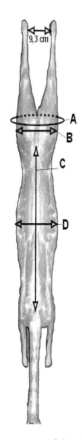

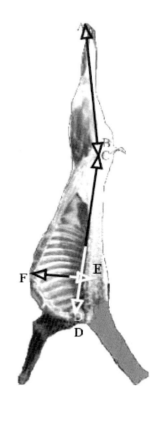

Figure 3. External measures of the alpaca carcass.   *Figure 4. Internal measures of the carcass.*

Results and discussion

Subjective classification of the carcass conformation

Results from the subjective classification of the rump demonstrates the presence of various differences among the three parameters studied (ie. profile of the leg, rump conformation and ischio-tarsal profile) (Table 1). With regard to the internal profile and conformation of the leg, the highest percentage of the carcasses was rated for Class 2 (28%), followed by Class 3 (28%) with 28% between Classes 2 and 3. One animal (corresponding to 4%) was assigned as Class 1, while 12% were rated between Classes 1 and 2. When classified according to rump conformation, the highest percentage was concentrated in Class 2 (68%), followed by Class 3 (20%) and 4% for Class 2-3, Class 1-2 and Class 1. With the classification based on the ischio-tarsal profile, most of the carcasses were rated as Class 2 (76%) with 20% in Class 3 and 4% between Classes 1 and 2. Considering all three classification parameters, no carcass showed "very good" or "excellent" conformation. (Class 4 and Class 5, see Table 1).

The average live weight of the animals was approximately 34.7 kg. When related to carcass conformation, the results are consistent with the report by Garnica *et al.,* (1993) for *Wacaya* alpacas of approximately the same age (18 months) and fed with natural pastures. These animals

Table 1. Proportion (%) of male alpaca carcasses classified by conformation and fat coverage, according to parameter and class (N =25, 16 months of age).

Parameter		Class (score)						
		1	1.5	2	2.5	3	4	5
Internal profile of the leg	(%)	4	12	28	28	28	0	0
Conformation of the rump	(%)	4	4	68	4	20	0	0
Ischio-tarsal profile	(%)	0	4	76	0	20	0	0
Fat coverage*	(%)	4	50	12	35	0	0	0

*Accumulated percentage

reached an average live weight of 45 kg. This information suggests that animals might exist with carcasses which have a better conformation, possibly rating as Class 4 or 5.

It has been observed that the internal profile of the leg of Class 1 alpacas is shaped as a closed "V" with straight branches which contrasts with the open shape typical of ovine livestock. Their legs appear to be stretched and the profiles of their muscles are straight; the rump is narrow, its length is bigger than its width and only slightly rounded, and, in general the thorax and the shoulders of the carcasses look stretched and narrow (see Figure 5).

In contrast, the two animals rated as Class 3 (good conformation) displayed rather short legs. They had a "Y"-shaped internal profile and slightly convex muscles; the rump tended to be square-shaped; the distance between the iliac spine and the ischium appeared shorter than the measurement between each hip joint (see Figure 5).

As for the ischio-tarsal parameter (originally designed for cattle), it was not possible to clearly determine the variation between the profiles of the buttock muscles and the legs. This is because the hind limbs of the bovine are shorter in comparison to the whole body and their profiles are better defined. In contrast, the hind limbs of llamas and alpacas are elongated, which makes it more difficult to evaluate the profile of the leg.

Subjective classification according to fat coverage

The average values based on fat classification demonstrated that 4% of the carcasses were generally rated as Class 1, 50% between Classes 1 and 2, 12% as Class 2 and 35% as between Classes 2 and 3. Likewise, no carcass showed the characteristics of Classes 1 or 2. These results are consistent with those of Foraquita and Bustinza (1993), who described 50% of 18-month-old alpacas as being deficient or regular, because the carcasses displayed a low or only incipient development of fat coverage.

The average scores for each region in the 25 carcasses used in this study are the following: 2.26 for the shoulder; 1.98 for the neck; 1.8 for the back; 1.8 for the ribs; 1.64 for the leg. These results demonstrate that the development of subcutaneous adipose tissue occurs most quickly in the shoulder with the slowest rate of occurring in the leg. Similarly, fat deposition tends to develop more quickly in the neck than in the loin. This result contrasts with the conclusion of Osorio (1995) that the back has a faster rate of development than the shoulder in Spanish breeds of sheep.

Class 1

Class 3

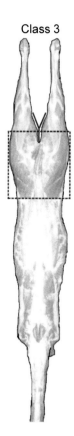

Figure 5. Alpaca carcasses classified according to the internal profile of the leg and rump.

By comparing the overall mean value obtained in the general qualification (1.89) and the average values of the neck (1.98), the loin (1.80) and the ribs (1.80), differences were shown not to attain statistical significance (P>0.05). In contrast, the differences are considerable when the overall mean is compared to the average values of the shoulder (2.26) and the leg (1.63). With regard to the correlation analysis, the loin is highly correlated to the general average (r = 0.86), followed by the ribs (r = 0.83), the shoulder (r = 0.82) and the neck (r = 0.81). In considering these results, it is concluded that the loin is the most representative region for the classification of alpaca carcasses. This statement is consistent with conclusion of Osorio (1995), that the loin is a good indicator for the tissue composition of ovine carcasses.

Indices of compactness

The analysis of the indices of compactness for the relationship of cold carcass weight to carcass length (WC/LC) resulted in values that varied between 0.23 and 0.37. Among these, the lower values are similar to those recorded for ovine livestock of the Spanish Merino and Manchega breeds (0.21). The indices of compactness for rump width and leg length (RW/LL) varied between 0.40 and 0.49. These values are lower than those of sheep, which ranged between 0.6 and 0.8 depending on the breed (Osorio, 1995). This difference can be explained by the greater length of leg of the alpaca as a function of body length, compared to that observed in sheep. This anatomical characteristic affects the index value, although it does not mean that alpacas, or camelids in general, are not valuable for meat production. In fact, Ruiz (2000) and Osorio (1995) indicated that long and (probably) poorly conformed carcasses may feature a higher or equal percentage of prime or lean joints when compared to sheep. Representative values, for alpaca carcasses, of 36.4%, 12.2%, and 21.7% for the leg, loin and shoulder respectively are higher than those reported for Spanish Merino sheep carcasses of 32.6%, 11.5% and 20.7%, although the latter are generally considered to be well conformed and compact.

Relationship between the Index of Compactness and the subjective classification

A low correlation coefficient (0.39) was obtained from analysis of the relationship between the compactness index for RW/LL and the subjective classification based on the internal profile and conformation of the leg. The relationships of this index with that for rump conformation and ischio-tarsal profile resulted in average correlation coefficients of 0.57 and 0.26 respectively. The relationship between the compactness index for WC/LC and the parameters for the classification based on the internal profile and the conformation of the leg is characterized by a low correlation (r = 0.35). Conversely, the same index when tested against rump conformation shows a high correlation (r = 0.61). No significant correlation was evident between the WC/LC index and the ischio-tarsal index. In all the cases the level of statistical significance is $P<0.05 \geq 0.32$.

As shown in Table 2, the parameter for the classification based on rump conformation is highly correlated to carcass weight and to the weight of such prime cuts as the loin and the leg. The parameter for the classification based on the internal profile of the leg shows a low correlation with the weight of the carcass, and even lower correlation values against both the weight of the loin and the ischio-tarsal profile. These results demonstrate that the rump conformation is closely related to the variation in the weight of the carcass, the leg and the loin.

Multiple regression analysis

The multiple regression analysis was conducted through the application of the Backward elimination method on the parts regarded as prime cuts (*e.g.* the leg and the loin). These cuts are important for their weight and for the possibility of obtaining additional quality minor cuts. Our interest focused on the prediction of their weight on the bases of other independent variables. The following most influential variables ($P<0.05$) were selected from the statistical procedure applied:

Table 2. Correlations between classification parameters and weight of prime cuts.

	Carcass weight	Leg weight	Loin weight	Internal profile, leg	Rump conf.	Ischio-tarsal prof.
Carcass weight	1					
Leg weight	0.96	1				
Loin weight	0.88	0.83	1			
Internal profile, leg	0.38	0.38	0.29	1		
Rump conformation	0.71	0.66	0.55	0.55	1	
Ischio-tarsal profile	0.23	0.24	0.23	0.30	0.25	1

Leg weight = -7.09 + 0.168 (rump width) + 0.007 (body length) + 0.471(F) + 0.087 (leg length) $r^2 = 0.91$ Eq. (1)

Loin weight = - 3.560 + 0.153 (rump width) + 0.474 (F) $r^2 = 0.67$ Eq. (2)

Carcass weight = -16.551 + 0.513 (rump width) + 0.274 (leg length) $r^2 = 0.85$ Eq. (3)

where: F = tarsus width

Equation (1) shows that for increases in each unit of measure for F (tarsus width) the weight of the leg increases by 471g. This value contrasts with the increase of 168g in the case of unit increases in the width of rump. Therefore, the most influential variable is F. Although they were not eliminated by the use of this method, the length of the leg and the width of thorax exerted no significant effect. The multiple determination coefficient shows that about 91% of the variation in the leg weight is caused by the independent variables.

In equation (2), the method selected two independent variables. Of these, F (tarsus width) is the most influential on the weight of the loin. Both variables explain 67% of the variation in the weight of the loin. In equation (3) the most influential factor in determining the weight of the carcass is the rump width, followed by the length of the leg.

Prediction parameters for prime cuts and the carcass

The multiple regression analysis of prime cuts, using the characteristics of the loin as independent variables, produced the following relationships:

Loin weight = -0.936 + 0.260 (B) + 0.051 (live weight) $r^2 = 0.91$ Eq. (4)

Leg weight = -0.383 + 0.228 (A) + 0.081 (live weight) $r^2 = 0.84$ Eq. (5)

Carcass weight = -1.67 + 0.080 (AB) + 0.24 (live weight) $r^2 = 0.90$ Eq. (6)

where: B = Thickness of the *Longissimus dorsi* at the level of the 12th thoracic vertebra.
A = Thickness of the *Longissimus dorsi* at the level of 9th thoracic vertebra.
AB = Area of the *Longissimus dorsi* at the level of the 12th thoracic vertebra.

It is concluded from equation (4) that the thickness of the loin (B) considerably affects its weight and that the latter increases by 260 g for each additional unit in the former. The equation also shows that approximately 91% of the variation of the weight of loin can be related to the independent variables live weight and thickness of the loin.

Equation (5) shows that the thickness of the loin (A) has the major significant effect on the weight of the leg, with each additional unit in loin thickness increasing the leg weight by 228g. The independent variables, live weight and thickness of the *Longissimus dorsi*, explain 84% of the variation in the weight of the leg. This conclusion is consistent with the report by Cadavez *et al.* (cited by Delfa, 2000), that 75-95% of the variation in the weight of the main parts of the ovine carcass is explained by live weight, depth of the *Longissimus dorsi* and thickness of the lumbar and sternal subcutaneous fat. Equation (6) indicates that the carcass weight variable is significantly affected by live weight so that each additional unit in the latter causes the former to increase by 240g.

Conclusions

The subjective classification based on conformation has demonstrated that within a sample population of twenty five 16-month-old alpacas there is only a small probability of finding all the conformation classes. The methods used to assess the carcasses are different and the results produced for each carcass do not always coincide. Thus, we obtained the highest percentage corresponding to those rated as "normal" for the three classification parameters. None of the carcasses used in this study was rated as Class 4 or 5, although some reports from Peru indicate the possibility of finding heavier animals which could fall into one of these classes.

Classification patterns for ovine carcasses were modified for application on camelids and were used to confirm the characteristics of the internal profile and the conformation of the leg. The following results were obtained for alpacas. Class 1 -"poor"- animals have a closed "V" shaped internal profile with the rump being long rather than wide. Class 3 -"good"- animals have a Y-shaped internal profile with slightly convex muscles and a square-shaped rump. The qualification based on the ischio-tarsal profile is more difficult due to the stretched morphology of the anatomy of the alpaca.

In general, the subjective classification based on fat coverage in five different regions of the carcass indicated the limited development of adipose tissue. Most animals were therefore rated as Class 1 – 2. No carcass presented the characteristics of Class 3, 4 or 5. These findings demonstrate that 16-month-old alpacas fed with natural pastures do not accumulate much sub-cutaneous fat. They thus produce lean meat, which is a characteristic typical of camelids.

In alpacas, the anatomical region showing the fastest development of adipose tissue is the shoulder, and the slowest is the leg. It is possible to consider the back region as the most representative for the subjective classification based on fat coverage, because of its very high correlation and similarity to the general average in the qualification of the entire carcass.

The analysis of indices for the compactness for the rump width and the length of the leg (RW/LL) demonstrates that the morphology of alpacas is less compact than that of Manchega and Merino sheep. This occurs even if the percentage of prime parts (leg, shoulder and loin) in alpaca carcasses is higher than in the carcasses of sheep classified as having a good morphology. The positive correlations among the weight of the prime cuts, the compactness indices and the subjective classification according to the rump conformation, demonstrate that this latter parameter must be considered in the classification of camelid carcasses.

Through multiple regression analyses using the method of Backward variable elimination, we obtained models that reveal the important influence of the tarsus width (F) and the rump width on the weight of the leg and the loin. As for the carcass weight, the rump width is more influential than the length of the leg. The thickness of the *Longissimus dorsi* is the prediction variable with the most significant effect on the variation of weights of the back and the leg, as important prime cuts. The most influential variable when considering carcass weight is shown to be live weight.

References

ASPA (Associazione Scientifica di Produzione Animale, IT), 1991. Metodología relative alla macellazione degli animali de interesse zootecnico e alla valutazione e dissezione della loro carcassa. Ismea- Roma, pp. 40-43.

Delfa, R., 2000. Predicción in vivo de la composición de la canal. Técnica de los ultrasonidos. In: V. Cañeque and C. Sañudo (ed.), Metodología para el estudio de la calidad de la canal y de la carne en rumiantes, Madrid ES. INIA – MCT. pp. 52–53.

Foraquita, S. and V. Bustinza, 1993. Inspección sanitaria y calificación de la carcasa. (Sanitary inspection and qualification of the carcass). In: V. Bustinza (ed.), Carne de Alpaca. Puno, PE, EPG-UNA, pp. 45–46.

Garnica, J. *et al.*, 1993. Carcasa y su rendimiento. (Carcass and their yield). In: V. Bustinza (ed.), Carne de Alpaca. Puno, PE, EPG-UNA, pp. 60.

IBNORCA (Instituto Boliviano de Normalización y Calidad), 1997. Clasificación de las canales de camélidos sudamericanos de matanza (alpacas y llamas). (Classification of the carcass of South American camelids of slaughter), Norma Boliviana, pp. 794–797.

Osorio, J. *et al.*, 1995. Estudio comparativo de la calidad de la canal en el tipo ternasco según procedencia, (Comparison of the quality of the carcas in the type ternasco according to origin). Rev. Bras. Agrociencia 1: 145–150.

Ruiz de Huidobro, F. *et al.*, 2000. Morfología de la canal ovina (Morphology of the carcass sheep). In: V. Cañeque and C. Sañudo (eds.), Metodología para el estudio de la calidad de la canal y de la carne en rumiantes, Madrid, ES, INIA – MCT, pp. 82–99.

A current perspective on the biology of fibre production in animals

H. Galbraith

School of Biological Sciences, University of Aberdeen, Hilton Place, Aberdeen, AB24 4FA, United Kingdom.

Abstract

South American camelids produce commercially important fibres from hair follicles. These are embedded in the skin and have close similarities to those of other fibre-producing animals. Two major types of follicle have been characterised. These are primary follicles, which are usually larger and produce fibres that are longer, have a greater diameter and are less commercially valuable than fibres from the more numerous secondary follicles with which they are associated anatomically. Animals with high ratios of secondary to primary follicles are generally favoured in breeding programmes. Follicles develop in the embryo from interactions of mesodermal and overlying ectodermal tissue, which extends into the dermis to form the hair follicle unit. Post-natal fibre is produced from division and subsequent differentiation of basal epidermal cell "keratinocytes" in the follicular bulb matrix under the influence of signals from the contiguous dermal papilla (DP). The volume of the DP affects the diameter of the hair fibre. The presence of pigmentation depends on the presence of melanocytes in the bulb matrix. The anatomical structure and activity of post-natal follicles vary according to species and location on the body. All follicles are characterised by cycles of fibre growth which differ in length of active growth (anagen) followed by no growth (telogen) subsequent to apoptosis (programmed cell death) of epidermal cells in the lower bulb matrix. Regeneration involves the production of a new bulb matrix with existing DP and frequently shedding of the previously produced fibre. Control mechanisms involve hormones including those mediating response to photoperiod and physiological state (*e.g.* pregnancy and lactation), and local growth factors and signalling pathways directing communication within and between dermis and epidermis. Fibre production also depends on the provision of nutrients including minerals, vitamins and sulphur-containing amino acids. This review considers the above topics with reference to information derived from a range of animal species and relates these to more limited information currently available from studies on South American camelids.

Resumen

Las especies camélidas en América del sur producen fibras de importancia comercial en sus folículos pilosos, los cuales están localizados en la piel y tienen una similitud cercana a aquellos de animales productores de fibra tradicionales. Dos tipos de folículo han sido caracterizados. Unos son los folículos primarios, los cuales son generalmente más grandes y producen fibras que son más largas, tienen un diámetro más grande y son de menor valor comercial que las fibras producidas por los muchos más numerosos folículos secundarios con los cuales están asociados anatómicamente. Aquellos animales con altas proporciones de folículos secundarios en relación a los primarios son generalmente deseados en programas de mejoramiento. Los folículos se desarrollan en el embrión a raíz de las interacciones entre los tejidos mesodérmico y ectodérmico que se extienden hasta la dermis para formar la unidad del folículo piloso. La fibra postnatal se produce de la división y subsiguiente diferenciación de la célula epidérmica base ("keratinocitos") en la matriz del bulbo folicular bajo la influencia de señales de la papila dermica contigua (PD). El volumen de PD afecta el diámetro del la fibra pilosa. La presencia de pigmentación depende de la presencia de melanocitos en la matriz del bulbo. La estructura anatómica y la actividad de los folículos postnatales varían de acuerdo con la especie y su localización en el cuerpo. Todos los folículos están caracterizados por ciclos de crecimiento de fibra que difieren de acuerdo a la longitud de tiempo del crecimiento activo (anágena) seguido de una de no crecimiento (telógena), subsiguiente a la apoptosis (muerte celular programada) de las células epidérmicas en la matriz de bulbo inferior. La regeneración consiste en la producción de una nueva matriz del bulbo con la PD existente y una frecuente perdida de la fibra producida anteriormente. Entre los mecanismos de control se encuentran varias hormonas incluso aquellas que intervienen en la respuesta al foto periodo y al estado fisiológico (Vg. preñez y lactación), factores de crecimiento local y las cascadas de señalización intracelular que controlan la comunicación entre la dermis y la epidermis. La producción de fibra también depende en la provisión de nutrientes incluyendo minerales, vitaminas y aminos ácidos que contienen azufre. Esta revisión considera los temas arriba mencionados con referencia a la información derivada de un rango de especies animales y los relaciona con información más limitada actualmente disponible de estudios que han sido llevados a cabo en especies camélidas de Sudamérica.

Keywords: hair follicle cycle, dermis, epidermis, nutrition, photoperiod

Introduction

The most important fibres produced by animals are those derived from secondary hair follicles with commercially value dependent on "fineness" as determined by smallness of diameter. The four species of South American camelids produce fibres with an approximate range in fineness as follows: Vicuña (<15μm), Guanaco (15-18μm), Alpaca (18-30μm) and Llama (>20μm) (Russel, 1993; SAC, 2005). Alpaca have a predominantly secondary follicle single coat and are domesticated as are the double-coated Llama while the two essentially undomesticated species are also double-coated. Annual yields of raw fibre have been reported as varying between 1.5 and 5.5kg for Alpaca, 1.5 to 2.0kg for Llama and 0.5-1.5kg for UK Guanaco (Moseley, 1995) with greater quantities of generally reduced fineness produced by males and associated with increasing age.

In keeping with practices used for other species such as sheep and goats (Russel, 1993) guard hair and kemp produced by primary follicles require to be separated from the fine fibre product of the secondary follicle prior to processing, and have limited value. The commercial value of fine fibres is optimised by high yields, long staple lengths, small diameters and minimal contamination with primary fibres and other materials. Those criteria are best met by animals with high ratios of secondary to primary follicles with the genetic potential for rapid growth while maximising fineness of fibres. Selection of alpaca in certain breeding programmes has included decisions based on measurement of follicular density in skin. Such an approach and its impact on fleece production is the subject of continuing discussion (*e.g.* Alpaca Journal, 2004). Other properties such as crimp and the presence of melanins and melanosomes as described by Cecchi *et al.*, (2004) are also important in defining the properties of fibres from individual species. Hair follicles are part of the mammalian integument that has many tissues with common properties. These include the production of keratinized end products in hair, skin, hoof and head horn. The growth of commercially valuable fine fibres by South American camelids may thus be considered in the context of general integumental tissue biology with precise morphology and behaviour specific to skin and hair follicle of the body site and genotype being considered. This review considers the development and basic structure of hair follicles, and identifies external and internal signalling mechanisms and nutritional and environmental influences which are centrally important to the regulation of hair fibre production.

Embryonic and post-natal development of hair follicles

Primary follicles are the first to form pre-natally in skin of animals in a window of development towards the end of the first trimester of pregnancy. Secondary follicles develop thereafter, in close association with primary follicles (Hardy, 1992). Hair follicles derive from epidermal (epithelial) and dermal tissues that originate from embryonic ectodermal and mesenchymal cells respectively. At the onset of follicle development, mesenchymal cells condense beneath the epidermal hair germ and release a signal (the 'first dermal message'), which stimulates the epidermal cells to form down-growths into the dermis. Subsequent development involves further interactions between epidermal and dermal cells to produce a structure, which has at its base a dermal papilla that is surrounded by a basement membrane and contiguous layer of epidermal cells in the matrix region of the follicle bulb. Such close proximity is important in facilitating the increasingly recognized "cross talk" between dermis and epidermis in skin and its appendages (Maas-Szabowski et al., 1999). This exchange of chemical signals provides regulation of morphological development and subsequent extracellular, cellular, and subcellular behaviour of both tissues pre- and post-natally. Development in the primary follicle gives rise to associated structures, the arrector pili muscle, the sebaceous and apocrine glands and hair canal. Development in the secondary follicle is confined to production of the sebaceous gland and hair canal only.

Dermal signals are considered to induce the production of a follicle with properties specific to location. The concentric arrangement of epidermal cells in the matrix region gives rise, on division and differentiation of epidermal basal cells, to the centrally positioned cortex and the increasingly peripheral cuticle and inner and outer root sheath layers. Numbers of follicles are considered to be genetically determined and fixed around birth with no new follicle production thereafter. However, changes in activity of individual follicles may occur during the lifetime of the animal. These include reductions in the activity of primary follicles in skin of animals such as Angora goats and Merino sheep with secondary follicle maturation for up to six months post-natally (Antonini et al., 2004). These workers have also described the activity of primary and secondary follicles in Peruvian Alpaca and Llama genotypes as affected by season and physiological development. The hair fibres present at birth are the product of proliferation, with progressive expression of cytoskeletal and other genes associated with differentiation, of cells of cortex and cuticle as the cells move towards the skin surface. Such genes include those for keratins and intercellular adhesion proteins, which are important in determining physical properties of fibre.

The dermal papilla and epidermal keratinocytes

While mechanisms responsible for the control of embryonic hair follicle development and post-natal behaviour are poorly understood, there is clear evidence that the dermal papilla plays a central role. It is formed from a stable population of specialized fibroblasts during embryonic development. These fibroblasts do not exhibit proliferative activity in vivo, but changes in volume of cytoplasm and composition of the extracellular matrix do occur in association with changes in physiological state of the hair follicle (Messenger, 1993). Macromolecular extracellular components of the dermal papilla include collagens and heparan sulphate and chondroitin sulphate proteoglycans. Along with collagens IV and VII and laminin these proteoglycans are important components of the basement membrane, which separates the dermal papilla and epidermal matrix of the hair follicle. In addition to a structural role, negatively charged proteoglycans may bind generally to positively charged molecules and in the case of transforming growth factor-β act specifically as a low affinity receptor, protecting it from degradation and promoting its binding to high affinity receptors (Bond et al., 1996). The basement membrane is considered to be largely continuous, although gaps have been described by electron microscopy during development. It appears that direct contact between dermal and epidermal cells is not required thereafter for successful interaction between the papilla and established follicular epidermis (Oliver & Jahoda, 1989). These workers also demonstrated the essential role of the dermal papilla in regulating postnatal epidermal cell proliferation and differentiation. Unlike the epidermal component of the follicle bulb region which is avascular the dermis is vascularised and hence directly exposed to systemic chemical signals and supply of nutrients. Cell-to-cell communication and exchange of nutrients in the follicle epidermis occur via gap junctions with well-recognised expression of connexin proteins. There is evidence to suggest that basal epidermal stem cells in skin do not actively express connexin-43 unlike those, which proliferate and commit to differentiation (Matic *et al.*, 2002). The volume of the dermal papilla is also a contributing factor in determining the numbers of basal epidermal cells (keratinocytes) lining the basement membrane and affecting directly the diameter of the hair cortex and subsequent diameter of the hair fibre.

The hair follicle growth cycle and composition of fibre

A characteristic property of the hair follicle postnatally is its cycle of active hair growth (anagen), proceeding through regression and shortening (catagen) to the resting (telogen) phase. Important features of anagen phase include the proliferation, in the bulb, of basal epidermal cells, one daughter cell of which moves suprabasally towards the skin surface to form the concentric layers of the hair shaft, the inner root sheath and the outer root sheath depending on original position in the follicle bulb (Ebling *et al.*, 1991). The other cell remains to divide again.

The growth of the hair fibre during anagen has been well described. In sheep wool, for example, a number of zones have been described which identify the changes occurring in hair follicle cells as they migrate from the bulb region to the skin surface (Orwin, 1989; Galbraith, 1998). The zones include basal epidermal cell mitosis after which daughter cells destined for differentiation move upwards undergoing progressive elongation and further keratinization, disulphide bond cross-linking and cortex dehydration, separation from inner root sheath and appearance at the skin surface. The hair product comprises largely of fully differentiated cortical cells containing large amounts of proteinaceous intermediate filaments (IFs) embedded in a matrix comprised of non-filamentous intermediate filament-associated proteins (Gillespie, 1991). The intermediate filaments, which are an important component of the cell cytoskeleton, are composed largely of combinations of the approximately twenty species of keratin proteins. These proteins are partly α-helical in structure and have a relatively low concentration of cystine (low-sulphur (S)-containing proteins). The intermediate filament-associated proteins (IFAPs) are composed of two molecular families one of cysteine-rich (high-S-containing) proteins and the other with high concentrations of glycine and tyrosine (high tyrosine proteins). Gillespie (1991) has summarized concentrations of sulphur S-containing amino acids in the low- and high-S and high-tyrosine protein fractions in a range of integumental tissues and products. Examples, expressed as S-carboxymethylcysteine residues per 100 amino acid residues, include values of 6.0 (Lincoln sheep wool: low S fraction), 18.0 (Merino sheep wool: high S fraction) and 6.0 (Merino sheep wool: high tyrosine fraction).

The hair shaft also contains a patterned cuticle outer layer with dimensions characteristic of the fibre type and, where present, a central hollow internal medulla. The presence of medullation in fibres arises from the loss of cells that lack the internal keratinous cytoskeletal structure and intercellular adhesion necessary to maintain their presence in the central cortex. Mechanisms responsible for the loss of structural integrity of such suprabasal cells arising from the tip of the dermal papillae are poorly understood. There are parallels with the presence of hollow tubules in cornified horn of bovine hooves.

Anagen-catagen-telogen changes and apoptosis

The duration of stages of anagen is a very important property for the production and maintenance of a fleece for animals in adverse environments and for its eventual harvesting. It is well recognized to be genetically determined for each hair type and anatomical location for a given species (Messenger, 1993). Anagen is followed by a short catagen phase involving the termination of mitosis of basal matrix cells and the elimination of the lower transient epidermis component of the hair follicle. The base of the follicle keratinizes to form the brush or club end, which rises to a position just below the sebaceous gland. The dermal papilla also moves higher in the dermis to remain in contact with epidermal cells of the hair germ. Certain epidermal cells in the bulb matrix are considered to undergo programmed cell death (apoptosis) following changes in expression of apoptosis-associated genes. In contrast, expression of apoptosis-inhibiting genes has been demonstrated to continue in dermal papilla fibroblasts, which remain viable throughout, although undergoing some morphological changes throuhout the follicle cycle (Galbraith, 1998).

Hair follicle cells of both the epidermis and dermis remain essentially inactive in the telogen (resting) phase which continues until the induction of a new anagen. The timing of shedding or moulting of the club hair during telogen or frequently at the beginning of a new anagen, is of considerable importance in domesticated wool-bearing animals, because of the commercial value of the fibre. Animals such as cashmere-bearing goats have a high degree of hair follicle cycle synchronization and combing the fleece or shearing with clippers are useful means of harvesting the fibre. Other species of animals such as guinea- pig and man, have a less orderly 'mosaic' pattern, with large numbers of both anagen and telogen follicles simultaneously present (Ebling *et al.*, 1991).

Development of new anagen and seasonal control

New anagen is considered to involve regeneration of the hair matrix from epidermal (stem) cells derived from the upper permanent "bulge" region of the follicle following receipt of signals from the dermal papilla (Rosenquist & Martin, 1996). The result is a new hair bulb with basal epidermal cells once again surrounding the dermal papilla following renewal of the proliferative component of the hair follicle. The new basal cells have a finite proliferative potential which, when achieved, results in mitosis ceasing and the follicle entering catagen. The characteristics of individual hair follicles are influenced by local and systemic control factors and intrinsic receptor expression which may be postulated to include molecular clock gene mechanisms (*e.g.* Badiu *et al.*, 2003).

There is particular interest in the development and cycling behaviour of follicles in hair-fibre-bearing ruminant or pseudo-ruminant animals that include South American camelids. Included among these are the fine wools from essentially single-coated breeds of sheep such as Merino and mohair-producing Angora goats and cashmere from the double coated Cashmere goat. Cashmere production is probably the most studied and is recognized to be subject to regulation by photoperiod with major fibre growth occurring between the summer and winter solstice with fibre shedding taking place as daylength increases in spring (McDonald *et al.*, 1987; Dicks, 1994). Growth of mohair by Angora goats is much less seasonally dependent although there is evidence to suggest from studies in the United States that up to 40% of follicles may be in catagen during winter (Margolena *et al.*, 1974). This value contrasts with 10% inactive follicles in summer although the contribution of other influences such as environmental temperature and nutrition is not clear. Studies in camelids include the observations of Russel and Redden (1995) that under northern UK conditions growth of undercoat and guard hair by adult male llama did exhibit seasonality although the periods of anagen and telogen extended beyond the normal time periods in any one calendar year. Length of undercoat was shown to be greatest in the period from July to December in a pattern similar to that typically observed in the Cashmere goat. Allain *et al.* (1994) have suggested that single-coated animals that have been genetically selected to produce fibre mainly from secondary follicles are essentially aseasonal and have long periods of anagen and unsynchronized patterns of telogen.

Systemic regulation of hair follicle growth and cycling

Studies in cashmere-bearing goats have provided a useful basis on which to assess the role of systemic and local signalling systems in the regulation of the hair follicle cycle. Such systems include the expression of seasonality in response to changes in photoperiod in which cashmere production is associated with increasing exposure of individual hair follicles to endogenous melatonin and reductions in prolactin concentrations in blood as daylength shortens. Following subsequent catagen and telogen, moulting of fibre and preparation for the next anagen is considered to be associated with increased prolactin concentration and reduced exposure to melatonin as daylength increases. A number of workers have studied hair follicle behaviour by treatment with, for example, melatonin implants (Dicks, 1994), pinealectomy or compounds, which affect circulating prolactin concentrations. Results obtained were frequently dependent on the timing of the treatment with respect to position in the follicle cycle and photoperiod. In addition, the presence of prolactin receptors in sites including dermal papillae and the apocrine sweat gland have been described in anagen wool follicles of Wiltshire sheep under New Zealand conditions (Choy *et al.*, 1995) and thus further implicating this hormone in the mediation of follicle cycle activity. The *in vitro* studies of Ibraheem *et al.* (1993, 1994) have also provided evidence for a direct effect of melatonin, although not prolactin, on caprine hair follicles. Dicks *et al.* (1996) did not detect the presence of melatonin receptors in such follicles. However, more recent reports have described expression of melatonin membrane receptors in the murine hair follicle and suggested that melatonin may affect directly regulation of the follicle cycle (Kobayashi *et al.*, 2004).

The presence of receptors for IGF-1 which may derive from systemic as well as local sources has also been described by Dicks *et al.* (1996) at different stages in the hair follicle cycle of Cashmere-bearing and Angora goats. Possible roles for a range of other hormones such as thyroxine and triiodothyronine, insulin, cortisol and other corticosteroids and oestradiol-17β (Messenger, 1993; Rhind & McMillen, 1995) have also been implicated in hair follicle regulation in species such as goats, mice and red deer. In addition, androgens such as 5α-dihydrotestosterone are recognised to influence hair growth activity in man and other mammals with particular effects dependent on site of follicle and presence of androgen binding receptors in dermis (*e.g.* scalp hair loss and stimulation of beard growth) (Messenger, 1993; Ando *et al.*, 1999) There is also evidence of a role of immune system activators and suppressors in determining hair follicle cycle behaviour (*e.g.* Christoph *et al.*, 2000).

Local regulation of hair follicle activity

Studies into the role of locally-produced activators and inhibitors of hair follicle activity have been assisted by techniques such as organotypic cell culture, whole follicle culture, localised *in vivo* studies, the detection of mitogens, morphogens and their receptors by techniques such as immunohistochemistry and polymerase chain reaction (PCR) methodology and investigation on animals with genetic mutations. The role of molecular mediators in the control of the adult hair follicle cycle has been recently reviewed in great detail by Stenn and Paus (2001). Mechanisms which may be paracrine, autocrine or intracrine are recognised to be involved in local control and these in turn may be subject to systemic regulation by hormones or individual nutrients.

These authors have highlighted the complexity of the epithelial-mesenchymal interactions, which determine follicle behaviour and its capacity for regeneration. They have considered particularly knowledge on families of genes responsible for producing molecular signals involved ubiquitously in a number of biological systems and which include fibroblast growth factors, transforming growth factor-β, the WNT pathway, Sonic hedgehog, neurotrophins and homeobox. These are in addition to their characterisation of signalling molecules such as epidermal growth factor, bone morphogenic proteins, hepatocyte growth factor, insulin-like growth factors and binding proteins, platelet derived growth factors and receptors, certain cytokines and enzymes including kinases which mediate cell cycle behaviour at particular sites within the follicle.

The recent review of Millar (2002) has also been helpful in defining current progress in understanding further the inter- and intra-signalling mechanisms that regulate hair follicle mitogenesis and morhogenesis. In particular, the WNT signalling pathway and its effect in inhibiting the degradation of β-catenin has been implicated in the production of the first dermal signal in embryonic skin as described above. The response of epithelial cells to this signal in the formation of a placode has been suggested to include WNT signalling. Fibroblast growth factors, the signalling molecule ectodysplasin and their receptor genes may also stimulate follicle formation, an effect which appears to be inhibited by certain members of the bone morphogenic family of molecular regulators. The first epithelial signals which stimulate formation of the dermal condensate are thought to include WNT and platelet-derived growth factor A. These along with the gene product Sonic hedgehog later in development are thought to regulate proliferation of epithelial cells and the development of the dermal papilla from the initial dermal condensate. Subsequent exchange of signals between dermis and epithelial components and their development is thought to be affected by expression of adhesion molecules such as β-integrin and α-catenin. The close proximity of the seven different layers of epithelial cells formed in the maturing hair follicle suggests tight regulation of commitment to individual cell phenotypes. Signaling molecules involved in the formation of the inner root sheath and hair shaft include WNT, Notch 1 and bone morphogenic protein 4 (BMP4). The inhibition of effects of BMP4 by the gene product Noggin has been shown to permit epithelial proliferation while impairing differentiation of hair shaft and cuticle. Another interesting transcription factor which suppresses differentiation while promoting expression of an acidic hair keratin gene is FOXN1. Mutation in WHN1 expression impairs the growth of hair as observed in nude mice.

Signalling molecules such as Sonic hedgehog and WNT are also thought to be involved in the regulation of the angle at which follicles are positioned in skin. These molecules, among a range of others (Sten & Paus, 2001) have been ascribed a role in the hair growth cycle where WNT signaling is important in both epithelia and in dermal papillae for epithelial activation and Sonic hedgehog supports subsequent epithelial proliferation in anagen. It is apparent that initial development of follicles in the embryonic skin and subsequent hair growth are the product of complex biological events, which include regulation at whole animal, follicle, cell and subcellular level. The mechanisms involved are incompletely understood for the intensively studied laboratory animals and even more so for animals such as camelids which have been the subject of limited research to date.

Nutritional influences on hair follicle activity

Energy, protein and amino acids

The development of feeding strategies for fibre-producing animal including those farmed under extensive husbandry where nutritional intervention may be difficult, requires consideration of the basic biology of hair production as outlined above. In addition, the effects of changes in physiological state such as body growth, lactation, pregnancy, heat or cold stress may increase competition and alter availability of nutrients for hair follicles. Provision of dietary nutrients may thus influence hair growth directly by providing components for cellular and extra-cellular structure, oxidisable energy, or catalysis in the follicle system or indirectly by altering concentrations of circulating hormones and numbers and receptivity of cellular receptors. Such homeostatic (short term) and homeorrhetic (longer term) mechanisms for maintaining the internal environment of the body (including the competition between other integumental and non-integumental tissues) would also be expected to apply. These direct or indirect influences, particularly in the context of fluctuating nutrient supply, may modulate the behaviour of the hair follicle cycle and alter fibre production by increasing rate of matrix cell proliferation and/or post-mitotic differentiation in anagen follicles. Examples of important nutrients (see also Galbraith, 1998; Reis & Sahlu, 1994) include amino acids, carbohydrates, lipids, minerals and vitamins.

In terms of fundamental biology, hair follicles require a range of nutrients to support tissue accretion and metabolic activity. Long chain fatty acids contribute to the structure of cellular membranes. A supply of oxidisable substrate is needed to provide energy for the synthesis of proteins and other molecules during proliferation and differentiation in the developing cortex, inner root sheath and supporting structures in follicular dermis and epidermis. The importance of dietary energy and protein and the high priority given to wool production in sheep was demonstrated in the early studies of Marston (1948). Other studies in fibre-producing goats have suggested that the magnitude of response to variation in energy and protein/nitrogen supply is dependent of factors such as genetic potential for fibre growth and competition between hair follicle and other body tissues according to physiological state. Reductions in fibre production typically associated with pregnancy and lactation have also been described (*e.g.* Galbraith *et al.*, 2000).

There is particular interest in the role of dietary protein and amino acids, particularly cysteine, in supporting synthesis of cells of the cortex and cuticle. Sheep wool with a value of approximately 9.0 residues cysteine per 100 total amino acid residues has a much higher concentration of this amino acid than skeletal muscle (1.1) as a competing tissue within the body (Galbraith, 2000). This concentration is also greater than that provided from rumen microbial protein (1.0) or extracted heat-treated soya bean meal (1.0), or white fish meal (0.9) as high quality protein supplements in diets for ruminants. Approximate concentrations of methionine in these sources are approximately 2.5, 1.4 and 3.0, respectively (Galbraith, 2000). Methionine supply is essential because of its in roles in polyamine synthesis, molecular methylation and initiation of polypeptide synthesis, in addition to provision of the sulphur moiety for cysteine synthesis, by transulphuration, in tissues such as liver and hair follicle of caprines (Galbraith, 2000; Souri *et al.*, 1997). The availability of amino acids from dietary protein sources for absorption at the small intestine depends on a number of factors which include, (a) for the dietary supply, ruminal degradability and digestibility of the undegraded fraction and (b) for ruminal microbial protein, its synthesis and post- ruminal digestibility.

The disparity in the concentrations of cysteine in hair fibre and other body tissues and dietary or ruminal microbial sources of sulphur amino acid supply are factors in the relatively low value (0.3) for conversion of metabolisable protein to hair follicle protein in animals such as sheep (AFRC, 1993). Shahjalal *et al.* (1992) obtained consistent increases in yields and diameter of raw fibre in response to protein supplementation in diets for Angora goats with evidence for a response to additional energy supply on low-protein diets used. Other authors (*e.g.* Russel, 1995) have suggested that secondary (cashmere-producing) follicles of Cashmere-bearing goats are insensitive to protein supplementation. In contrast, Souri *et al.* (1997) did obtain evidence of a response in cashmere but not primary follicle guard hair to dietary supplementation with protected methionine. Galbraith (2000) has suggested that variation in responses between studies may be explained by the adequacy of supply in meeting the requirements for uptake and utilization for individual follicles. Studies in the UK did not demonstrate responses by male llama (Russel & Redden, 1994) and male and female guanaco (Mosely, 1995) to supplementation of basal diets which contained above maintenance levels of nutrition. Evidence of responses in South American camelids to diets apparently deficient in energy and protein were also reviewed by Mosely (1995). There is, in addition, evidence for a particular signaling effect of amino acid supplementation on rates of mitosis and apparent keratin gene expression of epidermal basal cells, resulting in greater wool production in sheep (Hynd, 1989).

Vitamins and minerals

An adequate supply of vitamins and minerals is essential for normal function in skin (Sherertz & Goldsmith, 1991) and for hair follicles also. A range of deficiency symptoms have been described for vitamins which relate to their well understood roles in cell biochemistry. These include ribloflavin (epithelial atrophy in rat hair), niacin (impairment of hamster hair growth), pantothenic acid (hair loss in rats and dogs), folic acid (abnormal wool growth in lambs) and biotin (general effects on integumental tissues including caprine hair follicle (Tahmasbi *et al.*, 1996), vitamin A and retinoids (essential in proliferation).

Important minerals include magnesium, which has a role in energy-providing phosphate transfer reactions and DNA degradation and synthesis. Free calcium concentrations also provide an important mechanism of signal transduction in keratinocytes (Pruche *et al.*, 1996). Potassium deficiency in the rat causes hair loss and, in mice, production of a hair coat lacking in lustre. Manganese is essential for the synthesis of glycoproteins such as chondroitin, dermatan sulfate and collagen in the dermal extracellular matrix. Selenium has been shown to be important in the enzymatic de-iodination of thyroxine to triiodothyronine, which influences hair follicle activity in, for example, Cashmere goats (Rhind & Kyle, 2004). Zinc is an integral component of a large number of metallo-enzymes with important metabolic functions ranging from control of gene expression to metabolism of protein (includes zinc-metalloproteinases involved in tissue remodelling), fat and carbohydrate (Neldner, 1991), with cessation of wool growth with hair loss and dermatitis of the skin as symptoms of deficiency. Copper is also an essential micronutrient required for the formation and activity of important enzymes in the skin and its appendages (Danks, 1991). These include the production of cross-links in collagen and elastin fibres, the maintenance of melanin pigment production in hair follicles, the post-translational formation of disuphide bonds between cysteine molecules in hair proteins and a role in the formation of crimp and maintenance of structural strength in wool of sheep.

Conclusions

The production of fibre by South American camelids has similarities with hair production by primary and secondary follicles of other pseudo-ruminant, ruminant, and non-ruminant animals. These main types of hair follicle develop from dermis and epidermis in the embryo, with initial primary follicle production followed by formation of associated secondary hair follicles. Pre-natal development and the post-natal cycles of growth followed by quiescence and re-activation, specific to genotype and body site, are regulated by a variety of extrinsic and intrinsic influences including response to photoperiod and physiological status. It is influenced by systemic hormones, systemic and local growth factors, essential dermal-epidermal interactions and exchange of signalling molecules and intrinsic cellular sensitivities. Successful hair growth and optimal functioning of the hair follicle are dependent on an adequate supply of nutrients including energy from oxidisable substrates, amino acids including cysteine, at disproportionately high concentrations, and methionine and fatty acids, vitamins and minerals. Improved protein nutrition has been shown to increase wool growth in certain ovine and caprine genotypes but apparently not in South American camelids fed at supra-maintenance levels of nutrition. Further work is needed to characterize the precise hair follicle biology of these animals in order to optimise genetic selection, husbandry and successful commercial production of their fibre.

References

AFRC (Agricultural And Food Research Council), 1993. Energy and Protein Requirements of Ruminants. An Advisory Manual Prepared by the AFRC Technical Committee on Responses to Nutrients. Wallingford, CAB International.

Allain, D., R.G. Thébault, J. Rougeot and L. Martinet, 1994. Biology of fibre growth in mammals producing fine fibre and fur in relation to control by day length: relationship with other seasonal functions. In: Hormonal Control of Fibre Growth and Shedding. European Fine Fibre Network, J.P. Laker and D. Allain (eds.), Occasional Publication N° 2., Aberdeen, Macaulay Land Use Research Institute, pp 23-40.

Alpaca Journal, 2004. http://www.alpaca-journal.com/Category,category,Animal%20Selection.aspx (accessed 5 January 2005).

Ando, Y., Y. Yamaguchi, K. Hamada, K. Yoshidawa and S. Itami, 1999. Expression of mRNA for androgen receptor, 5 alpha-reductase and 17 beta-hydroxysteroid dehydrogenase in human dermal papella cells. British Journal of Dermatalogy 141: 840-845.

Antonini, M., M. Gonzales and A. Valbonesi, 2004. Relationship between age and postnatal skin follicular development in three types of South American domestic camelids. Livestock Production Science 90: 241-246.

Badiu, C., 2003. Genetic clock of biologic rhythms. Journal of Cellular and Molecular Medicine 7: 408-416.

Bond, J.J., P.C. Wynn and G.P.M. Moore, 1996. Effects of epidermal growth factor and transforming growth factor alpha on the function of wool follicles in culture. Archives of Dermatology 288: 373-382.

Cecchi, T., C. Cozzali, P. Passamonti, P. Ceccarelli, F. Pucciarelli, A.A. Gargiulo, E.N. Frank, and C. Renieri, 2004. Melanins and melanosomes from Llama (*Lama glama L.)*. Pigmentation Cell Research 17: 1-5.

Choy, V.J., A.J. Nixon and A.J. Pearson, 1995. Localisation of receptors of prolactin in ovine skin. Journal of Endocrinology 144: 143-151.

Christoph, T., S. Muller-Rover, H. Audring, D.J. Tobin, B. Hermes, G. Cotsarelis, R. Ruckert and R. Paus, 2000. The human hair follicle immune system: cellular composition and immune privilege. British Journal of Dermatology 142: 862-873.

Danks, D.M., 1991. Copper deficiency and the skin. In: Physiology, Biochemistry and Molecular Biology of the Skin, L.A. Goldsmith (ed.), New York, Oxford University Press, pp. 1351-1361.

Dicks, P., 1994. The role of prolactin and melatonin in regulating the timing of the spring moult in the Cashmere goat. In: Hormonal Control of Fibre Growth and Shedding. European Fine Fibre Network, J.P. Laker and D. Allain (eds.), Occasional Publication no. 2., Aberdeen, Macaulay Land Use Research Institute, pp. 109-125.

Dicks, P., C.J. Morgan, P.J. Morgan, D. Kelly and L.M. Williams, 1996. The localisation and characterisation of insulin-like growth factor-I receptors and the investigation of melatonin receptors on the hair follicles of seasonal and non-seasonal fibre producing goats. Journal of Endocrinology 151: 55-61.

Ebling, F.J.G., P.A. Hale and V.A. Randall, 1991. Hormones and hair growth. In: Physiology, Biochemistry and Molecular Biology in the Skin, L.A. Goldsmith (ed.), New York, Oxford University Press, pp. 660-696.

Galbraith, H., 1998. Nutritional and hormonal regulation of hair follicle growth and development. Proceeding of the Nutrition Society 57: 195-205.

Galbraith, H., 2000. Protein and sulphur amino acid nutrition of hair fibre-producing Angora and Cashmere goats. Livestock Production Science 64: 81-93.

Galbraith, H., B. Norton and T. Sahlu, 2000. Recent advances in the nutritionl biology of Angora and Cashmere goats. In: 7th International Symposium on Goats, L. Gruner and P. Chabert (eds.), INRA, France, pp. 59-65.

Gillespie, J.M., 1991. The structural proteins of hair: isolation, characterization and regulation of biosynthesis. In: Physiology, Biochemistry and Molecular Biology of the Skin, L.A. Goldsmith (ed.), New York, Oxford University Press, pp. 625-659.

Hardy, H., 1992. The secret life of the hair follicle. Trends in Genetics 8: 55-61.

Hynd, P.I., 1989. Effects of nutrition on wool follicle cell kinetics in sheep differing in efficiency of wool production. Australian Journal of Agricultural Research 40: 409-417.

Ibraheem, M., H. Galbraith, J.R. Scaife and S.W.B. Ewen, 1993. Growth and viability of secondary hair follicles of the Angora goat, cultured *in vitro.* Journal of Anatomy 182: 231-238.

Ibraheem, M., H. Galbraith, J.R. Scaife and S.W.B. Ewen, 1994. Growth of secondary hair follicles of the Cashmere goat *in vitro* and their response to prolactin and melatonin. Journal of Anatomy 185: 135-142.

Kobayashi, H., T.W. Dunlop, B. Tychsen, F. Conrad, T. Ito, N. Ito, S. Aiba, C. Carlberg and R. Paus, 2004. The murine hair follicle is a melatonin target. Experimental Dermatology 13: 583.

Maas-Szabowski, N., A. Shimotoyodome and N.E. Fusenig, 1999. Keratinocyte growth regulation in fibroblast cocultures via a double paracrine mechanism. Journal of Cell Science 112: 1843-1853.

McDonald, J.J., A. Hoey and P.S. Hopkins, 1987. Cyclical growth in Cashmere goats. Australian Journal of Agricultural Research 38: 597-609.

Margolena, L.A., 1974. Mohair histogenesis, maturation and shedding in the Angora goat. Technical Bulletin n° 1495, Beltsville, MD, United States Department of Agriculture.

Marston, H., 1948. Nutritional factors involved in wool production by Merino sheep. Australian Journal of Scientific Research 81: 362-375.

Matic, M. and M. Simon, 2003. Label-retaining cells (presumptive stem cells) of mice vibrissae do not express gap junction protein connexion 43. Journal of Investigative Dermatology Symposium Proceedings 8: 91-95.

Messenger, A.G., 1993. The control of hair growth: an over-view. Journal of Investigative Dermatology 101, Suppl. 1: 4S-5S.

Millar, S.E., 2002. Molecular mechanisms regulating hair follicle development. Journal of Investigative Dermatology 118: 216-225.

Moseley, G., 1995. Factors affecting the fibre yield and quality of domestic guanaco. In: European Fine Fibre Network, J.P. Laker and A.J.F. Russell (eds.), Occasional Publication N° 3., pp. 85-96.

Neldner, K.H., 1991. The biochemistry and physiology of zinc metabolism. In: Physiology, Biochemistry and Molecular Biology of the Skin, L.A. Goldsmith (ed.), New York, Oxford University Press, pp. 1329-1350.

Oliver, R.F. and C.A.B. Jahoda, 1989. The dermal papilla and maintenance of hair growth. In: The Biology of Wool and Hair, Q.E. Rogers, P.J. Reis, K.A. Ward and R.C. Marshall (eds.), Cambridge, Chapman and Hall Ltd., pp. 51-68.

Orwin, D.F.G., 1989. Variations in wool follicle morphology. In: The Biology of Wool and Q.E. Rogers, P.J. Reis, K.A. Ward and R.C. Marshall (eds.), Cambridge, Chapman and Hall Ltd., pp. 227-242.

Pruche, F., N. Boyera and B.A. Bernard, 1996. K + ATP channel openers inhibit the bradykinin-induced increase of intracellular calcium in hair follicle outer root sheath keratinocytes. In: Hair Research for the Next Millennium, D. Van Neste and V.A. Randall (eds.), Amsterdan, Elsevier, pp. 463-466.

Reis, P.J. and T. Sahlu, 1994. The nutritional control of the growth and properties of mohair and wool fibers - a comparative review. Journal of Animal Science 72: 1899-1907.

Rhind, S.M. and S.R. McMillen, 1995. Seasonal changes in systemic hormone profiles and their relationship to patterns of fibre growth and moulting in goats of contrasting genotypes. Australian Journal of Agricultural Research 46: 1273-1283.

Rhind, S.M. and C.E. Kyle, 2004. Skin de-iodinase profiles and associated patterns of hair follicle activity in cashmere goats of contrasting genotypes. Australian Journal of Agricultural Research 55: 443-448.

Rosenquist, T.A. and G.R. Martin, 1996. Fibroblast growth factor signalling in the hair growth cycle:expression of the fibroblast growth factor receptor and ligand genes in the murine hair follicle. Development Dynamics 205: 379-386.

Russel, A.J.F., 1993. Development of management systems. In: Alternative Animals for Fibre Production, A.J.F. Russel (ed.), Commission of the European Communities, Brussels, pp. 85-91.

Russel, A.J.F., 1995. Current knowledge on the effects of nutrition on fibre production. In: The Nutrition and Grazing Ecology of Speciality Fibre Producing Animals, J.P. Laker and A.J.F. Russell (eds.), European Fine Fibre Network Occasional Publication no. 3, Aberdeen, Macaulay Land Use Research Institute, pp. 3-21.

Russel, A.J.F. and H. Redden, 1994. Seasonal effects on camelid fibre production. In: European Fine Fibre Network, J.P. Laker and A.J.F. Russell (eds.), Occasional Publication N° 2, Aberdeen, Macaulay Land Use Research Institute, pp. 41-49.

SAC, 2005. http://www1.sac.ac.uk/management/External/diversification/Novstock/camelids.asp. (accessed 5 January, 2005).

Shahjalal, M., H. Galbraith and J.H. Topps, 1992. The effect of changes in dietary protein and energy on growth, body composition and mohair fibre characteristics of British Angora goats. Animal Production 54: 405-412.

Sherertz, E.L. and L.A. Goldsmith, 1991. Nutritional influence on the skin. In: Physiology, Biochemistry and Molecular Biology of the Skin, L.A. Goldsmith (ed.), New York, Oxford University Press, pp. 1315-1328.

Souri, M., H. Galbraith and J.R. Scaife, 1997. Comparisons of the effect of genotype and protected methionine supplementation on growth, digestive characteristics and fibre yield in Cashmere and Angora goats. Animal Science 66: 217-223.

Stenn, K.S. and R. Paus, 2001. Controls of hair follicle cycling. Physiological Reviews 81: 449-494.

Tahmasbi, A.M., H. Galbraith and J.R. Scaife, 1996. Development of an *in vitro* technique to investigate the role of biotin in regulating wool growth in sheep. Animal Science 62: 670.

Skin follicular structure in Bolivian llamas

T. Lusky[1], A. Valbonesi[1], T. Rodriguez[2], C. Ayala[2], Luan Weimin[3] and M. Antonini[4]
[1]Dipartimento Scienze Veterinarie, Università di Camerino, 62024 Matelica, Italy; alessandro.valbonesi@unicam.it
[2]Universidad Mayor de San Andres, La Paz, Bolivia.
[3]Jilin Agricultural University, Department of Animal Science, Chang Chun 1300118, Jilin Province, R.P. of China.
[4]ENEA C.R. Casaccia Biotec, Agro, S. M. di Galeria. 00060 Rome, Italy.

Abstract

Thirty llama kids of different sex and type "Q'aras" (or carguera) and "T'amphullis", born between January and April 1998 at the Patacamaya Experimental Station in Bolivia, were chosen for determining the age at which the hair follicles reach the maturity, as well as for comparing the skin follicular structure and activity among these different types of Bolivian llamas. Skin biopsies were taken from the right mid costal region at 2, 4, 6, 8, 10, 12 and 14 months of age in order to monitor four follicular parameters: 1) ratio of secondary to primary follicles (S/P), 2) percentage of active primary follicles (PAP); 3) percentage of active secondary follicles (PAS); 4) the percentage of medullated secondary (PMS) and 5) primary (PMP) fibres. The biopsies were immediately fixed in Bouin solution. Transverse sections of 7 μm were then cut with a rotary microtome and stained by using the Sacpic procedure, modified by Nixon (1993). As a general trend, the PAP parameter reached a maximum value at the 10th month and a minimum value at the 14th month, while the PAS parameter reached a maximum value at the 8th month and a minimum value at the 4th month. The mean value of the S/P parameter in the Q'aras type was found to be 5.35 with a maximum value reached at the 2nd month, while the minimum value was seen at the 6th month. In the T'amphullis type the mean value was found to be 5.45 with a maximum value at the 4th month and a minimum at the 12th.

Resumen

Trenta llamas crias de diferente sexo y tipo Q'aras (o carguera) y "T'amphullis", nacido entre enero y abril 1998 en la Estacion Experimental de Patacamaya en Bolivia, fueron escogido para determinar la edad de la maduridad de la estructura de los foliculos pilosos y a lo mismo tiempo para comparar la estructura y la actividad de lso foliculos entre los dos tipo de llama Boliviana. Cada dos meses (2, 4, 6, 8, 10, 12 y 14 de edad), del costillar medio se tomaron muestras de biopsias de piel para cuatros parametros de los foliculos: 1) relacion folicular (S/P); 2) porcentaje de los foliculos pilosos primarios (PAS); 3) porcentaje de los foliculos pilosos secundarios (PAP); 4) porcentaje de las fibras meduladas secundarias (PMS) y 5) primarias (PMP).La biopsias de piel fueron fijadas inmediatamente en solucion de Bouin. En las muestras, se realizaron cortes transversales de 7 μm y tincion Sacpic modificada por Nixon (1993). Como tendecia general la PAP llega a su maximo a 10 meses y el minimo a 14 meses de edad, cuando la PAS llega a su maximo a 8 meses y el minimo a 4 meses de edad. El valor medio del S/P en las Q es de 5.35, donde el maximo llega a 2 meses y el minimo a 6 meses de edad. El valor medio del S/P en las T es de 5.45 donde el maximo llega a 4 meses y el minimo a 12 meses de edad.

Keywords: llama, Bolivia, skin follicle

Introduction

South American camelids (SAC) population are dominated by alpaca (47%) and llama (44%); the former being mainly (91%) distributed in Peru and the latter (70%) in Bolivia (Bonavia D. 1996). The llamas population lives in Bolivian Andean regions, consisting of about 2 million heads (Martinez *et al.*, 1997). In Bolivian Andean Region, llamas breeding systems contribute to the economy of marginal production systems in extreme environments like the Andean area. Meat production is the main Bolivian llama "campesinos" strategy, while the fibre production is a secondary aim utilized for handcraft goods, because the llama, in contrast to alpaca fleeces, present coarser and more heterogeneous fibres. Alpaca is best used for fibre production because they present homogeneously fine, long and soft fleeces. Of the two types of alpaca, the "huacaya" is the more common and is characterized by compact, soft and highly crimped fibres, with blunt-tipped locks. The other type "suri" has straight, less-crimped fibres and locks with a "cork-screw" shape, very similar to those of Angora goat but not as bright.

In Bolivia, llamas are traditionally classified into two different types:
1. "Q'aras" (or cargera) is the typical double coated animal, with more guard hair (outer coat) and markedly less woolly fibres (undercoat) ranging from short to very short, and
2. "T'amphullis" (or woolly) which is a single-coated animal with soft, crimped secondary fibres but with a lowest quality fleece in respect to alpaca fibres, because many primary fibres are mixed within the secondary ones.

SAC fleeces are not, however, limited to the few types illustrated above, but are characterized by a great variability because of the interfertility occurring between llama and alpaca and the absence of selection programs in both Peru and Bolivia.

In any case Bolivian llama populations are unselected and show wide variability. In a Bolivian llama population, Martinez *et al.* (1997) analysed the followowing fleece parameters variability: percentage of medullated fibres (fragmented med, continuous med and kemp), fibre diameter of different fibre types (coarse, fine and medullated), average fibre length (undercoat and guard hair), clean fleece yields and crimps. In agreement to Martinez's results, the Bolivian llama fleece consist of a mix of heterogeneous types ranging from unmedulated to kempy fibres. Diameter and medullation vary according types of fleeces, age and types of animals. Quispe *et al.* (2003) studied llama population fleeces and fibre characteristics in the Potosì Department north and south Lipez Provinces. Quispe defined the T'amphullis (T) weight of commercial fleece higher then Q'aras (Q) type while there is no differences in weight of belly. He also showed no differences between the two types of llama in average fibre diameter (T 21.0 μm and Q 21.4 μm) and in percentage of medullation (T 13.9% and Q 15.4%), while the flock length of T (12.0 cm) was longer than Q flock length (10.4 cm).

The other parameter affecting the fleece and fibre production characteristics is the skin follicular structure and activity, but until now it has been scarcely investigated in SAC, despite the fact that the age at which alpaca and llama maximal fibre production is reached, is of fundamental importance for understanding the variability of Bolivian llama population, the correct SAC shearing management and the early animal selection. Aim of the present work is to fill this gap by determining the age at which all the secondary follicles reach the maturity in T and Q types, as well as to compare the skin follicular structure and activity among these different types of Bolivian llamas.

Materials and methods

Thirty llama kids of different sex (20 male and 11 female) and type (15 "Q'aras" and 16 "T'amphullis"), born between January and April 1998, were chosen for determining the age at which the hair follicles reach the maturity, as well as for comparing the skin follicular structure and activity. The experiment was performed at the Patacamaya Experimental Station in Bolivia, Department of La Paz, at 3,789 m above sea level (17°21'S, 67°45'W). The climate of this area is typical of high Andean Puna semi-arid ecosystem, with mean annual precipitation of 402 millimetres, with a coefficient of variation of 31%, and a mean annual temperature of 10.4 C degrees (Valdivia, 1996).

Skin biopsies were taken by means of suitable punch with 0.8 cm of diameter, from the right mid-side of each animal, *i.e.* approximately over the 10th rib, about half-way down the body (Frank *et al.*, 1989). This body area has been found to be more representative of fleece characters than other fleece regions (Martinez *et al.*, 1997). Skin biopsies were taken at 2, 4, 6, 8, 10, 12 and 14 months of age in order to monitor the llama skin follicular structure development. Skin samples were fixed in Bouin solution and stored in 80° alcohol for shipping to Italy. After storing, skin samples were dehydrated in a graded ethanol series and embedded in paraffin. Transverse sections of 7 μm were cut with a rotary microtome and stained by using the Sacpic staining procedure, modify by Nixon (1993). This dyeing method reveals the follicular Inner Root Sheath (IRS) and the cellular layer of the Other Root Sheath (ORS), which are the main histological structure for defining skin follicular activity. The level immediately under the sebaceous gland was defined as the most suitable depth for carrying out microscopical observations, because it contains the greatest number of detectable skin follicles and thus maximised the possibility of seeing the IRS.

The following parameters were recorded for each area: 1) total number of primary follicles; 2) active and inactive primary follicle number; 3) total number of secondary follicles; 4) active and inactive secondary follicle number; and 5) number of medullated fibres in both primary and secondary follicles. The recorded data were used to calculate:
* the ratio of secondary to primary follicles (S/P);
* the percentage of active secondary follicles(PAS);
* the percentage of active primary follicles (PAP);
* the percentage of medullated secondary fibres (PMS) calculated from the total number of secondary fibres;
* the percentage of medullated primary fibres (PMP) calculated from the total number of primary fibres.

Analyses of variance (ANOVA) were performed on S/P, PAS, PAP, PMS and PMP. As animals and age were not randomly chosen, a model I ANOVA was utilised, with fleece type, age, and sex considered as factors. All the statistical analysis were performed by means of SAS (1995) statistical software.

Results

Secondary to primary follicular ratio (S/P)

Among the two type of Bolivian llama no significant effect was detected in secondary to primary follicular ratio (Table 1.).

Moreover low variability was showed inside and between the both types of llama (minimum 5.18 and maximum 6.16). The result presents small differences with respect to the S/P 4.89

Table 1. Means ± SEM for skin follicle parameters for in two Bolivian Llama types.

Llama type	n.	S/P	PAS	PAP	PMS	PMP
T'amphullis	555	5.45 ± 1.53	49.59 ± 0.38	81.50[a] ± 0.37	15.06[a] ± 24.49	75.07 ± 41.28
Q'aras	920	5.35 ± 1.44	47.08 ± 0.37	74.76[b] ± 0.37	21.44[b] ± 26.25	82.35 ± 36.45

Means with different superscripts are significantly different. P≤0.05

value observed in a Peruvian woolly llama population (Antonini *et al.*, 2004). In the same study in Peruvian alpaca detected the following higher average value: 8.40 and 8.07 respectively in Huacaya Suri types. No significant sample period effect was detected in T and Q types (Table 2) while Antonini *et al.* (2004) in Peruvian llama recorded a S/P ratio highest value of 5.87 in the 4th months of age.

Percentage of active secondary follicles (PAS)

Present parameters showed no significant difference between sex and types of Bolivian llama while different age had a significant effect on PAS variation. The periods which corresponded to summer time (10 and 12 months old respectively 26.49 and 26.50% of the active secondary follicle) had the lowest value while intermediate seasons (8 months 62.69%, 2 months 51.49% and 14 months 57.93%) showed the highest PAS value. Finally, the medium secondary skin follicular activity value was expressed in winter period (4 months 44.25% and 6 months old 44.50%) (see Figure 1).

Table 2. Means for skin follicle parameters in Bolivian Llama types recorded at different ages.

Llama type	Age (months after birth)						
	2 (April)	4 (June)	6 (August)	8 (October)	10 (December)	12 (February)	14 (April)
S/P							
	5.49	5.46	5.34	5.35	5.33	5.48	5.45
Active secondary follicle %							
	51.49[c, d]	44.25[c]	44.50[c]	62.69[b]	26.49[a]	26.50[a]	57.93[b, d]
Active primary follicle %							
	71.31	80.60	75.83	84.58	77.70	54.95	74.80
T	100[a]	86.25[a, b]	80.82[a, b, c]	83.98[a, b]	88.02[a,b]	72.41[b, c]	71.42[c]
Q	64.28[c]	78.46[a, b]	71.21[b, c]	84.89[a]	70.83[b, c]	38.88[d]	77.76[a, b, c]
Medullated secondary fibre %							
	19.02	23.24	17.67	17.98	7.72	27.08	15.33
T	42.43[a]	17.83[b]	15.06[b]	14.82[b]	3.77[c]	5.98[c]	20.76[b]
Q	13.29[c, d]	24.38[b]	20.26[b, c]	19.62[c]	10.35[d]	46.50[a]	11.05[d]
Medullated primary fibre %							
	84.42[a]	89.78[a, b]	80.37[b]	76.12[b]	57.91[c]	75.61[b]	74.54[b]

Means with different superscripts are significantly different at P≤0.05.

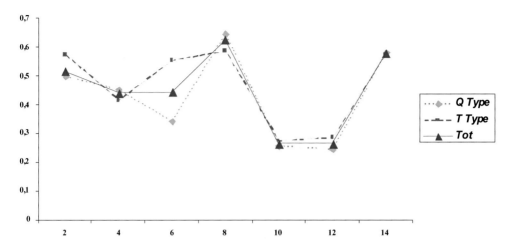

Figure 1. Percentage of active secondary follicles.

Percentage of active primary follicles (PAP)

T type primary follicular activities (81.50%) was significantly higher than Q types (74.76%). Different age had a significant effect on PAP variation. The trend was homogeneous in the two type of Bolivian llama in spite of the lower value for Q in respect to T type. The character had not a so clear season variation as in PAS, even if both types of Bolivian llama presented a reduction of activities during the summer time (10 and 12 months old). Moreover T at two months old presented the higher PAP value, but in this period only few skin samples had been taken (Figure 2).

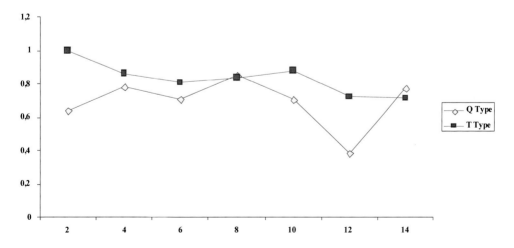

Figure 2. Percentage of active primary follicles.

Percentage of medullated secondary fibres (PMS)

T types present a significant lower PMS except at 2 months, probably due to the reduced number of samples in that period. Presented data vary significantly and homogenously in the both types of llama. Only in Q type at 1 year old, the medullation growth was out of the general trend (46.5%) with respect to T types (5.98%). If we do not consider the last data and T type at 2 month old result, the PMS character showed a reduction in the periods which correspond to summer time (Figure 3).

Percentage of medullated primary fibres (PMP)

There was no difference between types of Bolivian llama in terms of PMP, while the different sample period has significantly effect on the character. Also for PMP the data related to T at 2 months old is not reliable, because of the low number of samples. As for the PMS, PAS and PAP data, the PMP trend presented a significantly reduction at 10 months old which correspond to summer trial period (Figure 4).

Discussion

The present experiment showed no significantly differences between T and Q types in S/P and follicular activity, even if T types seem to show low medullated fibre and higher percentage of active primary follicle with respect to Q llama type. Among the two types of Bolivian llama no significant effect of sex was demonstrated in any of the parameters analysed and finally the ratio of secondary to primary follicles presented no variation in the different sampling periods (age of the SAC). In present study, both types of Bolivian llama differ only significantly for PAP and PMS in the sample period.

With respect to active secondary follicle, active primary follicle, medullated secondary and primary fibre it is possible to assert that the sample period presents a significantly effect: the values of all four characters generally decreased until August (6 month), increased in October

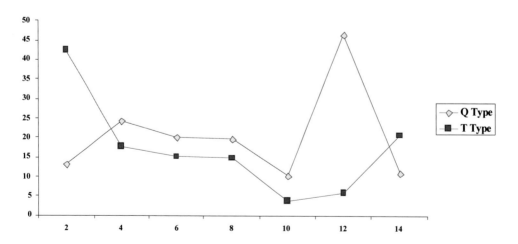

Figure 3. Percentage of medullated secondaty fibres.

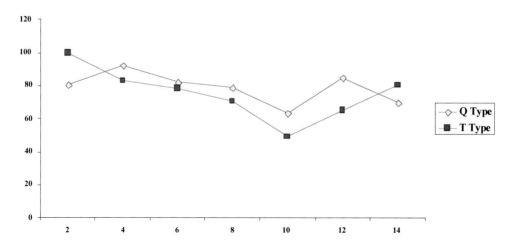

Figure 4. Percentage of medullated primary fibres.

(8 month), decreased again strongly in December – February period (10 – 12 month) and rised again at 14 month old (April period) to reach the same values described one year before.

A seasonal skin follicle activity, as well as in T llama type a positive correlation between this parameter and the percentage of medullated secondary and primary fibres, has been previously reported for a range of mammalian species and breeds, *e.g.* for different goat breeds (Nixon, 1993, 1991). Nixon (1993) reported that in animals selected for fibre production, like New Zealand Angora goats, activity of secondary follicles remained high from winter to spring, thus suggesting that the growth of fine, generally non-medullated fibres, was continuous or asynchronous.

The great genetic variability of the Bolivian llama (Lauvergne, 1994; Lauvergne *et al.*, 2001) could explain the morphological differences especially in fleece characters, which are strongly influenced by the environmental effect of the Bolivian breeding areas and by the absence of a concrete genetic selection plan.

Frank (2001) supported the present theory because he defined the genetic heredity of the double (Q) and simple coated (T) Argentinean llama obtained by an additive effect. Moreover results by Quispe *et al.* (2003) on fibre and fleece study on South Lipez Provinces Bolivian llama population seem to confirm our conclusion because they did not show any significant differences in fleece belly weight, average fibre diameter, and percentage of medullated fibres between T and Q llama types.

Antonini *et al.* (2004) described that in Peruvian Woolly Llama (*chacos*), the percentage of secondary active follicles underwent a progressive, significant decrease from its maximum value at 2 month old (February), until June, then increased again at 10 months old (October) but failed to reach the value attained in the second month. In alpaca Suri a consistently high percentage is maintained throughout the period of 10 months old, with a minimum at June and a maximum at October. Huacaya showed an "inverted" trend, as it exhibited maximum and minimum values at the June and October, respectively.

Conclusion

In the present study, S/P data demonstrate that Bolivian llama population gain a complete and mature skin follicle apparatus since birth, which remains constant throughout all the investigation period. Because S/P is a main parameter to define the follicular structure, present results demonstrated that T and Q types could be only a subjective classification due to the variation of the characters inside the same genetic population. Fibre analyses on fleece samples drawn at the same time of the skin sample collection, will have to be carried out in order to provide further and defined information on Bolivian llama types and to confirm the skin follicular structure data obtained in the present study.

All the parameters, *i.e.* the ratio of secondary to primary follicles, reached its maximum value at 2 month old, while Antonini *et al.* (2004) described that in the Peruvian llama the follicular parameter achieved their maximum value two months later. It was confirmed that SAC gained a complete and mature skin follicle apparatus at an early age. The observed variation seem to be in relation to the seasons. Thus, unfavourable environmental conditions might negatively affect both the physiological activity of the skin follicles, and, consequently, also medullation decrease. This would also enable the effect of any environmental factors on follicular skin activities (which are responsible of fibre production) to be investigated, potentially a very important consideration in the harsh conditions of the Andean plateau.

The data presented here may be exploited for a rational management of the "crias category". Particularly llama kid fleece is the most requested by the market, and fibre production is present in llama from an early age, producers should practise early selection based on fibre production potential, as well as an anticipated first shearing. This would increase producers' revenues by a value equivalent to one shearing over the productive life of an animal especially for meat production purposes. The present data must, however, be considered as merely indicative of the potential for Bolivian llama fleece production, since they are not based on complete analyses, few specimens, especially between 2 months and 12 months old periods, and on a single series of observations. In order to substantiate the present findings, further studies are required using larger numbers of animals and for periods of at least three years.

Acknowledgements

We are very grateful to Mr Enzo Stella (ENEA C.R. Casaccia) and Mr. Gianvincenzo Lebboroni (University of Camerino) for valuable help in histological processing of skin samples. Financial support for this research was provided by the European Union (EC DG XII – INCO Programme – ERBIC18CT960067 and) through the SUPREME (Sustainable Production of Natural Resources and Management of Ecosystems) project.

References

Antonini, M., M. Gonzales and A. Valbonesi, 2004. Relationship between age and postnatal skin follicular development in three types of South American domestic camelids. Livestock Production Science 90: 241-246.

Bonavia, D., 1996. Los Camélidos sudamericanos (Una introducciòn a su studio). Istituto Francès de estudios Andinos, Lima.

Frank, E.N., S.G. Parisi De Fabro, and T. Mendez, 1989. Determinacion de variables foliculares en cortes de piel de camelidos sudamericanos y su relacion con las caracteristicas de vellon. Rev. Arg. Prod. Anim. 9: 379-386.

Frank, E.N., 2001. Description y analisis de segregacion de fenotipos de color y tipos de vellon en llamas argentinas. (Segregation analysis and description of colour and fleece types phenotypes in Argentinean llamas). PhD Thesis Facultad de Ciencias Veterinarias. Universidad de Buenos Aires, 2001, pp. 151-153.

Lauvergne, J.J., 1994. Charactherization of domestic genetic resources of American Camelids: a new approach. In: M. Gerken and C. Renieri (eds.), Proceedings of the European Symposium of South American Camelids. Camerino, pp. 59-63.

Lauvergne J.J., Z. Martinez, C. Ayala and T. Rodriguez, 2001. SUPREME Project: Identification of a primary population of South American domestic camelids in the provinces of Antonio Quijarro and Enrique Baldivieso (Department of Potosì, Bolivia) using phenotypic variation of coat colour. In: M. Gerken and C. Renieri (eds.), Proggress in South American Camelids Research. Wageningen, EAAP publ. No 105, pp. 64-74.

Martinez Z., L.C. Iniguez and T. Rodriguez, 1997. Influence of effects on quality traits and relationships between traits of the llama fleece. Small Ruminant Research 24: 203–212.

Nixon, A.J., 1993. A method for determining the activity state of hair follicle. Biotech. and Histoch. 60: 316-325.

Nixon, A.J., M.P. Gurnsey, K. Betteridge, R.J. Mitchell and R.A.S. Welch, 1991. Seasonal hair follicle activity and fibre growth in some New Zealand cashmere bearing goats (*Capra hircus*). J. Zool. Lond. 22a: 589-598.

Quispe, J., T. Rodriguez, Z. Martinez and M. Antonini, 2003. Clasificacion y caracterizacion de fibra de llamas criadas en el altiplano sur de Bolivia (Fibre classification and characterization of llama bred in the South Plateau of Bolivian). Proceeding III° Congreso Mundial Sobre Camelidos, Primer Taller Internacional de DECAMA. 15-18 de Octubre de 2003 Potosì, Bolivia. Vol. II, pp. 651-655.

SAS, 1995. User's guide: Statistics, version 6.12. SAS Inst. Inc, Cary. NC.

Valdivia, C. and J. Christian, 1996. IBITA 181/technical report 49/sr-crsp 47/1996 - http://www.ssu.missouri.edu/SSU/SRCRSP/Papers/TR33/TR33_5.htm.

Yield and quality of the dehaired llama fibre of the community of Phujrata

N. Cochi Machaca

Avenida "E", N° 218, Villa Aroma, El Alto, Bolivia; nestorcochi@mixmail.com

Abstract

Fifty one llama fleeces were obtained from the community of Phujrata of the Department of La Paz, Bolivia. The fleeces were divided into two groups which were subjected either to complete or partial manual dehairing. Dehairing was achieved by means of the tactile-visual method based on fibre fineness and length. The dehaired fleeces were then classified into five categories of quality. These were AA (first), A (second), SK (briefs or claw), LP (locks/pieces) and CD (thick hairs). For each category, the diameter, medullation percentage and the fibre length were analyzed. The effect of separation into quality categories showed statistically significant differences (P < 0.01) for the characteristics studied. It was observed that the effects of the dehairing and the classificaton into quality categories were highly significant (P < 0.01) for diameter, medullation and length. However, comparisons between treatments indicated a statistically significant (P < 0.01) effect on the fibre diameter, but not length and medullation percentage. The average percentage yields of dehaired fleeces in the classified categories were 48.35, 20.27, 9.74, 13.28 and 6.37 for AA, A, SK, LP and CD respectively. Results for the complete dehairing process suggested greater losses of qualities AA and A than in partial dehairing. The dehairing of white fleeces was more time-consuming than that of the coloured fleeces. Partially manual dehairing was a quicker process and resulted in higher economic benefits than were obtained with complete manual dehairing of fleeces.

Resumen

De 51 vellones de la comunidad de Phujrata del Departamento de La Paz - Bolivia, se dividieron en dos grupos: en el primero se aplico un descerdado manual completo y al segundo un descerdado manual parcial, mediante el método tacto - visual en base a finura y longitud. Se clasificaron en cinco calidades: AA: Primera, A: Segunda, SK: Bragas, LP: Cortas/Garra y CD: Cerdas, en las que fueron analizados el diámetro, porcentaje de medulación y la longitud de mecha antes y después del descerdado. El efecto de las calidades mostró diferencias estadísticas (P<0.01) para los caracteres estudiados. En los vellones de Phujrata se observo que el efecto del descerdado y la clasificación en calidades fue altamente significativo (P<0.01) sobre el diámetro, medulación y longitud, y el efecto de tratamiento afectó estadísticamente (P<0.01) al diámetro de fibra excepto en el porcentaje de medulación y longitud. Los rendimientos de fibra descerdada y clasificada para vellones de Phujrata fueron de 48.35, 20.27, 9.74, 13.28 y 6.37% para las calidades estudiadas. Los vellones blancos de Phujrata fueron descerdados en mayor tiempo que los vellones de color. Vellones descerdados parcialmente reportan mayores beneficios que los que fueron descerdados completamente, en vellones de la comunidad de Phujrata.

Keywords: llama, fibre quality, dehairing

Introduction

South American Camelids constitute the main source of subsistence for residents of the high Andes and produce a fibre of high textile value. Bolivia has a fibre production of 850 TM/year approximately. This volume includes fine fibre and coarse hair.

The fleece of the llama contains thick fibre with a typical diameter of 28.1 μm and a high content (60.1%), of medullated hair which reduces the quality and commercial value of llama fibre.

Manual dehairing is a process which permits the removal of the thick hair of the fleece. It increases uniformity and inproves quality and commercial desirability since the dehaired fibre presents excellent qualities as a natural and specialist fine fibre. In considering the processing of considerable amounts of llama fibre, very little information exists on parameters such as: time for dehairing, manpower requirements and quality of the dehaired fibre.

The objectives of the present study were therefore to:
- determine the physical characteristics (diameter, length, medullation and yield) of the gross fibre and the dehaired fibre, subjected to different degrees of removal of thick hairs or hair;
- evaluate the yield of the llama fibre classified into different commercial qualities;
- determine the cost of manual dehairing.

Materials and methods

Geographical localization and climatic characteristics

The study was conducted with fleeces of llamas of the community of Phujrata of Pacajes county (La Paz), located 56 km to the northwest of the village of Patacamaya. Geographically this area is located between the parallel 17° 15' of southern latitude and 68° 17' of western longitude, at an altitude of 4,490 m. Annual rainfall averages 750 mm/year and the mean temperature is 7.2°C.

Dehairing and classification of the fibre by qualities

From the 51 llama fleeces selected in Phujrata, a group of 26 fleeces was completely dehaired and 25 fleeces were partially dehaired. Fibre samples were taken before applying the dehairing processes to determine the initial values of the characteristics studied. The actual dehairing process involved the manual extraction of hair or thick hairs from the llama fleece by means of the tactile-visual method.

In the partial dehairing, the only thick hairs that were removed were those that were apparent in the fleece. Almost all the thick hairs were removed in the complete dehairing. Simultaneously, a classification was made by the tactile-visual method into five categories of qualities: AA (first), A (second), SK (briefs or claw), LP (locks/pieces) and CD (thick hairs). Also, straw or vegetable residuals, garbage and earth were separated by shaking the fibre (Yujra & Chavez, 1997). This process was manually carried out by a trained teacher. Additional background information to the methodology used and previous studies may be obtained in the reports of Castro (1988), Mancilla (1988) and Martínez *et al.* (1993).

Determination of quality

The following parameters were taken into account in determining the quality of llama fibre. These were diameter μ (μm), fibre length (cm), medullation (as % of total fibre) and yield (%, applying quality criteria).

To determine the yield according to quality, the total (dirty) fleece of each llama was measured on an analytical weighing balance. After the dehairing and classification, five different qualities were obtained for each fleece. These were then separately weighed (Figure 1).

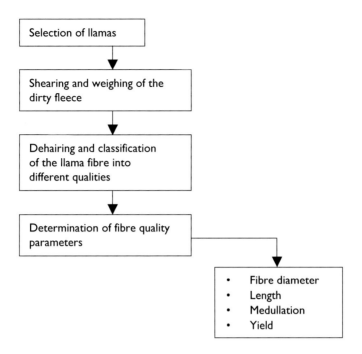

Figure 1. Diagram of the methodology used.

Statistical analysis of the factors influencing the variables studied

For the analysis of the factors that influenced the variation of the diameter, medullation and length, the following linear model (1) was used:

$$Y_{ijklmn} = \mu + \beta_i + T_j + E_k + C_l + G_m + K_n + e_{ijklmno} \tag{Eq. 1}$$

For the analysis of the weight of the dirty fleece (2) and the yield of the qualities (3), the model (1) was modified according to the effects of the main factors:

$$Y_{ijk} = \mu + T_i + E_j + C_k + e_{ijkl} \tag{Eq. 2}$$
$$Y_{ijkl} = \mu + \beta_i + T_j + E_k + C_l + e_{ijklm} \tag{Eq. 3}$$

The data were analyzed with the Statistical Analysis System package (SAS, 1992), version 6.04.

Results and discussion

Characteristics of quality of fleeces of the community of Phujrata

The complete and partial dehairing significantly affected (P <0.05) the fibre diameter of the fleeces of llamas of Phujrata (Table 1). These results may be considered in the context of the earlier observations that the fine fibre of llama is more uniform than that of the alpaca.

Table 1. LSQ-Means according to the effects influencing dirty fleece weight, diameter, medullation and fibre length of dehaired and classified llama fleeces of the community of Phujrata.

Main effects	N	Dirty fleece (g)	Diameter (μ)	Medullation (%)	Length (cm)
Treatment					
Complete dehairing	/206		28.22[a]	53.81	12.83
Partial dehairing	/200		29.21[b]	55.36	12.34
Type of animal					
T'amphulli	15[*]/120[**]	1,216.36[a]	27.39[a]	43.4[a]	13.94[a]
Q'ara	36/286	690.68[b]	30.04[b]	65.76[b]	11.24[b]
Age					
2 years	7/56	719.99[a]	27.79[a]	46.66[a]	11.44[a]
3 years	33/264	1,098.99[b]	28.58[ab]	58.59[b]	12.93[b]
4 years	11/86	1,041.57[bc]	29.78[c]	58.50[bc]	13.39[bc]
Colour					
White	10/80	953.24[a]	28.58[a]	61.11[a]	11.55[a]
Brown	17/128	1,072.62[ab]	28.91[ab]	53.01[b]	13.69[b]
Spotted	18/150	893.23[ac]	27.84[acd]	57.79[ac]	12.04[a]
Black	6/48	894.99[a]	29.53[ab]	46.43[d]	13.07[b]
Dehairing					
Before dehairing	/203		29.56[a]	57.48[a]	14.52[a]
After dehairing	/203		27.87[b]	51.68[b]	10.66[b]
Qualities					
AA (First)	/102		20.32[a]	39.84[a]	14.59[a]
A (Second-adult)	/100		23.13[b]	46.42[b]	14.09[ab]
SK (Briefs or claw)	/51		36.01[c]	67.64[c]	12.83[c]
CD (Thick hairs)	/51		42.60[d]	76.54[d]	13.38[bc]
LP (Locks/pieces)	/102		21.52[e]	42.48[abe]	8.03[d]

[a,b,c,d,e] LSQ-means within columns with no common subscript differ significantly (P <0.05); [*]: number of observations for dirty fleece; [**]: number of observations for diameter, medullation and length.

The effect of the manual dehairing was to significantly (P <0.01) influence the diameter of the fibre, so that the fibre without dehairing averaged 29.56 μ and after the deshairing 27.87 μ. These results are compatible with those of Parra (1999) who carried out a manual dehairing of fleeces of llamas of Potosí and Charaña and reported diameters of 24.90 and 23.55 μ, respectively.

The percentage of medullated fibre differed significantly (P <0.01) according to the type of animal. This observation is consistent with previous reports indicating the presence of 48.1% and 27.1% of medullated fibre in fleeces of Q'ara and T'amphulli llamas, respectively.

The length of fibre in the llamas of Phujrata was statistically influenced (P <0.01) by animal type. The highest lengths were obtained in T'amphulli llamas (13.9 cm) compared to Q'aras (11.2 cm). These results are superior to values of 7.39 and 12.3 cm reported for llamas of the types "vellocino" and "mechosa" (Parra, 1999). Cardozo (1995) has also described production characteristics in these two breeds of llama.

Fibre diameters determined on the basis of quality averaged 20.32 μ for AA, 23.13 μ for A, 36.01 μ for SK, 42.60 μ for CD and 21.52 μ for LP, respectively. The diameters obtained for the qualities AA, A and LP in the present study correspond to the qualities X and AA according to previously reported specifications.

The percentage of medullation was significantly different (P <0.01) for the different qualities of separated fibres. Values were 39.84% (AA), 46.42% (A), 42.48% (LP), 67.64% (SK) and 76.54% (thick hairs) repectively.

The dehairing and the classification into different qualities resulted in highly significant differences (P <0.01) for the lengths (cm) of the qualities AA (14.59), A (14.09), SK (12.83), LP (8.03) and CD (13.38) respectively.

A number of differences between individual means for the different parameters were also recorded. These included effects relating to increasing age of animal (increases in weight of dirty fleece, diameter, medullation and fibre length particularly between 2 to 3 years of age) and colour of fibres.

Determination of the yield of the qualities

The process of complete and partial dehairing applied to the fleeces of Phujrata was highly significant (P <0.01) for the yield of the qualities AA, LP and thick hairs (Table 2). It was observed that the percentage yield of the completely dehaired fibre, with respect to qualities, had smaller values for AA (38.51 vs. 58.51%) and A (18.05 vs. 22.50%) and greater values for SK (11.60 vs. 7.88%), LP (19.36 vs. 7.20%), CD (10.12 vs. 2.62%), earth (1.52 vs. 1.17%) and

Table 2. LSQ-Means according to the effects influencing yield of dehaired and classified fibre in llama fleeces of the community of Phujrata.

Main effects	Yield of dehaired and classified fibre (%)							
	N	AA	A	SK	LP	CD	Earth	Reduction
Treatment								
CD	26	38.51[a]	18.05	11.60[a]	19.36[a]	10.12[a]	1.52	0.83
PD	25	58.19[b]	22.50	7.88[b]	7.20[b]	2.62[b]	1.17	0.44
Type								
T'amphulli	15	51.12	22.93	10.73	9.39[a]	2.83[a]	1.96[a]	1.03 [a]
Q'ara	36	45.58	17.61	8.76	17.17[b]	9.92[b]	0.73[b]	0.23 [b]
Age								
2 years	7	41.73	20.84	9.40	16.71	9.67[a]	1.27	0.36
3 years	33	50.91	20.94	9.99	9.46	6.65[a]	1.49	0.57
4 years	11	52.42	19.05	9.83	13.67	2.79[b]	1.28	0.97
Colour								
White	10	53.92[a]	25.32[a]	8.77	5.53[a]	3.65[a]	1.74[a]	1.06
Brown	17	43.40[b]	18.99[ab]	9.65	18.10[b]	7.12[b]	1.97[ab]	0.77
Spotted	18	45.03[ab]	15.56[bc]	11.60	19.14[bc]	7.11[bc]	1.07[ac]	0.49
Black	6	51.05[ab]	21.22[ac]	8.94	10.36[ab]	7.60[ab]	0.60[acd]	0.22

AA: first; A: second/adult; SK: briefs or claw; CD: thick hairs or hair; LP: locks/pieces; CD: complete dehairing; PD: partial dehairing.
[a,b,c,de] LSQ-means within columns with no common subscript differ significantly (P <0.05).

reduction (0.83%), compared to the partially dehaired fibre. These data suggest that the complete dehairing process resulted in greater losses of qualities AA and A than partial dehairing. There were few differences in percentage yields associated with breed, age or colour.

Time for dehairing and classifying of the llama fibre

Measurements of the quantity of fibre processed in one day showed that 2.75 kg could be achieved with partial dehairment. This quantity compares with 1.28 kg for complete dehairment, which also involves the time taken to review the fleece on three or four additional occasions. However, the completely dehaired product is more uniform and free of thick hairs, except for the hair intermingled within the fleece.

The yield of processed fibre was smaller in coloured fleeces than in white ones. This fact can facilitate or hinder the dehairing and classification process of the fibre.

Economic analysis of the dehairing and classification process of the llama fibre

Comparisons of the relationship of benefit/cost (gross revenue/variable costs) showed superior returns for partial dehairing with a value of 1.39 compared with 1.21 for the completely dehaired system (Table 3).

Conclusions

The gross fibre in fleeces of llamas of the Phujrata community had an average diameter of 29.56 µ which was reduced to 27.87 µ after dehairing. Diameters (µ) in such fleeces in the present study, when separated for the qualities of AA (20.32), A (23.13) and LP (21.52), were smaller compared to the qualities for SK (36.01) and CD (42.6).

The gross fleece of the llamas, presented a high percentage of medullated fibre of 57.48%. After dehairing it was reduced to 51.68%. This reduction demonstrated that the extraction of the hair had a major influence on the quality of the product. The qualities AA and A had the lowest percentage of medullation of 47%.

The fibre length of the dehaired and classified fleeces attained the minimum conditions required by the textile industry.

Comparisons for percentage yields according to quality criteria and averaged for complete and partial dehairing were for AA (48.35%), A (20.27%), SK (9.74%), LP (13.28%), CD (6.37%),

Table 3. Economic analysis of the dehairing and classification process of the llama fibre, by treatment.

	Yield (Kg)	Value (Bs/kg)	Gross revenues (Bs)	Variable costs (Bs)	Net revenues (Bs)	B/C
Phujrata						
CD	596.80	25.00	14,920.00	12,300.00	2,620.00	1.21
PD	806.98	20.00	16,139.60	11,600.00	4,539.60	1.39

Yield: Yield of the qualities AA+A, IB: Gross revenues, CV: Variable costs, UN: Net utility, B/C: Relationship of benefit (gross revenue)/ variable costs, CD: complete dehairing, PD: partial dehairing, Bs: Bolivianos

earth (1.34%) and reduction (0.63%) respectively. Comparisons of the quantities of fleece that could be dehaired in one day indicated averages of 1.28 and 2.75 kg for complete and partial treatments.

The largest financial benefits were obtained for the fibre subjected to the partial dehairing, with a benefit/cost relationship of 1.39. This compares with a value of 1.21 for the complete dehairing system.

References

Cardozo, A., 1995. Tipification of the llamas K'aras and T'amphulli. In: Waira Pampa, A pastoral system Camelids - Sheeps of the Bolivian arid Highland. ORSTOM-CONPAC ORURO-IBTA, La Paz, Bolivia, pp. 65-72.

Castro. P.F., 1988. Analysis of the commercial fleece of the camelids alpaca and llama. Grade thesis to obtain Agronomist Engineer's title, Faculty of Agricultural and Cattle Sciences, UMSS. Cochabamba, Bolivia, pp. 34-61.

Mancilla, C.A., 1988. Physical characteristics of the fiber of llama type Ch'aco and K'ara of the C.E.C. La Raya. Grade thesis to obtain the title of Medical Veterinary Zootecnísta, UNA. Faculty of Veterinary Medicine and Zootécnia, Puno, Peru, pp. 23-54.

Martínez, Z. L. Iñiguez and T. Rodriguez, 1993. Characters of quality and determination of corporal areas of sampling more representative of the fleece of llamas. Bolivian Institute of Agricultural Technology (IBTA), series of works N° 4, Program of Cattle raising and Forages, Experimental Station of Patacamaya. La Paz, Bolivia, pp. 11-35.

Parra, P.I., 1999. Evaluation of the productive potential of the llama (*Lama glama*), in the fifth municipal section Charaña. Grade thesis to obtain Agronomist Engineer's title, Faculty of Agricultural and Cattle Sciences, UMSS. Cochabamba, Bolivia, pp. 54-98.

Yujra, E. and R. Chavez, 1997. Dishairing and technical classification of llama fiber. Foundation SARTAWI natural fibers. La Paz, Bolivia, 10 pp.

Variability of vicuña fibre diameter in two protected natural areas of Peru

J. Ayala[1], M. Lopez[2] and J. Chavez[3]
[1]Instituto Nacional de Recursos Naturales, calle Diecisiete Nº 355 El Paloma, San Isidro, Lima, Peru; juedayala@hotmail.com
[2]Universidad Nacional Agraria La Molina, Avenida La Universidad, La Molina, Lima, Peru.
[3]Instituto Interamericano de Cooperación Agrícola, Av. Jorge Basadre 1120, San Isidro, Lima, Peru.

Abstract

Fibre samples from one hundred two younger and older, male and female vicuñas were collected during the Chaku Festival 2000, in the National Reserve Pampa Galeras – NRPG and the National Reserve Salinas y Aguada Blanca – NRSAB, in order to analize the phenotypic variability of mean, standard deviation and coefficient of variation of the fibre diameter within and between populations. The Sirolan Laserscan (IWTO-12-95) method was used to measure the mean fibre diameter of vicuña samples, which were corrected for sex and age effects. Components of variance were estimated using a nested design. The results of mean diameter in both populations were 13.37 micron (NRPG) and 13.82 micron (NRSAB) (P<0.05), respectively. The component of variance between populations ($\sigma^2 p$) was lower than that within and between animals (σ^2_B and σ^2_w), for the mean ($\sigma^2 p = 0.09$ μ^2, $\sigma^2_B = 0.28$ μ^2 and $\sigma^2_w = 0.27$ μ^2) and the coefficient of variation of fibre diameter ($\sigma^2 p = 0.205\%^2$, $\sigma^2_B = 0.67\%^2$ and $\sigma^2_w = 26,9\%^2$). The component of variance between animals contributed 45% to all variance of the mean. On the other hand, the component of variance within animal contributed 96% and 97% to all variance of the standard deviation and the coefficient of variation of mean fibre diameter. The genetic variance and permanent environment contributed to the variation in fineness, and special environmental variance influenced the uniformity of vicuña fibre.

Resumen

Muestras de fibra de 102 vicuñas machos y hembras, juveniles y adultos de la Reserva Nacional de Pampa Galeras – RNPG y de Salinas y Aguada Blanca - RNSAB, fueron colectadas en las capturas del año 2000, por el Instituto Nacional de Recursos Naturales -INRENA, para analizar la variabilidad del promedio, desviación estándar y coeficiente de variación del diámetro de fibra entre población, entre y dentro de animal. Se empleó el método de Laserscan según la International Wool Organization (IWTO-12-95), para obtener las mediciones, las cuales fueron ajustadas por efecto de sexo y edad, utilizando un diseño de bloques al azar con submuestreo; y los componentes de varianza mediante un diseño jerárquico. La comparación entre los promedios de los diámetros corregidos entre las poblaciones RNPG y RNSAB, mediante la diferencia mínima significativa (dms = 0.2963), fue significativamente diferente (P< 0.05); y el componente de varianza entre población(σ^2_p) fue menor que entre y dentro de animal (σ^2_B y σ^2_W), para el promedio ($\sigma^2_P = 0.09\mu^2$, $\sigma^2_B = 0.28\mu^2$ y $\sigma^2_W = 0.27\mu^2$), desviación estándar ($\sigma^2_P = 0.0028$ μ^2, $\sigma^2_B = 0.020$ μ^2 y $\sigma^2_W = 0.54$ μ^2) y coeficiente de variación del diámetro de fibra ($\sigma^2_P = 0.205\%^2$, $\sigma^2_B = 0.67\%^2$ y $\sigma^2_W = 26.9\%^2$). La diferencia observada del promedio del diámetro de fibra corregido entre población fue de 0.45 micras (P< 0.0001), información que puede sustentar los certificados de origen de fibra de vicuña; y los rangos estimados son indicadores del grado de conservación de las poblaciones de vicuña en cada Área Natural Protegida. El componente de varianza entre animal contribuye en 45% a la varianza total del promedio, y dentro de animal en 96% y 97% a la varianza total de la desviación estándar y coeficiente de variación del diámetro de fibra, respectivamente. La varianza genética y de ambiente permanente contribuyen a la variación de la finura, y la variación de ambiente temporal influye mas sobre la distribución y uniformidad del diámetro de fibra de vicuña. Por lo tanto, se puede identificar y conservar en forma objetiva individuos que puedan mejorar y mantener la estructura de los vellones en las poblaciones de vicuña.

Keywords: fibre, genetic variance, vicuña

Results

Table 1. Means, standard deviation (S.D.) and coefficient of variation (%) of vicuña fibre diameter in Pampa Galeras and Salinas y Aguada Blanca.

Population	Pampa Galeras				Salinas y Aguada Blanca			
	n^c	Mean (μ)	S.D. (μ)	C.V. %	n^c	Mean (μ)	S.D. (μ)	C.V. %
Male[a]	18	12.95	3.8	29.4	7	13.87	3.6	25.7
Male[b]	18	13.53	3.5	25.9	1	13.63	4.4	32.2
Female[a]	12	13.17	4.1	30.9	8	13.78	3.8	27.2
Female[b]	33	13.83	3.4	24.7	5	13.83	4.1	29.9
Average		13.37	3.7	27.7		13.78	3.9	28.8

[a]Younger vicuña; [b]Adult vicuña; [c]n = number of animals.

Table 2. Variability of corrected mean fibre diameter (μ) by body region of vicuña populations in Pampa Galeras and Salinas y Aguada Blanca.

Body region		Pampa Galeras (n = 81)	Salinas y Aguada Blanca (n = 21)	Average of region
Shoulder middle	Mean (μ)	13.26	13.95	13.40[b]
	Range (μ)	11.52 -14.46	12.7 - 15.4	
	CV (%)			30.49[a]
Ribs middle	Mean (μ)	13.09	13.52	13.19[c]
	Range (μ)	11.42 - 15.06	12.3 – 14.9	
	CV (%)			27.76[b]
Thigh middle	Mean (μ)	13.76	13.93	13.79[a]
	Range (μ)	11.92 - 15.62	12.7 - 15.3	
	CV (%)			26.18[b]
Average		13.37[b]	13.82[a]	

[a,b,c]The values within a column or row with no common superscript differ significantly (P < 0.05); n = number of fibre samples tested.

Acknowledgments

The authors wish to express their appreciation to the Direction of Natural Protected Areas and Wild Fauna Wild of the National Institute of Natural Resources. Ours thanks are also due to Engineer Ricardo Tamaki, Mrs Patricia Luna and Mrs Jessica Amanzo for their colaboration in the field, and the laboratory of fibres of the Universidad Nacional Agraria La Molina, Lima.

Additional references

Ayala, J.E., 1999. Variabilidad del Diámetro de Fibra en Alpacas Huacaya Usando los Métodos de Microproyección y Análisis Óptico del Diámetro de Fibra – OFDA. Tesis para optar el grado de Magíster Scientiae. Escuela de Post Grado –UNALM. Lima –Perú.

Baxter, B.P., M.A. Brims and T.B. Taylor, 1992. Description and Performance of Optical Fibre Diameter Analyser (OFDA). Journal of the Textile Institute Vol 83,N4, 507-526 pp.

Becker, W.A., 1984. Manual of Quantitative Genetics. Fourth Edition. Washington Satate University. Published by Academic Enterprises. Pullman, Washington. 49-55 pp.

Caprio, M. and Z. Solari, 1981. Diámetro de Fibra en el Vellón de la vicuña Programa Académico de Zootecnia. Informes de trabajos de Investigación en Vicuñas UNALM - Lima. Vol I pp.77-93.

Carpio, M. 1991. La Fibra de los Camélidos. Capítulo V. Producción de Rumiantes Menores. Alpacas. Ed. Cesar Novoa - Arturo Flores. Convenio Universidad de California Davis - INIAA. RERUMEN, Apartado 110097, Lima 11, Perú. 297-356 pp.

Iman, N.Y., C. Le Roy Johnson, W.C. Russell and R,H. Stobart, 1992. Estimation of Genetic Parameters for Wool Fiber Diameter Measures. Journal Animal Science, Vol 70:1110-1115 pp.

Ralph, I.G. 1989. Componentes of variation in fibre diameter, Wester Australia. Department Agricultural. Seminar. Katanning, 15-18 pp.

SAS, 1995. SAS Users guide: Statistics, Statistical Analysis Systems Institute. Inc. Cary, North Carolina, USA.

Wheeler, J., M. Fernandez, R. Rosadio, D. Hoces, M. Kadwell and M. Bruford, 2000. Diversidad genética y manejo de población de vicuñas del Perú. Rev. Inv. Vet. Perú. Suplemento 1:170-183 pp.

Part VI
Short communications

Size and growth of the guanaco (*Lama guanicoe*) population at the Calipuy National Reserve

J.E. Ayala

Instituto Nacional de Recursos Naturales, calle Diecisiete Nº 355 El Paloma, San Isidro, Lima, Peru; juedayala@hotmail.com

Abstract

The first census of the guanaco populations was realized in June of 2002, 2003 and July 2004 in the Calipuy National Reserve and 462, 484 and 436 guanacos, respectively, were reported. These results indicate a population growth of 4.5%. On the average the population consisted of 40% adults, 6% juvenile, 5% crias, 19% single males and 30% male bachelor groups. The mean proportion of adult females in both census was 0.82, indicating that the number of adult females was 4 to 5 times higher than that of adult males. The monthly census realized in the strict protection area between February 2003 and September 2004, indicated that changes within the populations were observed in March and May in each year. These changes are related to the parturition and the migration of the guanacos. The results indicate that the guanaco population of Calipuy is mainly maintained by the contribution of the adult females and a minor contribution of offspring (5%). However, the bachelor male groups form a gene pool within the Calipuy National Reserve.

Resumen

Los primeros censos de la población de guanacos en la Reserva Nacional de Calipuy se iniciaron en junio del 2002, 2003 y julio del 2004, mediante la observación directa los cuales reportaron 462, 484 y 436 guanacos, respectivamente; que refleja un crecimiento poblacional de 4,5%. Respecto a la estructura social y sexos observados en los censos anteriores (excluyendo a los guanacos no identificados) en promedio los adultos representan el 40% de la población, los juveniles el 6%, los solitarios 5% y la tropilla el 19%. La media de la proporción de hembras adultas en la población de adultos para ambos censos fue de 0,82, lo cual indica que las hembras adultas fueron mas que los machos adultos, casi en una proporción de 5 a 4. Los censos parciales mensuales realizados en la zona de protección estricta en febrero del 2003 a septiembre del 2004 indican que los cambios cn la población se presentan entre marzo y mayo de cada ano, y esta relacionado con los picos de parición de guanacos y la migración de guanacos fuera de la zona de monitoreo. Estos resultados revelan que la población de guanacos de la Reserva Nacional de Calipuy se mantienen con un mayor aporte de hembras adultas y una menor contribución de crías de casi el 5% de la población. Sin embargo, cuentan con un pool de genes de las tropillas.

Keywords: guanacos, census, social structure

Introduction

The first national census of guanacos was realized in 1996 in Peru. From 3,810 guanacos found, 538 guanacos (14.12%) were living in the Calipuy National Reserve. This Reserve is located in the town of Santiago de Chuco, Province of La Libertad, at 3,400 m above sea level and covers 64,000 ha. The present research was conducted to 1) estimate changes in the size of the guanaco population and 2) estimate the contribution of the social groups to the growth of the guanaco population.

Materials and methods

The research was realized in the strictly protected zone of the National Reserve of Calipuy, during the guanaco census in July 2004. This result was compared with others census of 2002 and 2003. Observations were made at 14 places preferred by the guanacos. Binoculars were used to determine number and sex.

Results and discussion

The census of guanacos were realized within thirteen sectors of the Calipuy National Reserve. Between 2002 and 2004, the guanaco population increased by 4.5% and then decreased by 11%. These changes could be due to migration of the guanacos in search of pasture or water. The social structure of the guanaco population observed in each census averaged 8% adult males, 32% adult females, 6% juveniles, 5% crias, 19% single males and 30% male bachelor groups. These results suggest that the guanaco population in the Calipuy National Reserve (Table 1) is mainly composed of adult and single males.

The mean percentage of adult females in the census of 2002, 2003 and 2004 was 0.82 indicating that the number of adult females was 4 to 5 times higher than that of adult males. The male guanaco population consisted of 13% of adult males with territorial family groups, whereas single males and bachelor groups represent 87% of the male guanaco population. The monthly census realised between February 2003 to September 2004, indicated that the populations changed in March and May each year. These changes are related to the parturition and the migration of the guanacos (Figure 1).

Table 1. Census of guanaco population in the Reserve National Calipuy (48 km² of study area).

Date	Total	Adult Male	Adult Female	Juveniles	Cria	Single males	Bachelor male group	NR
June 2002	462	14	132	29	24	78	138	47
June 2003	484	41	138	17	20	144	124	-
July 2004	436	46	167	31	26	28	138	-

NR= not registered

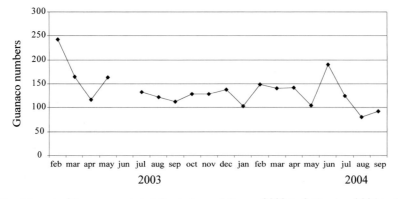

Figure 1. Development of the guanaco population between February 2003 to September 2004 in the guanaco protection area of the Calipuy National Reserve.

Determination of internal organ weights in llamas and alpacas

C.K. Cebra,[1] R.J. Bildfell[2] and C.V. Löhr[2]

[1]Department of Clinical Sciences, Oregon State University College of Veterinary Medicine, 105 Magruder Hall, Corvallis, OREGON 97371, USA; christopher.cebra@oregonstate.edu
[2]Department of Biomedical Sciences, Oregon State University College of Veterinary Medicine, 105 Magruder Hall, Corvallis, OREGON 97371, USA.

Abstract

To investigate whether there were species or gender differences among organ weights in adult camelids, the first and second gastric compartments (forestomach), third gastric compartment, liver, and pancreas were collected and weighed post mortem from 2 female and 6 male alpacas and from 16 female and 7 male llamas. These camelids all were euthanized for reasons independent of this project. Organ weights, and organ weights as a proportion of total live body weight or forestomach weight were calculated and compared between camelids of different species and genders. No gender differences were found. Llamas had proportionally larger forestomach and smaller third gastric compartments, livers, and pancreata than alpacas. The ratio of pancreatic weight to forestomach weight was also greater in alpacas, whereas the ratio of pancreatic weight to hepatic weight was not different between species. These findings suggest that the forestomach compartment makes up a greater proportion and soft tissue a smaller proportion of total body weight in llama. This may affect factors such as medication dosages. These findings also suggest that camelids have a similar pancreatic mass to ruminants, which makes it unlikely that lack of islet cell mass is responsible for the relative lack of insulin production in camelids.

Keywords: alpaca, llama, internal organ weights

Introduction

Compared to other domestic species, relatively little is known about New World camelids. Various investigations have demonstrated unique features about camelid anatomy and physiology, but much basic information is still lacking. Among the lacking information are normal organ weights from healthy camelids.

Some information exists concerning the weights of the gastric compartments (Fowler, 1998). However, how that information relates to modern management systems is unknown. The gastric compartments represent an important contribution to the weight of the camelids, but also represent the a sizeable fluid compartment which is separated from extracellular fluid by a semipermeable membrane. Given recent concerns about a possible difference in the extracellular fluid compartment between llamas and alpacas (Cebra et al., 2001), knowing the size of the gastric compartments in these two species would be helpful.

Another unique feature of camelids is their relatively poor insulin response to hyperglycemia (Cebra et al., 2001; Ommaya et al., 1995), which could relate to a lack of pancreatic islet cells, or poor function of those cells. The insulin response to glucose administration is smaller in alpacas than llamas receiving the same dosage. Camelids appear to have a normal population of islet cells per unit area of pancreas, and these cells stain with antibodies against the pertinent glucose transporter molecules, but the overall islet cell mass is unknown. Knowing the overall weight of the pancreas would help in this regard.

Liver weight has been noted in some clinical reports of illness, chiefly regarding hepatic neoplasms (Cebra *et al.*, 1995), but knowledge of normal liver weights is also lacking. The purpose of this paper was to establish reference values for the weights of various internal organs in llamas and alpacas, and to investigate whether these values gave insight into questions about extracellular fluid space, endocrine pancreatic function, or normal liver mass in New World camelids.

Materials and methods

Animals

This trial made use of 2 adult female and 6 adult castrated male alpacas, and 16 adult female, 5 adult castrated, and 2 adult intact male llamas that were euthanized under other protocols. Those protocols all were approved by the Institutional Animal Care and Use Committee of Oregon State University. All camelids were healthy and were allowed to eat grass hay up to the time of euthanasia.

Procedure

Camelids were weighed immediately prior to pentobarbital euthanasia. The liver, pancreas, first and second gastric compartments, and third gastric compartment were removed immediately after euthanasia, dissected free of connective tissue, and weighed. Gastric compartments were manipulated as little as possible to preserve contents. Raw weights were recorded, and the following ratios were calculated: liver, pancreas, first and second gastric compartments (the forestomach), and third gastric compartment each as a percentage of total live body weight, pancreas to liver, and pancreas to the first two gastric compartments.

Statistical analysis:

All raw weights and ratios were analyzed for differences between species and gender using 2-way ANOVA. Gelded and intact male llamas were pooled for the comparisons. Differences between mean values were detected by use of Tukey's test. Comparisons were considered significant at $P < 0.05$.

Results

All raw weights were significantly greater in llamas than alpacas (Table 1). Gender differences were not detected. Liver, pancreas, and the third gastric compartment made up a significantly greater proportion of live body weight in alpacas, and the first two gastric compartments made up a significantly greater proportion and body weight in llamas. Gender differences were not detected in these ratios.

There were no differences in the ratio of pancreas to liver based on species or gender, but alpacas had a significantly greater ratio of pancreas to the first two gastric compartments than llamas.

Discussion

Alpacas had a proportionally smaller forestomach, and proportionally larger other internal organs than llamas. Gender did not appear to play a role. The difference in the forestomach was greatest, where llamas had a forestomach that was proportionally about 50% bigger and represented a 4 to 5% greater contribution to total body weight. Comparable values appear in the literature (10 to 15% of body weight) but these suggest that the lower value is more applicable to alpacas

Table 1. Body weight (kg), liver weight (kg), pancreatic weight (g), forestomach weight (kg), and third gastric compartment weight (kg), the weight of each organ expressed as a percentage of total body weight (%), and the ratio of pancreatic to hepatic and forestomach weights in healthy adult llamas and alpacas. Data are mean ± SD. Different superscripts represent means that are significantly different (P < 0.05).

Variable	Alpacas		Llamas	
	Females (n=2)	Males (n=6)	Females (n=16)	Males (n=7)
Body Weight (kg)	53.6 ± 5.1[a]	67.8 ± 18.9[a]	135.2 ± 29.6[b]	127.6 ± 16.9[b]
Liver (kg)	1.02 ± 0.00[a]	1.13 ± 0.20[a]	1.98 ± 0.51[b]	1.92 ± 0.32[b]
Pancreas (g)	46.0 ± 0.7[a]	53.2 ± 8.7[a]	78.9 ± 19.0[b]	81.3 ± 13.4[b]
Forestomach (kg)	5.0 ± 1.1[a]	7.1 ± 1.8[a]	19.8 ± 3.5[b]	17.1 ± 3.9[b]
3rd Compartment (kg)	1.3 ± 0.0[a]	1.3 ± 0.3[a]	2.2 ± 0.5[b]	2.6 ± 0.9[b]
% Liver	1.9 ± 0.2[a]	1.7 ± 0.3[a]	1.5 ± 0.1[b]	1.5 ± 0.1[b]
min – max	1.8 – 2.0	1.3 – 2.1	1.3 – 1.7	1.4 – 1.7
% Pancreas	0.09 ± 0.01[a]	0.08 ± 0.02[a]	0.06 ± 0.02[b]	0.06 ± 0.01[b]
min – max	0.08 – 0.09	0.06 – 0.11	0.05 – 0.08	0.05 – 0.09
% Forestomach	9.6 ± 3.0[a]	10.6 ± 1.0[a]	15.1 ± 3.3[b]	13.4 ± 2.6[b]
min – max	7.4 – 11.7	9.6 – 12.5	9.4 – 20.8	11.1 – 17.8
% 3rd Compartment	2.5 ± 0.1[a]	2.0 ± 0.5[a]	1.7 ± 0.3[b]	2.0 ± 0.5[b]
min – max	2.4 – 2.6	1.6 – 2.7	1.2 – 2.5	1.5 – 2.8
Pancreas/Liver	45.0 ± 0.7 ×10⁻³	47.5 ± 8.1 ×10⁻³	40.6 ± 8.1 ×10⁻³	42.6 ± 7.0 ×10⁻³
	44.5 – 45.5	39.9 – 61.3	31.4 – 52.0	34.4 – 56.6
Pancreas/Forestomach	9.3 ± 1.9 ×10⁻³ᵃ	7.7 ± 1.3 ×10⁻³ᵃ	4.0 ± 1.0 ×10⁻³ᵇ	4.8 ± 1.0 ×10⁻³ ᵇ
	8.0 – 10.7	5.3 – 9.1	3.0 – 6.6	3.7 – 6.4

and the higher value to llamas. These data also suggest that estimates of the third compartment appearing in the literature (1 to 2%) are too low (Fowler, 1998). The differences may relate to the sample populations, diets, or other factors, and the current data may be more applicable to camelids raised in North America than in South America.

The approximately 4% difference in forestomach weight may account for the apparent difference in volume of distribution for glucose between llamas and alpacas (Cebra *et al.*, 2001). Glucose distributes readily through the extracellular fluid compartment, and the higher peak values seen in llamas may reflect that this compartment is smaller. The gastric compartments do not contribute to this fluid compartment. The estimate for this difference in size is about 7% of body weight, which is very similar to the difference in forestomach mass.

The difference in the other organs was smaller and the greater proportional weights in alpacas may simply have been due to the smaller contributions of their forestomachs to total body weight. Pancreatic weights were similar in llamas and alpacas and were also similar in proportional size to the pancreata of cattle and sheep (Schingoethe *et al.*, 1970). Coupled with histologic and immunohistochemical studies, these findings suggest that the poor insulin response to hyperglycemia seen in camelids is unlikely to relate to low islet cell mass, but rather is likely due to poor islet cell function. The difference in insulin production between llamas and alpacas also could not be explained by pancreatic mass, and the lack of difference in the pancreas:liver ratio between species supported that the two organs have the same relative importance in both species.

Hepatic weights have not been linked with the same sorts of questions as gastric and pancreatic weights. They are nonetheless important, particularly when trying to assess pathologic conditions. At 1.3 to 2.1% of body weight, the camelid liver may be proportionally slightly bigger than the ruminant liver (1 to 2%; Frandson *et al.*, 2003), but this also probably relates to the relatively smaller forestomach in camelids. Pathologic weights up to about 9% of body weight have been recorded.

References:

Cebra, C.K., F.B. Garry, B.E. Powers and L.W. Johnson, 1995. Lymphosarcoma in 10 New World camelids. J Vet Internal Med 9: 381-385.

Cebra, C.K., S.J. Tornquist, R.J. Van Saun and B.B. Smith, 2001. Intravenous glucose tolerance testing in llamas and alpacas. Am J Vet Res 62: 682-686.

Fowler, M.E., 1998. and surgery of South American Camelids: llama, alpaca, vicuña and guanaco. 2nd ed. Iowa State University Press, Ames, Iowa, 321 pp.

Frandson, R.D., W.L. Wilke and A.D. Fails, 2003. Anatomy and Phsysiology of Farm Animals. 6th ed, 327 pp.

Ommaya, A.K., I. Atwater, A. Yañez, M. Szpak-Glasman, J. Bacher, C. Arriaza, L. Baer, V. Parraguez, A. Navia, C. Oberti *et al.*, 1995. Lama glama (the South American camelid, llama): a unique model for evaluation of xenogenic islet transplants in a cerebral spinal fluid driven artificial organ. Transplant Proc 27: 3304-3307.

Schingoethe, D.J., A.D. Gorrill, J.W. Thomas and M.G. Yang, 1970. Size and proteolytic enzyme activity of the pancreas of several species of vertebrate animals. Can J Physiol Pharmaco l48: 43-49.

DECAMA-Project: Women of mountain in business of charque of llama

A. Claros Goitia, J. L. Quispe, A. Claros Liendo and J. Flores
Program Regional of South American Camelids (PRORECA), Sánchez Lima, No. 2340, La Paz, Bolivia; proreca@entelnet.bo

In the population of Lagunas, Oruro - Bolivia, five aymara ladies exist, dedicated to the charque elaboration. With this product they prepare typical plates as the "charquekan" that are marketed in the population of Tambo Quemado (border point with the republic of Chile). In this town, the technique of charque elaboration was precarious and unsanitary: the threaded meat was dried to the bleakness on wooden planks and sticks. To improve the quality of the charque, the Regional Program of South American Camelids (PRORECA) built eolic solar dryers taking advantage of materials of the area, being used as efficient tools for the dehydration of the meat. Starting from this novel construction and the tackle implementation and manual tools, the aymara ladies begin actions of technical and hygienic improvements in the process of charque elaboration. The main function of the solar dryers is to dehydrate the meat in relatively short time and to avoid any contamination of the meat with insects, dust and other agents with impact on the quality of the final product. The implementation of the dryers improved notably the quality of the product and its commercialization, and consequently the quality of the ladies' life.

Part VI

Referral service for South American camelids at the University of Bristol Veterinary School: A review of cases from 1999 to 2002

G.L. D'Alterio and K.J. Bazeley
Farm Animal Practice, University of Bristol, Langford, Bristol, BS40 5DU, United Kingdom

Abstract

South American or New World camelids are increasingly important species outside of South America. Britain is home to the largest population of New World camelids in Europe, estimated at 4-5,000 individual llamas and 7-8,000 alpacas. Knowledge in the field of veterinary medicine gained on the more common species, the llama and the alpaca could be transferred to the less common ones, the guanaco and vicuña. As a result of increasing numbers of camelids referred, and the rising number of inquiries from private veterinary practitioners, a new residency in camelid medicine was created, thanks to a joint effort between the University of Bristol and the Royal College of Veterinary Surgeons Trust. This paper provides a descriptive summary of clinical and surgical camelids cases (second opinion) referred to the Farm Animal Practice & Hospital of the University of Bristol, and the associated experience of first opinion cases as well as distance referrals.

Introduction

South American or New World camelids are increasingly important species outside of South America (Fowler, 1998). All four species belonging to the Family Camelidae-llama (*Lama glama*), alpaca (*L. pacos*), guanaco (*L. guanicoe*) and vicuña (*Vicugna vicugna*)-have been represented in zoos worldwide for a long time. More recently, relaxation on exporting restrictions from South American countries meant the establishment of breeding units of llamas, alpacas, and few guanacos in North America and Australia first, followed by New Zealand and several European countries afterward (Smith, 1993; Fowler, 1998). Britain is home to the largest population of New World camelids in Europe, estimated at 4-5,000 individual llamas and 7-8,000 alpacas (Newth, personal communication).

The llama, a key and multipurpose livestock species for rural economy in the high Andes (Wheeler, 1995), has had recreational purposes elsewhere (Smith, 1993). The alpaca, equally important in the Andes livestock industry, is farmed elsewhere primarily with the aim of establishing a high quality fibre industry. Both species are kept as companion animals, display and educational, and as a mean of diversification for the livestock industry in developed countries

Relatively little information is available on health issues affecting new world camelids, particularly at individual case level. As their popularity increases, so does the need for a better understanding of the pathological conditions that affect these species. This will enable clinicians to prevent and treat disease more effectively, with benefits both for the welfare of the animals and their economic performance.

In addition, knowledge gained on the more common species, the llama and the alpaca could be transferred to the less common ones, the guanaco and vicuña. The latter was only recently re-listed from CITES Appendix 1 to Appendix 2, thanks to the successful population recovery plans implemented in Peru` and Chile. Listing of the vicuña in CITES Appendix 2 should allow the sustainable commercial exploitation of this species, which produces what is currently the most expensive natural fibre in the world (Kadwell *et al.*, 2001).

This paper provides a descriptive summary of clinical and surgical camelids cases (second opinion) referred to the Farm Animal Practice & Hospital (FAPH) of the University of Bristol, and the associated experience of first opinion cases as well as distance referrals. The purpose of the paper is to allow zoo and livestock veterinary practitioners with an interest in alternative livestock species to anticipate the common clinical presentations encountered among camelid patients.

The Farm Animal Practice and Hospital of the University of Bristol

The Farm Animal Practice and Hospital primarily provides the practical teaching of undergraduate veterinary students. Among the various didactic activities, final year undergraduate veterinary students spend two weeks' clinical rotation based at FAPH. Students accompany members of staff during all clinical activities, including the out-of-hours service. Under close supervision, students are directly involved in dealing with all aspects of farm animal clinical work, from diagnostic work up to administration of treatment and monitoring of in-patients. The FAPH has over 120 registered first opinion clients, from large dairy cattle units to small holdings; five first opinion clients keep camelids. The case load is substantial also thanks to the location of the University campus in which the FAPH is based, the South-west of England, traditionally a region devoted to livestock farming. In addition to first opinion work, the FAPH receives referral cases from veterinary surgeons in private practice. Individual and on-farm referrals are accepted.

During the period August 1999 to July 2002 a total of 647 individual farm animal cases were admitted at the Farm Animal Hospital. It is important to notice that during this time, the Hospital closed for 10 months following livestock movement restrictions imposed as part of the steps taken to bring under control an outbreak of Foot-and-Mouth Disease.

The referral service for South American camelids

As a result of increasing numbers of camelids referred, and the rising number of inquiries from private veterinary practitioners, a new Residency in camelid medicine was created, thanks to a joint effort between the University of Bristol and the Royal College of Veterinary Surgeons Trust. The Residency promotes and expands the referral service, provides telephone advice for owners and veterinarians, and carries out herd visits and clinical research.

Camelids hospital and distance referrals

Seventy-four individual clinical cases, from February 1999 to December 2002 were admitted. Of this, sixty-five were alpacas (88 per cent) and 9 llamas (12 per cent). Out of the total number of cases referred, fifty-five were females (74 per cent) and nineteen were males (26 per cent). Crias (unweaned animals), often below 6 months of age) represented thirty-two per cent (24/74) of the animals admitted. Based on aetiology, body system and pathological process involved, thirty-one different diagnoses were formulated, the most common being mandibular/maxillary osteomyelitis (12/74). When considering the body system involved, the muscle-skeletal system was affected in twenty-seven cases (36 per cent), the digestive system in 8 cases, and the genito-urinary apparatus in 7, and all the other systems in sporadic presentations. A multi-systemic clinical condition was observed in 10 cases. A definitive diagnosis was not reached in twelve out of seventy-four cases. Table 1 summarises main presenting clinical signs and diagnoses for the hospitalised cases. Table 2 shows a breakdown of cases affecting juvenile camelids.

Table 1. Summary of camelid clinical cases referred to FAPH, from 1999 to 2002.

No. of cases	Main presenting clinical signs	Diagnosis
12	Mandibular/maxillary mass; discharging sinus	Mandibular/maxillary osteomyelitis
12	Various	No diagnosis
5	Inability to stand after birth	Failure to suckle
4	Infertility	Agenesis and or atrophy of ovaries and Fallopian tubes
4	Weight loss; diarrhoea; ill thrift	High endoparasitic infestation (often associated with nutritional imbalances)
4	Limb deformity	Angular limb deformity, particularly carpus valgus*
2	Weight loss; submandibular mass	Tuberculosis
2	Umbilical mass	Umbilical hernia
2	Lameness (sudden onset)	Bone sequestrum
2	Lameness	Fracture
2	Ocular lesion	Chronic keratitis and corneal ulcer
2	Testicle not in scrotum; scrotal oedema	Cryptorchidism
2	Various	Multifactorial condition (with poor genetic background and/or nutritional imbalance
1	Head down; stiff neck; ataxia	Cervical damage
1	Dyspnoea	Choanal atresia
1	Tail bending	Deformed tail
1	Head tilt; often recumbent	Cerebral abscess
1	Head tremors; torticollis; wide stance	Bacterial meningitis
1	Deformed skull	Maxilofacial dysgenesis
1	Recumbent; hypothermia; tachypnea	Extensive kidney damage of unknown aetiology
1	Lethargic; dysuria	Recto-genital abscess
1	Food material from nostrils	Fissure in the soft palate
1	Weight loss; lethargic	Lymphoid leukaemia
1	Recumbency; abnormal gait; stiffness	Hypovitaminosis D
1	Impaired mobility; tendons contractions; dyspnoea at handling	Hypovitaminosis E
1	Colic-like	Perforated stomach ulcers
1	Alopecia, crusting and scaling	Hypersensitivity to Chorioptes mite infestation**
1	Large subcutaneous mass	Fibroma
1	Heart murmur	Sinus arrhythmia ***
1	Hard udder and milk clots	Chronic *S. aureus* mastitis
1	Severe dyspnoea	Diaphragmatic paralysis
1	Recumbency; abnormal gait	Bilateral patellar sub-luxation

* Angular limb deformity was a frequent accidental finding in several other cases.
** Skin lesions were detected in several cases; the authors have extensively investigated this re-current clinical presentation and findings will be reported elsewhere.
*** Sinus arrhythmia was observed in other cases as an accidental finding.

Table 2. Summary of juvenile camelid clinical cases, 1999 to 2002.

N. of cases	Main presenting clinical signs	Diagnosis
5	Inability to stand after birth	Failure to suckle
2	Umbilical mass	Umbilical hernia
2	Various	Multifactorial condition (with poor genetic background and/or nutritional imbalance
2	Lameness (sudden onset)	Bone sequestrum
2	Lameness	Fracture
2	Limb deformity	Angular limb deformity, particularly carpus valgus
2	Various	No diagnosis
1	Deformed skull	Maxilofacial dysgenesis
1	Head tremors; torticollis; wide stance	Bacterial meningitis
1	Recumbency; abnormal gait; stiffness	Hypovitaminosis D
1	Impaired mobility; tendons contractions; dyspnoea at handling	Hypovitaminosis E
1	Head down; stiff neck; ataxia	Cervical damage
1	Heart murmur	Sinus arrhythmia
1	Severe dyspnoea	Diaphragmatic paralysis

The telephone advisory service and distance referrals have taken a significant portion of the time allocated by the authors to camelid medicine. Since records begun (March 2001), the referral service received thirty-nine different, individual and group clinical case reports from veterinarians in private practice, and a total of sixty-three general enquiries on camelids health. The latter are summarised in Table 3.

Selected clinical presentations

Mandibular ostemyelitis (tooth root abscess)

This was the most common clinical presentation at FAPH. Reported in the literature, it remains of unknown pathogenetic mechanism. The condition is associated to a firm swelling on the ventral aspect of the horizontal ramus of the mandible. More rarely, the rostral part of the mandible, or even the maxilla is affected. In chronic cases the swelling might involve the contra-lateral ramus and the mandibular profile is severely modified. Sinuses that discharge purulent material may also be observed. Cebra *et al.* (1996) reports hypersalivation, altered mastication and loss of condition, but in our experience the animals seem to tolerate the condition well, even when extremely progressed. The condition is likely to develop following disruption of the periodontal membrane of the molars, rather than from crown and pulp cavity disease (Anderson, 2002b). Radiographs of the affected area confirm the clinical diagnosis, although in the authors' experience it is sometimes difficult to identify the affected tooth/teeth radiologically.

Medical treatment only (long term antibiotics) might result in an initial improvement, but reoccurrence is common (Anderson, 2002b). Nearly all cases referred to FAPH received antibiotic treatment before referral, with little or no improvement.

Table 3. Telephone enquiries on camelids received from FAPH (starts 03/01).

Type of enquiry	Number alpacas	Number llamas
Vaccination protocol	2	
Mineral supplementation	1	
Legal case	2	
Field sedation	4	2
Development of questionnaire on FMD impact on camelids	1	1
Vitamin D3 deficiency and supplementation	4	
Skin disease (report of and advice on treatment	21	
Castration technique	3	1
Advice on pregnancy diagnosis	7	
Advice on microchip	1	
TB in camelids	1	
Possibility of Bracken (Pteridium aquilinum) poisoning	1	
Euthanasia		1
Use of GnRH	2	
Use of prostaglandins for treatment of metritis	1	
De-worming strategy	2	
Control of male libido		1
Use of phenylbutazone	1	
Sampling to assess mineral status	1	
Investigation of Johnes disease	1	
Presence of the bot fly in the UK	1	
Total	57	6

Tooth extraction is curative, and it should be performed under general anaesthesia and via a lateral, surgical approach. Anderson (2002b) strongly recommends not attempting repulsion of the affected molar/s by impact because of the risk of mandibular fractures.

Angular limb deformities

Angular limb deformities are commonly observed in camelids. We question the genetic merit of at least some of the founder members of the camelid herds outside of South America, which together with the small genetic pool available might account for the high prevalence of these defects.

Carpus valgus is prevalent, but several other defects are described (Fowler, 1998). For example, the authors have observed severe, bilateral inward bowing of the posterior fetlock in a mature female llama; surgical correction of the worst affected limb has been carried out and healing is in progress at the time of writing.

Regardless of the type of deformity, the clinician's concern is not aesthetic consideration, but the functional and long- term consequences of these defects. In our experience, detection of limb deformities was a common accidental finding, but it is worrying that only a few cases were referred to FAPH for further investigation. Even minor deformities warrant further investigation, first to establish whether the defect might be inherited or acquired, and second because early treatment is more likely to be successful. Radiology is fundamental in further assisting the diagnostic work up. Anderson (2002a) describes different techniques for the surgical treatment of carpus valgus.

Neonatal and juvenile clinical conditions

Approximately one third of the cases referred to FAPH were represented by juvenile camelids, often un-weaned. The neonatal phase is critical in respect to the subsequent development of a strong and healthy cria. Newborn camelids rely exclusively on passive transfer of immunity from colostrum, and its failure is an important predisposing factor to infections and determinant of neonatal morbidity and mortality (Garmendia *et al.*, 1987). Later in life, sub-optimal milk intake and therefore plane of nutrition will negatively influence development. Milk intake from the dam should be regularly monitored, including regular checks of the udder and the dam`s body condition. Weaning should not take place at a pre-determined time, but should vary according to the latter assessment. In at least three of the four cases of fracture and bone sequestrum seen at FAPH we suspected an underlying pathological process, including sub-clinical hypovitaminosis D (Parker *et al.*, 2002). Among the congenital abnormalities, some were compatible with life (*e.g.* umbilical hernia), others were not (maxillary dysgenesia). Irrespective of the outcome, the advice is not to breed from these patients. Post-mortem and histopathological examination confirmed a case of bacterial meningitis caused by *Salmonella Newport* in a newborn alpaca cria presented with neurological signs, such as torticollis, head tremors and paddling (D' Alterio *et al.*, 2003).

Mineral and vitamin imbalances

As in other species, camelids are susceptible to nutritional imbalances. Under European farming conditions, absolute deficiency in energy and protein is less likely to occur than in South America. Mineral and vitamin imbalances could be more likely on the fertilised and planted pastures of Europe, compared to the large range of plant species grazed in the native environment. Juveniles seem more prone to display clinical signs related to nutritional imbalances. We came across several overweight adult animals, but we have also observed lactating, pregnant females that progressively lose body condition, probably due to an energy deficit.

In late winter, a 5 months old male alpaca was referred to FAPH. History and clinical examination findings were abnormal gait, tendency to recumbence and poor growth. Radiographs of selected joints showed no obvious changes in bone density, but irregular growth plates were observed. The animal was found to be markedly hypophosphatemic. A suspect diagnosis of hypophosphatemic rickets was formulated. The animal responded very well to repeated parenteral administration of ADE supplement (Duphafral ADE Forte; Fort Dodge Animal Health), at the dose rate of approximately 800 iu/kg BW of cholecalciferol at bi-weekly intervals. Response to treatment should be monitored based on amelioration of clinical signs and increase of phosphorus blood levels. Vitamin D analysis is expensive and not readily available from veterinary laboratories.

In the authors' experience, crias born in late summer-early autumn are predisposed to develop hypophosphatemic rickets due to reduced sun exposure; crias born late in the breeding season should therefore receive vitamin D supplementation.

Myonecrosis (white muscle disease) due to vitamin E deficiency (without Se deficiency) was diagnosed in a 4 months old male llama, showing severe ataxia, contractions of the flexor tendons and severe dyspnoea during handling. Clinical signs greatly improved following treatment with vitamin E, at the dose rate of 20 mg/kg BW of alpaha-tocopheryl acetate (Vitenium; Novartis Animal Health UK Ltd).

Ill thrift due to multiple vitamins and trace elements imbalances was formulated in the case of two juvenile alpacas referred to FAPH. In one case a poor genetic background was also thought to be a contributing factor to the condition. Both animals were sampled for blood and found deficient in copper, vitamin E, Selenium, Iodine and cobalt. Interestingly, a follow up herd visit and blood sampling found several adult and sub-adult sub-clinically deficient in copper and cobalt.

The above case suggests how trace elements clinically deficient juveniles represent a window to detection of sub-clinically deficient adult animals.

Conclusions

Although the caseload so far is relatively small, a remarkably wide variety of clinical presentations have been seen. This is a feature of second opinion clinical work, where cases of particular complexity tend to be referred. First opinion work is generally more routine, and often concerned with prophylactic, rather than curative, measures. The telephone enquiries from first opinion practitioners reflect this difference (see Table 3). No diagnosis was made in several cases, despite extensive diagnostic and pathological investigation. This suggests that we do not yet know enough about these species, and highlights the need for clinical research and reporting. For example, no validated reference ranges are available for certain trace elements and vitamins, although diseases that respond to supplementation are commonly seen or suspected. Also, there are no medicines that are licensed for use in these species in the UK, and dosages have often been empirical. Clostridial vaccines are commonly used in camelids, following the protocol formulated for sheep, but the efficacy of this practice has not been validated. Owners of these animals (that may be very valuable) should be made aware of this.

The review of clinical records reported here has enabled the authors to retrospectively formulate hypotheses about the likely diagnoses for several of the unsolved cases, so that periodic review of clinical cases is likely to be a valuable exercise.

The advisory service has gained valuable information on camelids health in the UK, through direct clinical experience at individual and group level, and through contact with veterinary practitioners and camelid owners.

Acknowledgements

We would like to thank all colleagues and support staff at FAPH. Development of the referral service relied substantially on the expertise of clinical and surgical specialists at the University of Bristol Veterinary School. Undergraduate students are a constantly challenging source of inspiration to learn more and do better. We would also like to thank all the referring veterinarians. G.L. D'Alterio is recipient of an RCVS Trust Millennium Production Animal Scholarship.

References

Anderson, D.E., 2002a. Angular limb deformity. In: Proceedings of the course in camelid medicine, surgery and reproduction for veterinarians, D.E. Anderson (ed.), The Ohio State University, Columbus, Ohio, pp. 31-33.

Anderson, D.E., 2002b. Tooth root abscess. In: Proceedings of the course in camelid medicine, surgery and reproduction for veterinarians, D.E. Anderson (ed.), The Ohio State University, Columbus, Ohio, pp. 87-88.

Cebra, M.L., C.K. Cebra and F.B. Garry, 1996. Tooth root abscesses in New World Camelids: 23 cases (1972-1994). Journal of the American Veterinary Medical Association 209: 819-822.

D'Alterio, G.L., K.J. Bazeley, G.R. Pearson, J.R. Jones, M. Jose and M.J. Woodward, 2003. Meningitis associated with *Salmonella* Newport in a neonatal alpaca (*Lama paco*) in the United Kingdom. Veterinary Record 152: 56-57.

Fowler, M.E., 1998. Medicine and surgery of South American Camelids: llama, alpaca, vicuña and guanaco. 2nd ed. Iowa State University Press, Ames, Iowa, 321 pp.

Garmendia, A.E., G.H. Palmer, J.C. Demartini and T.C. McGuire, 1987. Failure of immunoglobulin transfer: a major determinant of mortality in newborn alpacas (*Lama pacos*). American Journal of Veterinary Research 48: 1472-1476.

Kadwell, M., M. Fernandez, H.F. Stanley, R. Baldi, J.C. Wheeler, R. Rosadio and M.W. Bruford, 2001. Genetic analysis reveals the wild ancestors of the llama and alpaca. Proc. R. Soc. Lond. B 268: 2575-2584.

Parker, J.E., K.I. Timm, B.B. Smith, R.J. Van Saun, K.M. Winters, P. Sukon and C.M. Snow, 2002. Seasonal interaction of serum vitamin D concentration and bone density in alpacas. American Journal of Veterinary Research 63: 948-953.

Smith, B.B., 1993. Major infectious and non-infectious diseases in the llama and alpaca. Veterinary and Human Toxicology 35: 33-36.

Wheeler, J.C., 1995. Evolution and present situation of the South American Camelidae. Biol. J. Linn. Soc. 54: 271-295.

Morphological variation of Italian alpaca population

G. De Fidelibus[1], A. Vecchi[1], G. Minucci[1], G. Lebboroni[1], M. Antonini[2] and C. Renieri[1]
[1]Department of Veterinary Science, University of Camerino, Via Circonvallazione 93/95, 62024 Matelica (MC), Italy; carlo.renieri@unicam.it
[2]ENEA CR Casaccia, BIOTEC AGRO, Via Anguillarese 301, 00060 S. M. di Galeria, Roma, Italy.

Abstract

In order to describe the Italian alpaca population and establish the technical documents for the ITALPACA Association, a research on 153 adult animals, 47 male and 106 female, bred in 17 farms located in Centre and North Italy was carried out. No variation has been observed for skin colour, hoof colour, ears length and type, forehead and nose profile. Thirty one percent of the animals are full white, 69% pigmented. Thirty two percent are red with black extremities, 32% black, 18% wild, 10% reddish brown, 1% black and tan. Five percent of the animals present an unreadable pattern. Twenty nine animals are diluted or grey. Twenty percent of animals are spotted. Forty seven percent of animals are Huacaya, 8% Suri and 44% intermediate. Ninety four percent of the animals present an intermediate fleece extension, 5% of animals have fleece covering heads and legs and only 1% of animals have uncover heads, neck, legs and belly. Only the age affects significantly withers height, chest height and rump width. The measures are significantly positively correlated.

Keywords: alpaca, morphological characteristics, colour

Introduction

Alpaca breeding started in Italy in 1992. First animals have been imported from France, Germany and Switzerland. At present various hundred animals are bred in about 40 farms located in North and Centre Italy.

In 1996 the Italian Breeding Alpaca Association (ITALPACA) was established. The Technical Committee was charged to organized the selection plan, to implement the Stud Book, to define the standard of population and to propose a lincar system for morphological evaluation. In order to collect available data, an inquiry in the population was carried out. Present paper shows the result of the morphological evaluation.

Materials and methods

The research was carried out on 153 adult alpaca, 47 male and 106 female, breed in 17 farms located in Centre and North Italy. With regards the population typology, the analysed characters are shown in Table 1. Fleece colour has been classified according to Lauvergne *et al.*, (1996), Renieri *et al.* (2002) and Frank *et al.* (2002). The fleece structure has been classified according to Antonini *et al.* (2001, 2004). All the intermediate animals have been classified in a single category. All the other characters have been classified according to the direct observation carried out by the authors on the Italia alpaca population.

According the decision of the breeders, biometrical characteristics recorded have been: rump width, withers and chest height. The records have been carried out with the sheep Lydtin stick adjusted to South American Camelids. A factorial model with age (7 levels), sex (2 levels), farm (4 levels) and South American origin (3 levels, Bolivia, Chile and Peru) as factors has been used

Table 1. Morphological characteristics analysed.

Character	Variable	Description
Coat colour	Pigmentary pattern	Non agouti
		Black and tan
		Badger face
		Mule stripe
		Wild (vicuña or guanaco type)
		Red with black extremities
		Red
		Other
	Type of eumelanins	Black
		Dark brown
		Light brown
	Alteration of pigmentation	Full white
		Dilute
		Grey or greying
		Other
	Spotting	Presence
Coat structure	Type	Huacaya
		Suri
		Intermediate
	Extention	Covered
		Medium covered
		Uncovered
Skin colour		
Outer ear	Length	Normal
		Short
		Atrophic
	Type (Direction)	Normal
		Banana type
		Left direction
		Right direction
	Tonicity	Erect
		Horizontal
		Atonic
Forehead-nose region	Profile	Convex
		Right
		Concave
	Direction	Left direction
		Right
		Right direction

for the analysis of variance by SAS GLM procedure. Phenotypic correlations of measures have been estimated by the SAS CORR procedure (SAS, 1996).

Results and discussion

No variation has been observed for skin colour (black), hoof colour (black), ears length and type (in both case normal), forehead and nose profile (straight). For the coat colour, 31% of the animals are full white, 69% pigmented. For the pigmentation patterns, 32% alpaca are red with black extremities, 32% black, 18% wild, 10% reddish brown, 1% black and tan. Five percent of the animals present an unreadable pattern. Eighty percent of the animals present black eumelanin, 18% brown eumelanin. The alteration of the pigmentation exists on 29 animals: 44% fleece colour are diluted and 56% grey. Twenty percent of animals are spotted, with a variable extension of spotting.

About the type of fleece, 47% of animals are classified as "Huacaya", 8% as "Suri" and 44% are intermediate. About the extension, 94% of the animals present an intermediate fleece extention, 5% of animals have fleece covering heads and legs and only 1% of animals have uncover heads, neck, legs and belly.

Only the age affects significantly the three measures variation. The estimated means for the factor are presented in the Table 2.

Withers height, chest height and rump width are significantly positively correlated. The values of correlation are very high: 0.543 for withers and chest height, 0.375 for withers height and rump width, 0.457 for chest height and rump width.

Conclusions

A large variation exists in Italian alpaca population on both coat colour and coat characteristics. This variation can be explained by both the different origins of imported animals and the absence of an univocal trend on the animal selection during the importation. The large variation has been taken into account by the Official Standard of ITALPACA population, approved by all associated breeders. About the coat colour, in fact, all variables can be accepted with a different score. The selection is for full white and uniform patterns and against spotting. Alteration like grey or dilution can be accepted only in specific lines. About fleece, the complete separation of Huacaya and Suri has been decided. Intermediate animals may to be used only in the Huacaya program. The effect of age being largely foreseen, no variation seems exist among the animals on biometric measures. The withers height is an important indicator for the influence of llama in some alpaca lines, but in the Stud Book, no selection criteria have been taken into account on biometric characteristics.

Table 2. Estimated means for morphological characteristics by age of the animals.

Age	Withers height x ± sd	Chest height x ± sd	Rump width x ± sd
2	82.318 ± 2.351	32.597 ± 1.162	19.345 ± 1.304
3	85.065 ± 2.427	34.232 ± 1.199	21.965 ± 1.346
4	84.057 ± 3.238	34.387 ± 1.600	21.212 ± 1.796
6	89.670 ± 2.485	36.871 ± 1.228	22.798 ± 1.378
7	83.087 ± 2.271	35.386 ± 1.122	22.057 ± 1.260
8	84.598 ± 3.226	34.567 ± 1.594	22.319 ± 1.789
9	76.057 ± 4.662	31.387 ± 2.304	18.212 ± 2.586

References

Antonini, M., M. Gonzales and A. Valbonesi, 2004. Relationship between age and postnatal skin follicular development in three types of South American Camelids. Livestock Production Science 90: 241-246.

Antonini, M., F. Pierdominici, S. Catalano, E. Frank, M. Gonzales, M.V.H. Hick and F. Castrignanò, 2001. Cuticular cell mean scale frequency in different type of fleece of domestic South American Camelids (SAC). EAAP Publ. 105, pp. 110-116.

Frank, E.N., C. Renieri, M.V.H. Hick, V. La Manna, C.D. Gauna and J.J. Lauvergne, 2002. Segregation analysis of Irregular Spotting and Full White in llama. 7th WCALP, communication no. 12-17.

Lauvergne, J.J., C. Renieri and E.N. Frank, 1996. Identification of some allelic series for coat colour in domestic camelids of Argentine. Proc. 2nd Eur. Symp. South American Camelids, Camerino, pp. 39-50.

Renieri C., E.N. Frank, M.V.H. Hick, V. La Manna, C.D. Gauna and J.J. Lauvergne, 2002. Segregation analysis of coat colour phenotypes in llama. 7th WCALP, communication no. 12-16.

SAS, 1996. User's guide: Statistics, Version 6.12 Edition, SAS Inst. In., Cary, NC.

Breeding and/or handling problems? Causes of death in camelids

I. Gunsser[1], T. Hänichen[2] and C. Kiesling[1]
[1] *Römerstr. 23, 80801 München, Germany; ilona.gunsser@t-online.de*
[2] *Einsteinstr. 65, 81675 München, Germany*

Abstract

Supported by the „Institute of Veterinary Pathology of the University of Munich", *post-mortem* examinations in 179 camelids were performed between 1993 and 2003. In addition, the results of 43 *post-mortem* examinations of other institutes and laboratories were added which were supplied by camelid owners. These evaluations include 144 llamas, 63 alpacas, 5 guanacos, 1 guanaco-llama mix, 5 dromedary camels, 3 Bactrian camels and one vicuña. Most pathological changes were found in the thorax (62.2%). The reason for this is in most cases a pulmonary edema which develops as a consequence of cardiovascular failure during the process of dying. Apart from this the most frequent problems were found in the liver (48.6%), the digestive system (43.2%), and the abdomen (36.0%), caused by endoparasites, infections, and chronic feeding faults. Further pathological problems were found in the urinary tract (31.5%), the head (26.6%), the spleen (23.0%) and the skin (22.5%). The reasons were degeneration of parenchymas, teeth problems, spleen reactions and mites or infections, respectively. Less frequent were pathological situations in the genitalia (14.0%), the neck (10.4%), the bones (9.0%) and the limbs (6.3%). Since in many cases more than one pathological finding was diagnosed in an animal, the summed percentages exceed 100%. The death reasons found in this investigation were primarily infectious diseases (22.5%), euthanasia (17.1%), emaciation (9.5%), and fatty degeneration of the parenchyma (9.0%). In 14.4% of the cases the diagnosis was inconclusive.

Keywords: camelids, *post-mortem* examination, causes of death

Introduction

During the last 10 years the number of owners of camelids has grown constantly. However, due to lack of information about the handling, feeding and breeding of these animals numerous problems have arisen, which in some cases even caused the death of these animals. Supported by the "Institute of Veterinary Pathology of the University of Munich", *post-mortem* examinations could be offered to the owners with the aim to collect detailed information about the causes of diseases and deaths in camelids.

Material and methods

In the years between 1993 and 2003, *post-mortem* examinations were performed on 179 camelids. The technique of choice was dissection and histology. In addition, the results of 43 *post-mortem* examinations carried out by other institutes and laboratories were added, which were supplied by camelid owners. These evaluations include 144 llamas, 63 alpacas, 5 guanacos, 1 guanaco-llama mix, 5 dromedary camels, 3 Bactrian camels and one vicuña. The pathological problems were separated as diseases of the head, the neck, the thorax (heart and lung), the abdomen, the digestive system, the urinary tract, the spleen, the genitalia, the skin, the limbs and the bones. In addition, based on the various pathological observations, the ultimate causes of death were determined.

Results

Numbers and types of camelids

During the first years of this investigation the amount of llamas in the camelid population in Europe was much higher than the amount of alpacas. For this reason relatively more lamas died and were sent in for pathological investigations. During the last years the numbers of alpacas have increased step by step, however, the exact numbers were not registered. The number and types of camelids which were dissected are shown in Table 1.

Location of pathological findings

Most pathological changes were found in the thorax (62.2%). The reason for this is in most cases a pulmonary edema which develops as a consequence of cardiovascular failure during the process of dying. Apart from this the most frequent problems were found in the liver (48.6%) and the digestive system (43.2%). The distributions of further pathological problems are shown in Table 2. Since quite frequently two and more findings in one animal had been registered, the summed percentages will exceed 100%.

Causes of death

The reasons of death found in this investigation were primarily infectious disease (22.5%), euthanasia (17.1%), emaciation (9.5%) and fatty degeneration of parenchyma (9.0%). In 14.4% of the cases the diagnosis was inconclusive. Further causes of death are presented in Table 3.

Discussion

Reasons of pathological findings

In the following the pathological findings are listed, ordered by frequency (see Table 2) in the various body regions and indicate the reasons for the findings (in parentheses) based on the results from pathology and histology, blood and fecal tests, and the interview of the owner:
1. *Thorax:* besides the heart and lung changes while dying (pulmonary edema), degeneration of myocardium (selen deficiency), malformation (defect of septum, overriding aorta), pneumonia (mykotic, bacterial), hydrothorax, metastases, and parasites (lungworm) has been found.

Table 1. Number and types of animals from own post-mortem examinations ("own") and from other institutions ("foreign").

	own	foreign	Sum
Alpaca	44	19	63
Llama	121	23	144
Guanaco	5	0	5
Vicuña	1	0	1
Llama/Guanaco	1	0	1
Dromedary camel	4	1	5
Bactrian camel	3	0	3
Sum	179	43	222

Table 2. Locations of pathological problems from own post-mortem examinations ("own") and from other institutions ("foreign").

	own	in %	foreign	in %	sum	in %
Thorax	112	62.6	26	60.5	138	62.2
Liver	88	49.2	20	46.5	108	48.6
Digestive tract	69	38.5	27	62.8	96	43.2
Abdomen	58	32.4	22	51.2	80	36.0
Urinary tract	58	32.4	12	27.9	70	31.5
Head	52	29.1	7	16.3	59	26.6
Spleen	46	25.7	5	11.6	51	23.0
Skin	46	25.7	4	9.3	50	22.5
Genital organs	24	13.4	7	16.3	31	14.0
Neck	16	8.9	7	16.3	23	10.4
Bones	19	10.6	1	2.3	20	9.0
Extremities	13	7.3	1	2.3	14	6.3

Table 3. Causes of deaths, from own post-mortem examinations ("own") and from other institutions ("foreign").

	own	in %	foreign	in %	sum	in %
Infection	40	22.3	10	23.3	50	22.5
Euthanasia	29	16.2	9	20.9	38	17.1
Unclear	27	15.1	5	11.6	32	14.4
Cachexia	17	9.5	4	9.3	21	9.5
Fatty degen. of parenchymas	17	9.5	3	7.0	20	9.0
Peritonitis	10	5.6	5	11.6	15	6.8
Abortion	13	7.3	1	2.3	14	6.3
Cardiac insufficiency	7	3.9	2	4.7	9	4.1
Liver cirrhosis	6	3.4	1	2.3	7	3.2
Trauma	6	3.4	1	2.3	7	3.2
Meningitis	2	1.1	1	2.3	3	1.4
Parasitosis	2	1.1	0	0.0	2	0.9
Tumor	2	1.1	0	0.0	2	0.9
Urinary calculi	0	0.0	1	2.3	1	0.5
Poisoning	1	0.6	0	0.0	1	0.5
Sum	179	100.0	43	100.0	222	100.0

2. *Liver*: fatty degeneration (metabolic problems), cirrhosis (fasciola hepatica, dicrocoelium dentriticum) and cumarin poisoning could be seen.
3. *Digestive tract*: Problems were caused by parasites (*Strongylides, Trichuris, Nematodirus, Eimeria, Moniezia*), stomach ulcus, compartment 1-3, with or without perforation, infection (*Clostridium perfringens, E. coli, Corona virus, Helicobacter* (serology)), sand, and torsion of small intestine.

4. *Abdomen*: Findings recorded were ascites, caused by parenchyma degeneration (wrong nutrition, anemia as result of parasitosis), peritonitis as result of ulcus perforation (wrong nutrition, stress), uroperitoneum after accident or stones in the urethra (rupture of the bladder), infection of lymph node (tuberculosis, *Corynebact. pseudotuberculosis*), furthermore tumor metastases and hernia umbilicalis.

5. *Urinary tract*: changes resulted in parenchyma degeneration of the kidney (metabolic problems), malformation, hypoplasia, renal cysts, calcinosis, infection (*Cl. perfringens, E. coli*).

6. *Head*: tooth infection, stomatitis, eye problems (traumatic), brain parasites (*Elephastrongylus*), malformation (wry face, choanal atresia, teeth) and otitis (parasites, insects) has been seen.

7. *Spleen*: atrophy, hyperämie, tumor and granuloma (parasites) were found.

8. *Skin*: Skin-problems in camelids were caused by infection (*Staphylococcus, Trichophyton*), mites (*Chorioptes mange*), further subcutaneous edema and myiasis.

9. *Genital organs*: endometritis (bred in short intervals), hypoplasia, cryptorchism, malformation, penis necrosis (caught in long fibre while breeding) was diagnosed.

10. *Neck:* problems were seen by infection of the esophagus mucosa, obstipation of the oesophagus (wrong food) and esophagus parasites (Sarcoporidia)

11. *Bones/extremities*: fractures, rickets (lack of vitamin D), malformation and the sagging of limb in the fetlock joint (breeding), were found.

Reasons of death

The interpretation of pathological findings as reason of death is generally problematic. In *post-mortem* examinations different pathological problems can be found. Those which were finally the causes of death are explained, ordered by frequency from Table 3:

1. *Infections* which finally caused the death of Camelids were *Escherichia coli*, Clostridium perfringens, Corynebact. pseudotuberculosis, Corona virus, Klebsiella, Mycobacterium bovis, *Pseudomonas, Fusobacterium necrophorum, Candida* and *Clostridium chauvoei*.

2. *Euthanasia*, the reasons were incurable leg problems, traumatic injuries, chronic diseases, malformations, neurological problems, berserk problems.

3. *Unclear*, there was no obvious cause of death.

4. *Cachexie* was the reason of death after chronic infection, chronic parasitosis and chronic feeding problems (lack of milk).

5. *Degeneration of parenchymas* which were incurable after chronic feeding problems or unclear intoxication.

6. *Peritonitis* as result of compartment 1-3 ulcus perforation or enteritis necrotica.

7. *Abortion*, the reason was mostly unclear, in some cases malformation of the foetus existed, some *Chlamydia* infections were found.

8. *Cardiac insuffiency* was the result of cardiomyopathy, selen deficiency or malformation.

9. *Liver cirrhosis* was caused by *Fasciola hepatica* or *Dicrocoelium dendriticum*.

10. *Tauma* resulting from fights among non-castrated males.

11. *Meningitis,* the reasons were parasites (*Elephastrongylus*), infections (*E. coli*, cocci). In some cases no specific reason could be found.

12. *Parasitosis*, in these cases to many *Stronylides, Trichuris, Nematodirus, Coccidia* and *Moniezia* has been found.

13. *Tumors* (carcinoma).

14. *Urinary calculi,* the reason of development was unclear, probably dietetic genesis.

15. *Poisoning* with cumarin has been seen.

Conclusions

Better informed owners can avoid health problems in llamas and alpacas. This information should include the techniques of correct feeding, which means hay, grass and minerals. Too much energy (from grain or vegetables) resulted in acidosis of the stomach. Furthermore, the analysis of the worm burden is crucial to enable to use the appropriate medicaments against the parasites which caused the problems in the individual animal. Knowledge about the correct breeding technique to avoid harming the uterus or strangulation of the penis in the wool of the female is imperative. Very helpful also is the education of the owners how to train their camelids, so that they are able to get into close contact with them and are therefore put in a position to control their health.

References

Hänichen, T. and H. Wiesner, 1995. Erkrankungs- und Todesursachen bei Neuweltkameliden. Tierärztl. Praxis 23: 515-520.

Hänichen, T., H. Wiesner, and E. Göbel, 1994. Zur Pathologie, Diagnostik und Therapie der Kokzidiose bei Wiederkäuern im Zoo. Verh. Ber. Erkrg. Zootiere 36: 375-380.

Fowler, M.E., 1998, Medicine and Surgery of South American Camelids, Iowa State University Press, Ames, Iowa, USA.

Gunsser, I., T. Hänichen and J. Maierl, 1999. Leberegelbefall bei Neuweltkameliden. Tierärztl. Prax. 27 (G): 187-192.

Gunsser, I., H. Wiesner and T. Hänichen, 2002. Krankheiten. In: M. Gauly (ed.), Neuweltkameliden. Parey Verlag, Stuttgart.

Hertzberg, H., C. Wenker, J.-M. Hatt, P. Ossent, T. Hänichen, A. Brack and E. Isenbügel, 1997. Dirocoeliose bei Neuweltkameliden (Fallberichte und Therapievorschlag). Verh. Ber. Erkrg. Zootiere 38: 399.

Wenker, C., J.-M. Hatt, H. Hertzberg, P. Ossent, T. Hänichen, A. Brack and E. Isenbügel, 1998. Dikrozöliose bei Neuweltkameliden. Tierärztl. Prax. 26 (G): 355–361.

Wiesner, H., T. Hänichen and J. Hector, 1997. Tuberkulose bei Säugetieren im Münchner Tierpark Hellabrunn im Zeitraum von 1972 bis 1996. Verh. Ber. Erkrg. Zootiere 38: 299-303.

Effect of oestradiol on embryo mortality in llamas

W. Huanca[1], J. Palomino[1] and T. Huanca[2]
[1]IVITA, Faculty of Veterinary Medicine, University of San Marcos, P.O. Box 03, 5034 Salamanca, Lima, Peru; whuanca@appaperu.org
[2] ILLPA, INIA, Puno, Peru

Embryo mortality is an important factor that affects the reproductive performance of camelids under Peruvian conditions. Maternal recognition of pregnancy would be an important factor involved in the embryo mortality of camelids. The objective of this study was to evaluate the effect of application of oestradiol and progesterone on day 8 and 9 after mating on the embryo survival from day 20 to 35 in llamas.

80 adults female llamas were used. Animals were evaluated with an ultrasound equipment ALOKA SSD500 and probe of 7.5 MHz to determine presence of a dominant follicle $\geq$ 7.0 mm and then mated with males. After mating the animals were assigned randomly to the following groups G1 (n = 20, Control): 2 ml IM saline solution on day 8 and 9; G2 (n = 20): 0.2 ml, oestradiol day 8 and 9 after mating; G3 (n= 20): 15 mg progesterone day 8 and 9 after mating; G4 (n = 20): 0.2 mg oestradiol and 15 mg progesterone day 8 and 9 after mating. Ultrasound examinations were performed on days 0, 3, 9, 20, 25, 30, and 35 to determine occurance of ovulation (day 3), corpus luteum diameter (CL (day 9) and presence of embryonic vesicle or embryo (between day 20 to 35). Male acceptance, ovulation and conception rate was compared by Chi Square analysis between groups.

Acceptance rate to males was 100% in all groups and ovulation rate was 95, 100, 95 and 100% in G1, G2, G3 and G4, respectively (P > 0.05) and no differences were detected between groups in CL diameter. Conception rate on day 20 was 57.9, 75.0, 63.1 and 55.0% to G1, G2, G3 and G4 (P > 0.05). Embryonic loss rates from day 20 to day 35 were not significantly different between groups. These results suggest that oestradiol has a positive effect on embryonic survival and would be a factor involved in the maternal recognition of pregnancy. More research is necessary to determine the role of oestradiol and its effect on embryo mortality in llamas.

Additional references

Aba, M., M. Forsberg, J. Sumar and L. Edqvist, 1995. Endocrine changes after mating in pregnant and non pregnant llamas and alpacas. Acta Vet. Scand. 36:489–496

Fernandez-Baca, S., W. Hansel and C. Novoa, 1970. Biological Reproduction 3:243–251

Skidmore, J., W. Allen and R. Heap, 1997 Maternal recognition of pregnancy in the dromedary camel. Journal of Camel Practice and Research 4 (2):187–192

Embryo transfer in camelids: Study of a reliable superovulatory treatment in llamas

W. Huanca[1], M. Ratto[2], A.Santiani[1], A.Cordero[3] and T. Huanca[4]
[1]*IVITA – Faculty of Veterinary Medicine, University of San Marcos, P.O. Box 03 – 5034, Salamanca, Lima, Peru, whuanca@appaperu.org*
[2]*WCVM, University of Saskatchewan, Canada*
[3]*Faculty of Zootechny, UNALM, Peru*
[4]*ILLPA, INIA, Puno, Peru*

Genetic improvement in camelids under Peruvian highland conditions is limited by the long gestation period, by traditional schemes of breeding in communities and by the breeding season from January to March. The use of reproductive biotechnology such as embryo transfer would be an interesting alternative for genetic improvement in camelids but the development of protocols of super ovulation according to the special characteristics of camelids is required. The objective of the present study was to evaluate the response to super ovulation, embryo recovery rate and pregnancy in synchronized recipients.

Female llamas 4–6 years of age were used. Animals were evaluated with ultrasound to determine presence of a follicle ≥ 7.0 mm previous to a treatment of synchronization of the follicular wave. After 12 days ovulation was induced (Day 0) and super ovulation was induced with 1,000 UI of eCG on Day 2. Prostaglandin was used on Day 5 and animals were mated on Day 6 with males of probe and on the same day females with the presence of follicles ≥ 7.0 mm were synchronized as recipients. Non-surgical embryo collection was utilized on Day 7 after mating with a 2-way balloon catheter passed through the cervix and flushing each horn with 250 ml of medium. Embryos were evaluated and transferred to recipients. Pregnancy was evaluated with ultrasound on day 20 and 30 after embryo transfer.

Number of follicles as response to the treatment of super ovulation was of 11.2 ± 2.4 follicles and the recovery rate was of 8.7 ± 1.5 embryos with a good quality and development, but with differences in size. Embryo transfer was successful with a 42.8 of pregnancy determined by ultrasound on day 20 and 30 after transference. Results obtained suggest that is possible to use embryo transfer as an alternative in a program of genetic improvement in camelids.

Additional references

Bourke, D.A., C.E. Kyle, T.G. McEvoy, P. Young and C.L. Adam, 1995. Superovulatory responses to eCG in llamas (Lama glama). Theriogenology 44:255-268

Correa, J.E., M.H. Ratto and R. Gatica, 1997. Superovulation in llamas (*Lama glama*) with pFSH and equine Chorionic Gonadotrophin used individually or in combination. Animal Reproduction Science 46:289-296

Ratto, M.H., J. Singh, W. Huanca and G.P. Adams, 2003. Ovarian follicular wave synchronization and fixed-time natural insemination in llamas. Theriogenology 60:1645-1656

Part VI

DECAMA-Project: Characteristics of the supply and demand of charqui

N. Pachao

DESCO - Centro de Estudios y Promoción del Desarrollo. Málaga Grenet 678, Umacollo, Arequipa, Peru; nadescauna@hotmail.com

Abstract

The article analyzes the Peruvian market of charqui with especial focus on the market of Arequipa. The analysis had three objectives: These were (1) to identify volumes and production, (2) to determine characteristics of demand and (3) to identify commercialization systems and financial margins. The results indicate that the production of charqui is focused in southern Peru and in particular provinces of Azangaro, Yunguyo and Carabaya (Puno), Canchis and Espinar (Cuzco) and Arequipa (Arequipa). The total production is at least 443.4 MT/year. The main characteristics of the charqui are weight losses (65% as "offal") in its production from fresh meat and its sophistication, including further processing, according to the market to which it goes. The demand is diverse, because the main consumption is found in 11 cities distributed across coast, mountain range and tropical rainforest. It also derives from urban and rural types of population with consumptions of 30% and 60% respectively. Diverse systems of commercialization are evident according to origin and destiny of the charqui. The system of Arequipa is most complex. This is due to the importance of the city, firstly as a center of wholesale distribution for the production of Azangaro and Arequipa, and secondly, as supplier to other provinces which include Mollendo, Ica, Ilo and Moquegua y Tacna.

Resumen

El artículo analiza el mercado de charqui en el Peru, centrándose en el caso de Arequipa, por lo que se plantean tres objetivos: 1. Identificar volúmenes y zonas de producción del charqui, 2 Determinar las características de la demanda del charqui en Arequipa y 3. Identificar el sistema y márgenes de comercialización en Arequipa. Los resultados indican que la producción de charqui está focalizada en departamentos sureños y en algunas provincias, como son: Azangaro, Yunguyo y Carabaya (Puno), Canchis y Espinar (Cuzco) y Arequipa (Arequipa). La producción de los tres departamentos, asciende a 443.4 TM/año. Entre las características del producto se tiene: elevados costos de producción debido al porcentaje de merma (65%), la no realización de actividades secundarias en la zona rural y su sofisticación de acuerdo al mercado al que se dirige: rural o urbano. La demanda es diversa, por lo que su consumo principal es en 11 ciudades distribuidas en costa, sierra y selva, abarcando dos tipos de población: urbana y rural, cuyos consumos son 30% y 60% respectivamente. De acuerdo a origen y destino del charqui, se articulan diversos sistemas de comercialización, sin embargo el de Arequipa es el más complejo, debido a la importancia que tiene la ciudad como centro de distribución mayorista, al comercializar la producción de Azangaro y Arequipa para satisfacer el consumo de departamentos del sur del Peru.

Keywords: alpaca, llama, charqui, commercialization system, margins

Introduction

The alpaca and llama species of South American camelids (DSAC) were the most representative livestock of pre-Columbian Peru. At the time of the virreinato there were approximately 7 million alpaca and 23 million llama, numbers which have decreased to little more than 3 and 1 million, respectively. Alpaca and llama are nowadays located mainly in southern Peru, in the localities of Arequipa, Cuzco and Puno, and their provinces which include (with proportion of national population of DSAC) Caylloma (6%), Canchis (5%) and Azangaro (15%).

The keeping of DSAC is important for the human population of the high Andean zones. DSAC are animals which best resist the geographical adversities. In addition, the production of the DSAC sub-products introduces the population to commercial activity arising from the sale of fiber, meat and leather. The production of charqui, which is DSAC meat subjected to a process of salting and drying, deserves special mention, because it is important for consumption by the local population. In addition, the migration of people from the high Andes to the agricultural areas of the coast or rain forest, and urban areas (in particular Lima) has linked the centers of production and consumption, where nutritional customs have an important role.

The technique of storage and the process of drying of meat dates from pre-Columbian times at around 6.000 years BC and is still maintained at present. It represents a rich protein source for rural populations. Selected background to population patterns, methodology of DSAC meat preservation and the marketing system is provided in references (Caldentey & Gomez, 1993; INEI, 2002; Instituto Canto, 2002; Kallpa Proyecto; Mendoza, 1987; Proyecto Alpacas, 1989, 1990).

Material and methods

The geographical regions of study were Puno, Cusco and Arequipa. The primary data were gathered by surveys and interviews. These were designed to determine the production volume, prices and production costs of charqui and to identify system, channels and agents participating in the commercialization process. The data correspond to the year 2003.

The information of the volume of commercialized charqui in the rural area was derived for the market fair of the Camaná district which was chosen as a reference because it is the main agricultural zone of Arequipa. Consumption characteristics were determined from the surveys conducted in the locality of "The Carmen", which functions as a center for the hiring of the agrarian labor force, Two markets, "The Inkas" and "The Altiplano", were also studied in Arequipa city in order to determine the same parameters in the urban area.

The prices and margins of commercialization were analyzed from the survey results for each agent. These provided prices and costs of production and commercialization. We used Two concepts were developed. These were (1) Gross Margin of Commercialization (GMC) calculated as the difference between the price paid by the consumer and the price paid to producer, expressed in percentage:

$$GMC = \frac{Pc - Pp}{Pc} \times 100$$

And (2) Net Margin of Commercialization (NMC) calculated as the the percentage of the final price that perceives the intermediation as net beneficiary, after having subtracted the commercialization costs.

$$NMC = \frac{Pc - Cost\ of\ commercialization}{Pc} \times 100$$

Where: Pc = Price paid by the consumer
  Pp = Price paid to the producer

Results

Production of charqui

The research shows that southern Peru makes a significant contribution to national production of charqui (Table 1). The order of production (MT/month) is Puno (23.4) > Cuzco (9.3) > Arequipa (4.1). It is therefore concluded that the estimated total national production of charqui is at least 36.8 MT/month.

The three major-producing provinces of Puno contribute 61.4% of the national production. The most important province is Azangaro which includes districts such as Azangaro and José Domingo Choquehuanca (JDCH). These provide 80.8% of Punenian charqui and 51.4% of the national production.

Cuzco has two provinces of importance which contribute, in total, 25.3% of the national production. Production in Canchis (Sicuani) accounts for 54.8% and that of Espinar, 45.2% of the Cuzco region. The production of the Arequipa region is concentrated in a single province: Arequipa, which provides 11.1% of the national total.

The provinces identified as charqui producers, especially those in Puno and Cuzco regions, are also those which have the greatest population of DSACs. This explains the direct relationship between the potential of provinces for meat production and actual production of charqi. The greater availability of meat supply also helps to create small, medium and large production centers for charqui. The presence of important markets for general food transaction has facilitated the creation of marketing centers for charqi. Examples include the two food markets (see below) in Areqipa which have made possible the organised sale of charqi even although the raw material (DSAC meat) is transported the 260km distance from the locality of Caylloma-Chichas.

Volumes of urban consumption

The monthly urban consumption in Arequipa is calculated to amount to 2,669 kg. Of this quantity, 1,821 kg derives from the two markets "The Inkas" and "The Altiplano", where 89% is sold to the retailers and the rest directly to consumers. The second means of charqui supply, where the product is directly sold by small producers to the consumer, bypasses these markets. This system contributes 31.8% of the city consumption.

The commercial-wholesaler character of Arequipa is sustained both by the volume of sales in the main markets and that destined for sale outwith the province which amounts to 6.5 MT/month. The main external destinations (with quantities per month) are Moquegua-Ilo (2.7 MT), Tacna (1.8 MT), Mollendo (0.33 MT) and Lima at 1.55 MT/month.

Table 1. Average monthly production (MT) of charqui, by geographical location.

Region	Puno					Cusco		Arequipa
Province	Azangaro		Yunguyo	Carabaya	Others	Canchis	Espinar	Arequipa
City	Azangaro	JDCH	Unicachi	Crucero		Sicuani	Espinar	Arequipa
MT/month	16.0	2.9	3.5	0.2	0.8	5.1	4.2	4.1

Characteristics of consumption

The first important result is the relatively small percentage of the urban population (27.4%) which purchases charqui. This indicates a product with low demand which is confirmed by the small average purchase weight of only 136 g (Table 1).

Charqui, as sub-product of alpaca meat, has a low incidence of consumption due to perceptions of bad taste associated with smell and flavor. In the case of the fresh meat of alpaca this reason was given in 38.7% of the answers, although for charqui it was lower at 24.6% (Table 2). The reasons for the absence of consumption can be divided into two categories. These are firstly, total ignorance of the product (21.8%) and secondly, straight rejection (8.7%) because of dislike and distrust of the processing methods used for its production. It is notable that the urban population does not know of the product and that, in comparison with other meat products, it is considered difficult to find.

We can therefore conclude that the consumption is low in the urban area. The last reason is related to the limited variety of dishes that can be prepared with charqui. For example, in Arequipa only one dish (*olluquito* with charqui) was cited in 50.0% of the answers, with 34.2% awareness of other preparations including soups.

Volumes of rural consumption

The consumption in the rural areas is mainly by the agrarian labor force. The Majes irrigation in Arequipa utilizes the water of the Colca river. It extends to the coastal zone of the Arequipa province and has created conditions for agraricultural activities which produce the resulting demand for labor. Most of these workers who come from the high Andean zones of Arequipa, Puno and Cuzco are settled in Camaná, Majes and the valley of Arequipa. In Camaná, the workers acquire the charqui at a local market which takes place on Sundays. The total consumption is only about 1.87 MT/month in Camaná. This derives from purchases from the market and fair in Camaná (765 kg/ month) and those via the "The Altiplano" (900 kg/ month) and "The Inkas" (200 kg/ month) wholesale fairs. Additionally, a further 779 kg/month are consumed from supplies purhased at local fairs in the remainder of the valley of Arequipa.

Characteristics of the rural consumption

The present survey reveals that the consumption in the agricultural-rural area was evident in 62.3% of the responses obtained. The average quantity and frequency of consumption is 1.0 kg/ week per family (Table 3).

Table 2. Characteristics of the urban consumption of charqui.

Region	Average of purchase (g)	Reasons of no consumption (%)				
		No custom	Dislike	Not known/ sold	No trust in hygiene meat/ elaboration	Others
Arequipa	136.53	36.2	24.6	21.8	8.7	8.7

Table 3. Characteristics of the rural consumption of charqui.

Region	Average of purchase (kg/week)	Reasons of consumption (%)			
		Preservation	Taste/ custom	Nutritious/ healthy	Others
Arequipa- Camaná	1.0	42.4	27.1	10.2	20.3

In considering the reasons given for consumption of charqui, the most important related to the fact that it can be preserved effectively. It is necessary to mention that the rural income, for example in Camaná, ranges between 54.9 and 85.7 $/month, which means that the consumer has to be careful with expenditure. In addition, Charqui represents an alternative source of nutrition for the rural populations. A second reason is appreciation of the the product`s taste and a third, the longstanding custom of consumption which is related to the origin of the consumers from high Andean zones. This pattern suggests a strong relationship between the geographical origins of the human population and charqui consumption. It is also important to note that consumers are generally not aware of the high content of protein, since only about 10% consider it as nutritious and healthy.

Similarly, as in the urban population, non-consumption in the rural population is ascribed to doubts about meat quality, processing methods and dislike of the taste of the meat. Differences in preferences for types of product were also evident. In the urban area the preferred charqui is white, without bone and is required flayed or crumbled. In the rural area, the charqui is consumed black, with bone and fat. The reason for this divergence is the greater income generated by urban economic activity and the finance available to purchase a more processed and presentable form of charqi. This situation contrasts with that in the rural area, where the agricultural activity is the main source of employment and provides lower income.

Systems and margins of commercialization

The basic elements of the comercialization system and the supply chain include the producer, wholesaler, retailer and consumer of charqui. The system requires the coordination of the activities between these interdependent components to create a functional network between production and eventual consumption.

The charqui-producing cities (6 provinces in 3 regions) are identified in Table1. Table 4 shows the locations of main demand of charqui (7 urban and 4 rural, where 3 are in the rain forest territory) and which result in the different market systems.

Table 4. Average monthly consumption of charqui, by city.

Region	Lima	Madre de Dios	Arequipa				Cusco		Tacna	Moquegua	Others
Province	Lima	Puerto Maldonado	Arequipa	Camana	Islay	Others	Cusco	Quilla-bamba	Tacna	Ilo	
City	Lima Metrop.	Several	Arequipa Metrop.	Camana	Mollendo	Others	Cusco Metrop.	Several	Several	Several	
TM/month	11.40	9.60	2.66	1.87	0.33	0.84	0.69	4.56	1.84	2.73	0.37

The southern Peru system is detailed in the following analysis (Figure 1). The 9.14 MT/month commercialized in the city of Arequipa derives from Arequipa and Azangaro via different commercialization channels. The first channel is formed by the sales made in markets on the periphery of Arequipa where charqui arrives directly from the producer to the consumer and representing 0.43% of the total. The second is made up by the producers of Puno and Arequipa whose production is directly bought by retailers resulting in commercialization of 7.2% of the charqui. Considering additionally the volume sold by the retailers, who purchase the charqui from the wholesale markets, the Arequipenian retailers handle 21.8% of the southern consumption.

In the city, "The Inkas" and "The Altiplano" markets cover 90.6% of the total commercialization of charqui which is supplied from Arequipa and Azangaro. The producers sell 83.7% of the commercialized charqui in both markets, whereas the wholesalers (Puno) handle 16.3%. A series of further channels is formed from this supply and includes the third channel which is made up of the direct sales to the Arequipenian consumers and represent 1.8%. The fourth channel constitutes the wholesaler-distributors of Moquegua-Ilo, who handle 24.5% of the volume commercialized in both markets. In the fifth channel there are the wholesaler-distributors of Tacna who handle 16.5%. Other destinations such as Mollendo, Ica and Lima join to form the sixth channel which handles 17.1%. Aproximately 58.1% of the charqui commercialized in the city of Arequipa in the latter three channels is thus destined for consumption in external provinces.

The seventh channel is represented by the consumption of the agricultural valleys of Arequipa and represents 17.8%. The main supply (16%) derives from purchases made in the two major markets.

In considering financial margins for the supply chain for charqui, the channel chosen was the system which involved Arequipenian producers wholesalers, retailers and consumers. The price

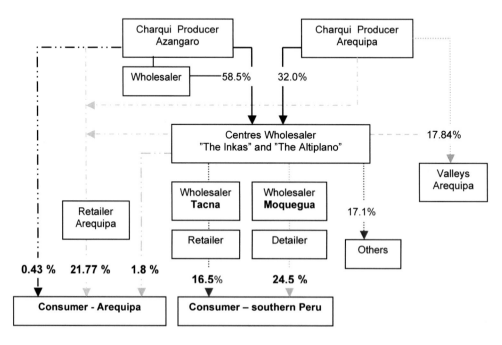

Figure 1. Channels of commercialization of charqui: Arequipa and Azangaro, Southern Peru.

(per kg) of charqui set in the "The Inkas" and "The Altiplano" markets at the time of sampling was $2.29 for the wholesaler (paid to the producer) and $2.86/kg ($0.29 per 100 g) received from the consumer for the retailer with $4.29 ($ 0.43 per g) required from the consumer for crumbled charqui.

Figure 2 shows the GMC for both types of charqui. In the first case, for each $1.00 spend for the end-purchase of charqui, $0.80 is for the producer, to cover production and other costs with $0.20 for retailer. In the second case for crumbled charqui, the margin is larger, with the GMC of the retailer increasing from $0.20 to $0.46. The last result arises from the retailer adding value to the product by making charqui available and more desirable (utility of place and form respectively) and thus motivating the demand by consumers.

The NMC is an alternative method of analysis in considering the gain generated by the commercialization process, and its division to each one of the agents (see information on costs given below). The results indicate that from the gain generated by the sale of conventional charqui, 17% accrues to the producer and 83% to the retailer. The difference is accentuated when crumbled charqui is analyzed, where the gain partitions 6% to the producer and 94% to the retailer.

The previous results are based on information on the costs of each agent. The Arequipenian producer buys DSAC meat with its price fixed in the city, not on the farm. The major costs (per kg) of charqui production when related to the purchase cost of $2.16 fresh meat are losses in offal of $1.19 (55% by weight of fresh material) and labor, transport and salt, contributing $0.06(3%). The remaining 42% of value of the original meat is the residue after the removal of water in the drying process and which is the material in the final charqui product. This reflects the typical coefficient of conversion by weight of raw meat to charqui of 0.45. A further comparison can be made in terms of the relationship of weight of charqui to live weight of animal which approximates to 0.23.

Discussion

Published information related to the commercialization of charqui is limited with existing data approaching the subject in partial form (see bibliography below). There are, in addition, the technical studies on DSAC produced in 1980 by Ing. Enrique Ampuero and Dr. Virgilio Alarcón. These indicate that greater numbers of charqui-producing localities were present two decades

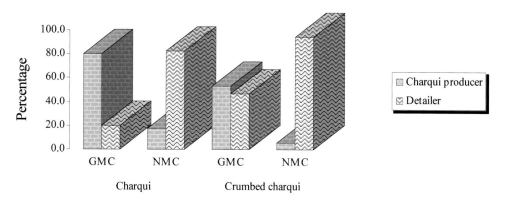

Figure 2. Participation of the producer of charqui in the final price to consumer.

ago in centers of production such as in Sicuani and Yauri (Cusco), Santa Rosa, Lampa and Azangaro (Puno). They were all located in departments in Southern Peru. In three documents (4;7) the importance of the contribution of costs of the purchase of meat, in the total costs, is highlighted. This contribution ranges between 73% (charqui with bone) and 85%. In the present investigation the costs of fresh meat and losses constitute 94.4%. Such results depend on the characteristics assumed in the process of charqui production and the efficiency of conversion of fresh meat to charqui that ranges between 40 and 46%.

These results suggest limited possibilities of reducing production costs by increases in economies of scale because remaining factors only represent 10 to 20% of the total costs.

In terms of demand for the charqui product there are currently no publications which apply to economic concepts. However, it is important to note the identification by the NGO "Kallpa" of a main market in the agricultural population settled in the Valley of Convención and Lares (Cuzco). This population has characteristics similar to those living in the agricultural valleys of Arequipa. It was also concluded that both custom and the absence of necessity for refrigeration have had very important roles in supporting consumption of charqui.

Conclusions

1. The volume of production of charqui is at least 36.95 MT/month. The main centers of production are located in seven cities belonging to three departments: Puno (4), Cusco (2) and Arequipa (1).
2. The market of charqui includes urban and rural areas, with significant differences between them. The urban consumption is represented by, at the most, 30% of the population, whereas the percentage is 64% in the rural area.
3. The amount and frequency of purchase for consumption differs between both types of populations. In the urban areas purchase, usually once monthly, does not surpass 200g which contrasts with the rural purchase of around 1,000 g at intervals of one week.
4. The high demand in the rural area in comparison with the urban one is due to the conservation characteristics of the product (no need for refrigeration), the taste and the custom of consumption.
5. Reasons for the decision not to consume charqui are common for both types of population. Among these are the absence of custom, dislike and distrust of meat quality and the process of production, and concern about hygiene.
6. There are seven charqui-producing and eleven charqui-buying cities. They form diverse systems of commercialization. They connect the supply from the high Andean areas with the rural and urban consumption of the three natural regions of Peru which are coast, mountain range and rain forest.
7. The regions of Puno, Arequipa, Areqipa, Southern Peru, have seven channels of commercialization. The main ones supply regions such as Arequipa, Tacna and Moquegua.
8. The charqui producer does not have a high participation in the final price to the consumer since the NMC does not surpass 18% due to the high production costs.

Acknowledgements

I wish to thank the producers, wholesalers and consumers of charqui who helped me do this investigation, mainly for maintaining their own customs and technologies, and for letting me know a part of the Peruvian economic reality. I wish to thank the EU-Project DECAMA and the NGO DESCO for giving me the opportunity to investigate this interesting topic.

References

Caldentey, P. and A.C. Gómez, 1993. Economía de los mercados agrario. (Economy of the markets agrarian) Ediciones Mundi Prensa.

INEI, 2002. Proyecciones de la población por años calendario según departamento, provincia y distrito (1990-2005). (Projections of the population per years calendar by department, province and district). Boletín especial N ° 16, 2002.

Instituto Cuanto, 2002. Peru en Números. (Peru in Numbers). R. Webb and G. Fernández (eds.).

Kallpa Proyecto. Procesamiento y comercialización de charqui y carne fresca de camélidos s/f. (Processing and commercialization of charqui and fresh meat of camelids).

Mendoza, G., 1987. Compendio de mercadeo de Productos agropecuarios. (Compendium of farming Product trade) IICA.

Proyecto Alpacas, 1989. Alternativas de comercialización a la carne de alpacas. Evaluación de los avances del proyecto de elaboración y comercialización de charqui. (Alternatives of commercialization to the meat of alpacas. Evaluation of the advances of the project of elaboration and commercialization of charqui) Informe técnico N° 8.

Proyecto Alpacas, 1990. Procesamiento de la carne de camélidos para charqui. (Processing of CSAD meat for charqui) Informe técnico N° 9.

11,000 years of camelid use in the Puna of Atacama

D. Olivera[1] and H. Yacobaccio[2]
[1]*Instituto Nacional de Antropología y Pensamiento Latinoamericano, CONICET y Universidad de Buenos Aires, José P. Varela 3016, CP 1417, Buenos Aires, Argentina; deolivera@movi.com.ar*
[2]*Consejo Nacional de Investigaciones Científicas y Técnicas (CONICET) y Universidad de Buenos Aires; Concordia 3590 2° F, CP 1419, Buenos Aires, Argentina*

The aim of this paper is to show the changes in the use of camelids through time by analyzing the different modes of relationships between people and camelids (hunting, domestication and pastoralism).

Two associated processes have been determined between 11,000 and 3,000 years B.C. These are: 1) a process of intensification in the use of camelids, and 2) the domestication of the guanaco since *ca.* 4,800 B.C. The analysis of the intensification process involved the study of 25 sites in the Southern Andes, allowing us to discern a pattern of increasing utilization of camelids through time. Camelids account for 49% of the utilized species between 11,000 and 8,400 B.C., but rise to 83% in the 8,400-5,000 B.C. period, and increase more between 5,000 and 3,000 B.C. accounting for 85-100% of the identified species in the archaeological sites. In this last time-span osteological, fiber and contextual evidence showing protection and domestication of camelid populations appear. This process was independent to that occurred in the Central Andes and happened as a consequence of deep social modifications in the hunting-gathering societies of the late Holocene.

After 3,000 years B.C. began a process of consolidation of the economies based on pastoralism and agriculture, together with the development of settlement strategies with a more important sedentariness. The camelids represent around 85-95% of the NISP in the sites, in both permanent residential bases ("villages") and transient sites of specific activities. Pastoralism seems to be the most important strategy with the use of a generalized llama type useful at the same time for meat, fiber and as pack animal.

Between 1,400 to 500 years B.C. the domesticated camelids turned to more specialized types. Perhaps, in coincidence with the increment of caravan traffic and fiber exploitation, related to an important process of social and political complexity in the Puna societies. Nevertheless, two things are remarkable: 1) hunting of wild camelids remained as a very important activity during all the period between 3,000 to 500 years B.C., including Inca's times, and 2) the important cultural changes did not modify the use of different environmental patches in an optimization model aiming at reducing environmental risks.

Paternity testing using microsatellite DNA in alpacas (*Vicugna pacos*)

J. Rodriguez[1], *C. Dodd*[2], *R. Rosadio*[1], *J.C. Wheeler*[1] *and M.W. Bruford*[2]
[1]*CONOPA, Los Cerezos 106, Lima 03, Peru; webmaster@conopa.org*
[2]*School of Biosciences, Cardiff University, Cardiff, U.K.*

Abstract

Ten alpaca and llama microsatellites (Lang *et al.*, 1996; Penedo *et al.*, 1998) were used to evaluate paternity in 47 alpacas registered at the IVITA Research Station in Maranganí, Canchis Province, Cusco, Peru. The microsatellites were amplified in three multiplex reactions and were polymorphic for all samples. Allele numbers varied between 4 and 20, and both allelic frequencies and exclusion probabilities were calculated using Cervus 2.0. All loci, except two, were within the size range published by Lang *et al.* (1996) and Penedo *et al.* (1998). The cumulative exclusion probability for the ten loci was 0.9998, and for each multiplex greater than 0.90. The results confirmed paternity in 18 parent-offspring pairs, but in a further 4 cases different adults were identified as parents than those recorded in the registry, demonstrating the necessity of DNA testing to ensure accurate recordkeeping and guarantee the parentage of registered animals.

Resumen

Diez microsatélites para alpacas y llamas (Lang *et al.*, 1996; Penedo *et al.*, 1998) fueron usados para evaluar parentesco en 47 alpacas (18 crías, 18 madres y 11 padres) registradas de la Estación Experimental IVITA - Maranganí, provenientes de la provincia de Canchis (Cusco - Peru). Los microsatélites fueron amplificados en tres reacciones de PCR múltiple y fueron polimórficos para todas las muestras. El número de alelos varió entre 4 y 20, las frecuencias alélicas y probabilidad de exclusión fueron calculadas utilizando Cervus 2.0. Todos los loci, a excepción de dos, estuvieron dentro de los rangos publicados por Lang *et al.* (1996) y Penedo *et al.* (1998). La probabilidad de exclusión acumulada para los 10 loci fue 0.9998 y mayor a 0.90 para cada PCR múltiple. Los resultados confirmaron la paternidad en 18 casos, sin embargo, en 4 casos diferentes adultos fueron identificados como padres comparando con los registros, demostrando la necesidad de pruebas de ADN para asegurar la certeza de los registros para garantizar el parentesco de los animales registrados.

Keywords: alpaca, microsatellite, paternity

Introduction

Accurate methods for parentage determination are very important for domestic animal breeders, as well as for the development of breed registries. For many years, conventional methods such as blood groups and biochemical polymorphisms were the only tools available for parentage determination, but during the last decade, with development of the polymerase chain reaction (PCR), many DNA based techniques have been developed and are now commonly used. Microsatellites, or short tandem repeats (STRs), are repeat sequences of up to six base pairs (Hancock, 1991). When amplified by PCR and observed through gel electrophoresis, microsatellite alleles can be distinguished based on differences in the number of repeats.

In most countries, including the U.S.A., Canada, United Kingdom and Australia, verification of parentage using DNA analysis, normally microsatellites, is required to place an animal in a breed registry. In Peru, however, registries are based entirely on observation and controlled breeding. In the present study, use of DNA analysis to certify the accuracy of one such registry has revealed shortcomings.

Materials and methods

Genomic DNA from 47 alpacas (*Vicugna pacos*) (11 fathers, 18 mothers and 18 offspring) registered at IVITA's Maranganí Research Station in Canchis Province, Cusco, Peru, was extracted from blood using the ULTRACLEAN DNA BloodSpin Kit® (Mo Bio Inc).

Ten microsatellite loci (LCA19, LCA22, LCA5, LCA23, YWLL08, YWLL29, YWLL36, YWLL40, YWLL43, YWLL46) isolated from llama (*Lama glama*) and alpaca (*Vicugna pacos*) (Lang *et al.,* 1996; Penedo *et al.,* 1998) were amplified in three multiplex PCR reactions with the QIAGEN Multiplex PCR Kit® (QIAGEN) and 10 µl made up of 25 ng genomic DNA, 2mM of each primer (one labelled at the 5' end with a fluorochrophor) and 5 µl of QIAGEN Multiplex Kit Master Mix (1U HotStar Taq® DNA polymerase, 10X Multiplex PCR Buffer, 3.0 mM $MgCl_2$ and 0.2 mM of each dNTPs). PCR followed, using the protocol: 1 cycle of 95°C for 15 minutes; 25 cycles of 94°C for 30 seconds, 61°C (for the first multiplex) and 59°C (for the second and third multiplexes) for 90 seconds and 72°C for 60 seconds; with a final extension of 60°C for 30 minutes in a thermocycler model 9700 GeneAmp® (Perkin Elmer).

PCR products were diluted 1:5 in a mix of formamide (AMRESCO), Genescan-350 TAMRA® (ABI Prism) and ABI Gene Scan Loading Buffer® (ABI Prism) (5:1:1) and denatured for 2 minutes at 95°C. PCR products were separated in a 4.25% polyacrylamide gel (Gene Page® - AMRESCO) at 1,000 V in TBE 1X (89 mM Tris base, 89 mM Boric Acid and 2 mM EDTA) using an ABI 377 DNA sequencers® (Applied Biosystems). Allele sizes were determined using ABI Genotyper® v.2.5 (Applied Biosystems).

Paternity assignment was determined by exclusion, observing incompatibility between father, mother and offspring allele sizes. The power of the test was established for exclusion probability (individual and accumulated) using the method of Jamieson and Taylor (1997). Individual and accumulated exclusion probabilities, heterozygosity and the polymorphic information contents (PIC) were calculated using Cervus 2.0 Parentage analysis ©.

Results and discussion

Because microsatellite markers are very accurate (close to 100%) they are commonly used for parentage determination in many domestic species. In camelids, Sasse *et al.* (2000) and Sam *et al.* (2001) have demonstrated the feasibility of paternity testing in camels (*Camelus dromedarius*), while Skidmore *et al.* (1999) used microsatellite markers to determine paternity of camel x guanaco hybrids.

In the present study, we utilized ten previously published South American camelid microsatellites (Table 1) to develop paternity tests in alpacas and llamas. All microsatellites amplified and were polymorphic. Allele number varied between four for the loci LCA22, YWLL46 and 20 for the locus YWLL08, with an average per locus of 9.1, similar to data published by Lang *et al.* (1996) and Penedo *et al.* (1998). Slight variations were observed in loci YWLL08 and YWLL36, probably due to the different origins of the animals (Table 1).

The exclusion probability for each locus varied between 0.174 (YWLL46) and 0.844 (YWLL08) with an accumulated exclusion probability of 0.9998 for ten microsatellites. Furthermore, four

Table 1. Allele size, allelic frequency and exclusion probability (EP) for the ten microsatellites analyzed.

Microsatellite	Allele	Frequency	EP	Microsatellite	Allele	Frequency	EP
LCA 19	85	0.1170	0.548	YWLL 08	134	0.1489	0.844
	89	0.0213			138	0.0106	
	91	0.0106			140	0.0638	
	93	0.0319			142	0.0532	
	95	0.0213			148	0.0426	
	99	0.1277			150	0.1170	
	101	0.4787			152	0.0745	
	103	0.0638			154	0.0319	
	105	0.0319			156	0.0532	
	107	0.0213			158	0.0319	
	111	0.0638			162	0.0106	
	117	0.0106			166	0.0745	
YWLL 29	214	0.0957	0.634		168	0.0319	
	216	0.2128			170	0.0426	
	218	0.0957			176	0.0745	
	220	0.3191			178	0.0851	
	222	0.0745			180	0.0106	
	224	0.1277			182	0.0213	
	226	0.0532			184	0.0106	
	228	0.0106			186	0.0106	
	234	0.0106		LCA 23	132	0.0106	0.621
YWLL 40	180	0.2340	0.408		136	0.0851	
	182	0.0106			138	0.0532	
	184	0.0957			142	0.2128	
	186	0.1489			144	0.0106	
	188	0.5106			146	0.2447	
YWLL 46	97	0.8085	0.174		148	0.2660	
	103	0.0106			150	0.0106	
	105	0.0745			156	0.0106	
	109	0.1064			158	0.0106	
YWLL 43	128	0.0326	0.452		160	0.0851	
	138	0.0109		LCA 22	113	0.6383	0.294
	142	0.2826			115	0.2340	
	144	0.0761			117	0.0532	
	146	0.4674			119	0.0745	
	148	0.0217		YWLL 36	149	0.2340	0.733
	150	0.0978			151	0.1277	
	156	0.0109			153	0.0638	
LCA 5	188	0.2021	0.575		157	0.1596	
	190	0.0851			159	0.0745	
	192	0.0957			165	0.0851	
	194	0.0638			169	0.0532	
	202	0.3617			171	0.0319	
	204	0.1809			173	0.0213	
	206	0.0106			175	0.1277	
					179	0.0213	

microsatellites were found to be highly informative based on their higher individual exclusion probability (YWLL08, YWLL36, LCA23 and YWLL29).

The three multiplex PCR reactions obtained a high cumulative exclusion probability (>0.90). The cumulative and individual exclusion probabilities, heterozygosity and polymorphic index contents (PIC) for the ten microsatellites and for the three multiplex PCR reactions are detailed in Table 2.

It was possible to resolve all 18 paternity cases (Figure 1). Four errors (22%) in parentage attribution were documented relative to the breeding records of the research station (Table 3).

Table 2. Individual and accumulated exclusion probability (EP and EPa), heterozygosity and polymorphic information content (PIC) for the different microsatellites used in alpacas.

Multiplex	Microsatellite	EP	EPa	$H_{(o)}$	PIC
M 1	LCA 19	0.548		0.702	0.709
	YWLL 29	0.634		0.787	0.786
	YWLL 40	0.408		0.660	0.604
	YWLL 46	0.174	0.9191	0.383	0.307
M 2	LCA 23	0.621		0.638	0.779
	LCA 22	0.294		0.511	0.477
	YWLL 36	0.733	0.9286	0.894	0.852
M 3	YWLL 43	0.452		0.304	0.639
	LCA 5	0.575		0.809	0.744
	YWLL 08	0.844	0.9637	0.915	0.918
M1 + M2 + M3	10 Microsatellites		0.9998		

In one case it was not possible to determine the true father because he was not among the samples provided. These results clearly demonstrate the necessity of DNA testing to ensure accurate recordkeeping and guarantee the parentage of registered animals.

Conclusions

Analysis of ten microsatellites (LCA19, LCA22, LCA5, LCA23, YWLL08, YWLL29, YWLL36, YWLL40, YWLL43 and YWLL46) permits paternity determination with an exclusion probability of 0.9998 in alpacas.

Analysis of six microsatellites (YWLL43, LCA5, YWLL08. LCA 22, LCA 23 and YWLL 36) in two multiplex PCR reactions determines paternity with an exclusion probability of 0.9637.

Acknowledgments

Special thanks to Juan Olazábal (IVITA, Maranganí), Rosa Dávalos (IVITA, Huancayo), Matilde Fernandez and the authorities and workers at the IVITA, Maranganí for help with sample collection and logistics. Support for this research was supplied by CONOPA (Peru), Cardiff University (UK) and the Darwin Initiative for the Survival at Species (UK).

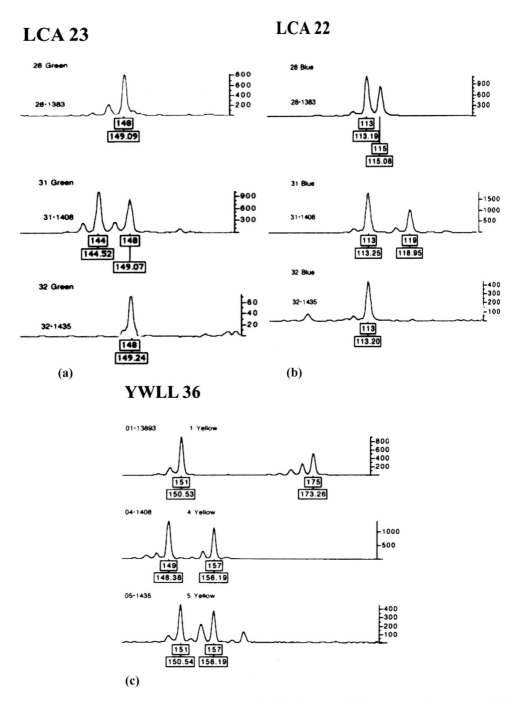

Figure 1. Electropherograms (in Genotyper v.2.5) (ABI PRISM). (a) Locus LCA23 labeled with fluorochrophor TET. The samples correspond to: 1383 (Father), 1408 (Mother) and 1435 (Offspring). (b) Locus LCA22 labeled with fluorochrophor FAM. (c) Locus YWLL36 labeled with fluorochrophor HEX. Y axis indicates amplification intensity in fluorescence units.

Table 3. Result the paternity testing with Cervus 2.0 Parentage analysis ©.

Case	Offspring	Mother	Aspirant father	Record data
01	1356	1352	1385	ERROR
02	1357	1353	1349	CORRECT
03	1358	1354	1385	ERROR
04	1359	1355	1348	ERROR
05	1416	1389	1374	CORRECT
06	1428	1401	1377	CORRECT
07	1429	1402	1380	CORRECT
08	1430	1403	1350	CORRECT
09	1434	1407	1383	CORRECT
10	1435	1408	1383	CORRECT
11	1438	1411	1385	CORRECT
12	1415	1388	1348	CORRECT
13	1418	1391	1379	CORRECT
14	1419	1392	1348	CORRECT
15*	1420	1393	????	ERROR
16	1422	1395	1374	CORRECT
17	1426	1399	1379	CORRECT
18	1433	1406	1386	CORRECT

*Case 15 belongs to one animal whose father was undeterminable because the true father was not found in the samples analyzed.

References

Hancock, J., 1991. Microsatellites and other simple sequences: genomic context and mutational mechanisms. In: Microsatellites. Evolution and Applications. D. Goldstein and C. Schlötterer (eds.), Oxford University Press, Oxford, pp. 1-9.

Jamieson, A. and C.S. Taylor, 1997. Comparisons of Three Probability Formulae for Parentage Exclusion. Animal Genetics 28: 397-400.

Lang, K., Y. Yang and Y. Plante, 1996. Fifteen Polymorphic Dinucleotide Microsatellites in Llamas and Alpacas. Animal Genetics 27: 285-294.

Penedo, C, R. Caetano and I. Cordova, 1998. Microsatellite Markers for South American Camelids. Animal Genetics 29: 398-413.

Sam, C., J. Schaaf and C. Lauk, 2001. Improved Breeding in Camelids through Molecular Techniques: Development of STR Multiplexes for Identification and Parentage Verification. In: Progress in South American Camelids Research. Proceedings of the 3rd European Symposium and SUPREME European Seminar. EAAP publication N° 105, M. Gerken and C. Reniere (eds.), Göttingen, pp. 296-302.

Sasse, J., M. Mariasegaram, M. Jahabar Ali, S. Pullenayegum, R. Babu, J. Kinne and U. Wernery, 2000. Development of a Microsatellite Parentage and Identity Verification Test for Dromedary Racing Camels. Abstracts of the 27th Conference of the International Society of Animal Genetics (ISAG), Mineappolis, EE.UU.

Skidmore, J, M. Billah, M. Binns, R. Short and W. Allen, 1999. Hybridizing Old and New World Camelids: *Camelus dromedaries* x *Lama guanicoe*. Proceedings of the Royal Society London B. 266: 649-656.

Biology of *Eimeria macusaniensis* in llamas

S. Rohbeck[1], C. Bauer[1] and M.Gauly[2]
[1]*Institute of Parasitology, Justus Liebig University Giessen, Rudolf-Buchheim-Strasse 2, 35392 Giessen, Germany; christian.bauer@vetmed.uni-giessen.de*
[2]*Institute of Animal Breeding and Genetics, Georg August University, Albrecht-Thaer-Weg 3, 37075 Goettingen, Germany*

Introduction

Eimeria macusaniensis is a parasite of the small intestine of South American camelids. Knowledge on its biology is very scanty. The present studies were performed to obtain first substantial data on the epidemiology and life cycle of this protozoon.

Material and methods

Course of natural infection

Faecal samples were taken from llamas and alpacas on a farm in central Germany in monthly intervals during a 12-month period (Oct. 2000 to Sep. 2001). Faecal examinations were performed using McMaster method with sugar solution (specific gravity 1.3).

Prepatent and patent period

Six llama crias reared parasite-free were orally infected with 20,000 (n = 5; one month old) or 100,000 (n = 1; two months old) sporulated oocysts of *E. macusaniensis*. Re-infection of crias (n = 4) with 20,000 or 50,000 oocysts were done 2 to 3 weeks after ending of patent period.

Sporulation time

To estimate sporulation time incubation of oocysts were performed in petri dishes at different temperatures.

Results

Course of natural infection

Crias started oocyst shedding in the 2nd month of life showing highest prevalence (67–71%) and maximum oocyst counts per gram faeces (OPG) (400–450 OPG in mean) 2–3 months *post partum*; from month 8 *post partum* less than 5% of the foals shed oocysts. In dams and male yearlings the prevalence of the infection (16% and 26%, respectively, in maximum) and oocyst counts (24 OPG and 45 OPG, respectively, in mean) was distinctly lower than in crias during the study.

Prepatent and patent period

The prepatent period was 32–36 days and the patent period 39–43 days. The total oocyst output (500–1300 OPG in maximum) was $3.3–10 \times 10^6$ and the mean reproductive rate 1:310 after the initial infection. Reinfections with 5×10^4 oocysts 2 or 3 weeks after the end of the first patency resulted in a longer prepatent period (37–40 days), a shortened patent period (20–23 days) and reduced oocyst output (mean reproductive rate 1:125) indicating a immune response to the first infection. The individual total daily oocyst output of the infected animals in shown in Figure 1.

Sporulation time

E. macusaniensis oocysts were isolated from fresh faecal samples and incubated in 2% potassium dichromate solution at temperatures of 6–7, 18–19, 25 or 30°C. Maximum sporulation rates (92–94%) were obtained at 18–19 or 25°C on day 21 and at 30 C on day 13 (Figure 2). The oocysts did not sporulate at 6–7°C.

Total daily *E. macusaniensis* oocyst output in individual crias:

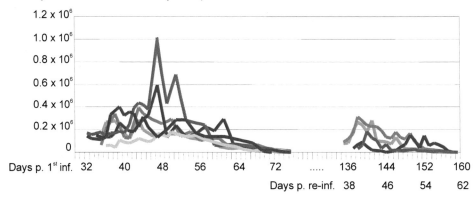

Figure 1.Total daily oocyst output of animals infected with Eimeria macusaniensis.

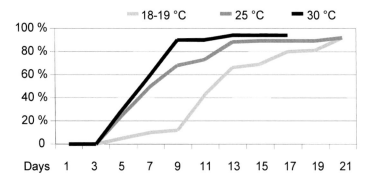

Figure 2. Sporulation times of E. macusaniensis oocysts at different temperature.

Changes in testicular histology and sperm quality in llamas (*Lama glama*) following exposure to high ambient temperature

A. Schwalm[1], G. Erhardt[1], M. Gauly[2], M. Gerken[2] and M. Bergmann[3]
[1]*Institute of Animal Breeding and Genetics, University of Giessen, Oberer Hardthof 25, 35398 Giessen, GERMANY, anja.schwalm@agrar.uni-giessen.de*
[2]*Institute of Animal Breeding and Genetics, University of Göttingen, Albrecht-Thaer-Weg 3, 37075 Göttingen, GERMANY*
[3]*Institute of Veterinary Anatomy, University of Giessen*

High environmental temperature can cause heat stress and infertility in male llamas. Reduced fertility of males during the summer has been described by Fowler (1998). Even under middle European climatic conditions seasonal changes in sperm parameters were detected. Sperm concentration was negatively correlated with the environmental temperature (Gauly, 1997). The objective of the present study was to demonstrate the changes in testicular histology and in mitotic activity of spermatogonia following a constant high ambient temperature under experimental conditions, corresponding to the changes in semen parameters in llamas.

For this study three fertile male llamas were castrated and were kept as control group. Ten fertile male llamas were housed in heated stables. After one of week acclimatisation period, the ambient temperature was elevated up to 30°C for 4 weeks. Afterwards the animals were allowed to recover for at least 7 weeks at 20°C. Semen was collected once a week from 4 animals using an artificial vagina and a phantom as mounting partner. Standard semen parameters were recorded each time. Six of the experimental animals were castrated at different times during the experiment (directly, two, four and 6 weeks after the heat period) to evaluate the histological changes in the testes. A staining with haematoxillin and eosin for microscopic study was performed. Using monoclonal antibodies (MIB-1) against the proliferation marker Ki-67 protein, the quantitative distribution pattern in the seminiferous epithelium of llamas was studied, in order to investigate the mitotic activity of the spermatogonia at different times after heat exposure.

The testes of animals castrated directly after the heat-period showed a general disorganization of the seminiferous epithelium consisting of the absence or reduction in number of various generations of germ cells, resulting in an increase of tubules cross sections where no stage could be determined. Six weeks after the heat-period the histological findings were still different compared to the control group. In the control group 36.9±7.6% of the spermatogonia were Ki-67 positive, *e.g.* were proliferating. Within the experiment the percentage of proliferating spermatogonia decreased to 8.2±5.8% two weeks after the heat-period. Four weeks after the heat-period the number of proliferating spermatogonia recovered to the levels of the control group. The sperm concentration declined significantly in all animals after the heat-period, the minimum level being reached 4 weeks after heat exposure with the percentage of abnormal sperms decreasing simultaneously.

High ambient temperatures can lead to significant changes in testicular histology and sperm quality in llamas and can result in infertility of the males.

References

Fowler, M.E., 1998. Medicine and Surgery of South American Camelids. Second Edition, Iowa State University Press.

Gauly, M., 1997. Saisonale Veränderungen spermatologischer Parameter und der Serumkonzentration von Testosteron, Oestradiol 17β, Thyroxin sowie Trijodthyronin männlicher Neuweltkameliden (*Lama glama*) in Mitteleuropa. Inaugural Dissertation, Justus-Liebig-Universität Giessen.

An evaluation of the growth and change in body dimensions from birth to maturity of the llama (*Lama glama*) and the huarizo (crossbred camelid) in the Bolivian Andes

B.R. Southey[1], T. Rodriguez[2] and D.L. Thomas[1]
[1]Department of Meat and Animal Science, University of Wisconsin, Animal Science Building 1675, Madison, Wisconsin, USA
[2]Faculty of Agronomy, Major University of San Andrés, Heroes del Acre street 1850, La Paz, Bolivia; rodriguezct01@hotmail.com.

The development of 25 llamas and 5 huarizos (llama - alpaca cross) was evaluated for body weight, height, length, chest circumference and staple length from birth to 900 days of age, at the Patacamaya Experimental Station in La Paz, Bolivia. The llamas were significantly heavier than the huarizos at all ages. The overall growth pattern for both genotypes was described as a nonlinear mixed growth curve model and as a quadratic function of age (Fitzhugh, 1976). The llamas tended to be taller with larger chest circumference than the huarizos, but they were not always significantly different. There appears to be no difference in body length. The huarizos produced significantly longer staple length than the llamas. The results from the llamas were similar to other South American studies but smaller than those obtained in North American studies (Escobar, 1982).

References

Escobar, R.C., 1982. Producción y mejoramiento de la alpaca. Fondo del Libro, Banco Agrario del Peru, Lima, Peru, 334 pp.

Fitzhugh, H.A., 1976. Analysis of growth curves and strategies for altering their shape. Journal of Animal Science 42: 1036-1051.

Seroprevalence of *Neospora caninum* und *Toxoplasma gondii* in South American camelids

D. Wolf[1], M. Gauly[2], W. Huanca[3], O. Cardenas[3], C. Bauer[1] and G. Schares[5]

[1]*Institute of Parasitology, Justus Liebig University Giessen, Rudolf-Buchheim-Strasse 2, 35392 Giessen, Germany; christian.bauer@vetmed.uni-giessen.de*
[2]*Institute of Animal Breeding and Genetics, Georg August University, Albrecht-Thaer-Weg 3, 37075 Goettingen, Germany; mgauly@gwdg.de*
[3]*Laboratorio de Reproducción Animal, Facultad de Medicina Veterinaria, Universidad Nacional de San Marcos, P.O. Box 03-5034, Salamanca, Lima - 3, Peru*
[4]*Estación Experimental ILLPA, Instituto Nacional Investigación y Extensión Agraria (INIEA), Rinconada de Salcedo s/n, Puno, Peru*
[5]*Institute of Epidemiology, Federal Research Institute for Animal Health, Seestrasse 55, 16868 Wusterhausen, Germany; gereon.schares@wus.bfav.de*

Abstract

Our results indicate that at least a low proportion of SAC living in this particular region of Peru has been exposed to *N. caninum*. The higher seroprevalences of both *N. caninum* and *T. gondii* in adult SAC than in foals suggest a predominance of postnatal routes of infection. The clinical significance of these findings remains to be investigated.

Introduction

Neospora caninum is a cyst-forming sporozoan parasite that has the dog as a definitive host and probably a wide range of animals as intermediate hosts. It is currently regarded as one of the most important causes of abortion in cattle worldwide (Dubey & Lindsay, 1996). However, nothing is known about its possible presence in South American camelids (SAC) until now. The aim of the present study was to develop and evaluate various diagnostic tools to examine *N. caninum* infections in SAC serologically. These tools were used to obtain information on the presence of *N. caninum* in alpacas (*Lama pacos*), llamas (*L. glama*) and vicuñas (*L. vicugna*) in Peru and in Germany. In addition, the SAC were serologically examined for *Toxoplasma gondii* infection.

Materials and methods

Experimental infection

A llama yearling being seronegative for *N. caninum* and *T. gondii* before inoculation was inoculated *i.v.* with 4.8×10^6 cell culture *N. caninum* tachyzoites. Serum samples were collected twice a week after inoculation and examined by western blot for antibodies against immunodominant *N. caninum* tachyzoite antigens (IDAs) of 17, 29, 30, 33 and 37 kDa molecular weight.

Seroprevalence survey

Serum samples were collected from a total of 869 young (3–6 months of age) and adult SAC of two farms in Peru (Department Puno) and from 32 SAC of a farm in central Germany. The sera were examined by western blot for antibodies against *N. caninum* and *T. gondii*.

Results

Experimental infection

The llama developed antibodies against two of the IDAs as early as 12 days *p.i.* and against all five IDAs later on. In contrast, the control animal inoculated with cell culture medium remained serologically negative.

Seroprevalence survey

The results from Peru are presented in Table 1. On the German farm one of 13 foals and 14 of 19 adult alpacas and llamas were seropositive for *T. gondii*, but no reactions with *N. caninum* IDAs were observed in any of these animals.

Acknowledgements

The authors want to thank the staff of Mallkini and Quimsachata for there technical and scientific support. Especially M.V.Z. Teodosio Huanca Mamani, M.V.Z. Rómulo Sapana Valdivia and Ignacio Garaycochea.

Table 1. Seroprevalence (immunoblot) of N. caninum and T. gondii in SAC from Peru.

			Total	Seropositive for N. caninum		Seropositive for T. gondii	
Origin	Host	Group	n	n	%	n	%
Mallkini	Alpaca	Foals	195	5	2.6	2	0.5
Quimsachata	Alpaca	Foals	161	1	0.6	1	0.6
		Adults	319	10	3.1	17	5.3
	Llama	Adults	81	1	1.2	7	8.6
	Vicuña	Adults	113	0	0	3	2.7

References

Dubey, J.P. and D.S. Lindsay, 1996. A review of *Neospora caninum* and neosporosis. Vet. Parasitol. 67: 1-59.

DECAMA-Project: Technological and nutritional parameters of fresh meat of Argentinean llamas (*Lama glama*)

A.P. Zogbi, E.N. Frank and C.D. Gauna
Facultad de Ciencias Agropecuarias, Universidad Católica de Córdoba, Cno. Alta Gracia Km 10,5017, Córdoba, Argentina; anazogbi@campus1.uccor.edu.ar

The objective of this work is the evaluation of color and texture parameters and the quantification of fatty acids and conjugated linoleic acids of fresh meat of Argentinean llamas (*Lama glama*). The first stage of the research work has been performed with 12 male animals of 21 months old, fed for some time previous to this study in an experimental field located in the La Pampa Province. They were slaughtered according to the methodology used for cattle. Samples of shortloin (*Longisimus dorsi* muscle), "Peceto" (*semitendinus* muscle), "cuadril" (*gluteous medium* muscle) were obtained. For the quantification of fatty acids and conjugated linoleic acid (CLA) two grams of fat were extracted from each loin (*Longisimus dorsi* muscle) following the technique described by Folch *et al.* (1957) and using a methanol/chloroform solution.

The average values obtained from the samples show that the Llama meat (LD) contains 47.80 mg of CLA/100 g of meat, SD ± 8.51. The following fatty acid profile was found in the samples: myristleic acid: 1.93%, pentatonic acid: 0.94%, palmitic acid: 21, palmitoleic acid: 2.97%, hexadecadienoic acid: 1.01%, heptadecenoic acid: 0.43%, estearic acid: 19.4%, vaccenic acid: 4.38%, oleic acid: 24.38%, linoleic acid: 8.37%, 9 cis,11-trans-linoleic conjugate CLA, 1.41%, linolenic acid: 3.72%, arachidic acid: 0.30%, eicosadienoic acid: 0.13%, arachidonic acid: 3.44%. These results are similar to the ones obtained by Patkowska-Sokola *et al.* (2002) in other ruminant species. The color was measured 48 h *post-mortem* using a spectrophotometer Minolta CM-508 d using iluminant D65 and a 10° standard. Nine color determinations were made in each samples following the recommendations of the American Meat Science Association for color measurements in meat. Low reflectance glasses were placed between the samples and the equipment. Color parameters were analysed as lightness (L*), redness (a*) and yellowness (b*) and the colour differences (ΔE) of the CIELAB color space. The reflectance spectra, every 10 nm, between 400 and 700 nm were also obtained. The average value of the meat samples indicate that the luminosity (L*) of the "Peceto" was of 34.52 ± 3.61 of the "Cuadril" 31.06 ± 2.40 and of the shortloin 31.82 ± 2.40. The coordinate a* of the "Peceto" was 6.98 ± 1.61, of "Cuadril" 7.35 ± 1.78 and of shortloin 6.73 ± 1.78. The average values for the coordinate b* in the "Peceto" was 6.99 ± 1.32, in the "Cuadril" 6.4 ± 1.64 and in the shortloin 6.22 ± 1.64. The color differences (ΔE^*) for the "Peceto" were 12.76 ± 2.4, "Cuadril" 12.76 ± 2.44 and shortloin 12.29 ± 2.44. The texture was measured using a TZ-TA-TX2 (texture analyser according to Warner-Bratzler). Each sample was then cut into 10 pieces of 10 x 10 mm x 15 mm (wide x long x thick) parallel to the longitudinal axis of the sample. They were cooked wrapped in aluminium foil in a revolving oven at 200°C till the inner sample temperature reached 70°C. The following average shearing force values were recorded (in pounds): "Peceto" 14.03 SD ± 0.80, cuadril 11.74 ± 1.30 and shortloin 13.11 ± 1.70. In the second stage of this research work (August 2004) meat samples of 25 animals were analysed. A comparison with the previously mentioned values will be carried out, among 25 animals located in Cieneguillas (Jujuy), considering that these animals are under different feeding and productive conditions than in the Province of La Pampa. Thus, it will be possible to evaluate how much these conditions affect the colour and texture of the llama meat.

References

Patkowska-Sokola, B., D. Jamroz, R. Bodkowski, A. Cwikla and T. Wertelecki, 2002. Fatty acids profile content of conjugated linoleic acid of meat fat from young cattle, lambs and kids. Animal Science and Reports 20 Supplement 1: 63-73.

Folch, J., M. Lees and G.H. Sloane-Stanley, 1957. A simple method for isolation and purification of total lipids from animal tissue. J. Biol. Chem. 226: 497-507.

Roundtables

Management of alpaca populations outside of South America

Co-ordination: *M. Gauly (Germany)*
Chairpersons: *A. von Baer (Chile), G. Berna (Italy), U. Lippl (Switzerland), M. Trah (Germany) and J. Vaughan (Australia)*
Reporter: *A. Riek (Germany) and M. Gerken (Germany)*

The roundtable was opened with a short description of the situation in the different countries by each Chairperson.

G. Berna (Italy): Italy has about 500 alpacas. The Italian alpaca association was founded in 2000 and has 36 associated members of whom four are large breeders (30 to 60 animals) and the rest are small breeders (5 to 20 animals). The main focus is on the creation of an alpaca breeding registry in co-operation with the University of Camerino and a laboratory for DNA analysis. The registry has been divided into the following five sections: Section I, the cria up to three months from birth; Sections II and III, the cria (males and females) evaluated at one year of age and selected for improvement; Sections IV and V, the present existing animals, selected in the absence of basic genetic defects. The wool committee organised the processing of about 300 kg of fibre in 2004.

M. Trah (Germany): In Germany, 5,000 to 10,000 camelids are mainly kept as pet animals. There are two breeding associations. There is no fibre industry for alpaca wool. Small "cottage industries" making products of handicraft quality have been developed in the last five years.

U. Lippl (Switzerland): The number of camelids is estimated as between 3,000 and 4,000, of which half are alpacas. The breeder association numbers 300 members. The animals are mainly kept as pets in groups of 10 to 20 animals.

J. Vaughan (Australia): Australia currently has 55,000 to 60,000 Alpacas kept with an average herd size of 20 animals. In addition, there are 5,000 to 7,000 llamas. A strict selection programme for fibre quality has been established using certified males. The breeding aim is to decrease the present fibre diameter of 26 to 20 µm. Alpaca fibre is in competition with the very fine wool of 100 millions of Australian sheep.

A. von Baer (Chile): Only a small breeder association of 12 members exists in the South of Chile in a very different environment to the Altiplano. The register of the Association is not recognised by any other association. There is only a marginal development of the fibre market and animals are mainly kept as pets.

The main focus of the following discussion was concentrated on the use of Alpacas outside of South America. While there is a strong emphasis on selection for fibre quality in Australia, there are, until now, no proper breeding or selection criteria applied in European countries. In contrast, alpacas in the USA are mainly kept as pet animals and are also successfully marketed as such.

In his remarks, *J. Gaye (UK),* explained his experiences in alpaca breeding in the UK, where two studbooks have been established. The major problem that arises, as in any other European country, is that the marketing of the raw material cannot compete with other international wool markets, for example, in South America. The reasons are, firstly, because of the small numbers of animals which have small yields of wool and secondly because of the differences in labour costs between countries such as England and Peru.

W. Huanca (Peru) compared the Peruvian and European situation. Peru has about three millions of alpacas and a well established fibre industry. It is clear that the small number of European animals cannot compete with the quality and quantity of Peruvian fibre. However, Peru is disadvantaged by lack of a national breeding programme and the present registration system includes only 500 animals.

From that point, the question then arose as to the most appropriate way forward for Europe in terms of breeding strategies and selection criteria. Should selection be based on fibre quality, like the Australians or should the alpaca in Europe be considered less as a farm animal, and mainly as a pet animal?

In general, the following objectives in establishing breeding programmes were identified:
- development of objective criteria for description and evaluation of breeding animals;
- provision of information on quantitative genetic parameters (*e.g.*, heritabilities);
- establishment of a central studbook for the different European breeding associations.

There was agreement that breeding goals will be very different between individual breeders. It was suggested that a first step towards a proper breeding strategy would be the establishment of a studbook in which all animals that fulfil certain selection criteria are registered. This stud book would allow the breeders to choose animals according to their respective breeding aims. *Gianni Berna (Italy)* advanced the suggestion of a European Federation of the National Associations with a Committee and a Secretariat held, for example, at a University. Such a Federation could start with a small number of breeding associations and would be open for all interested participants.

The roundtable was concluded by supporting the idea of a solution for Europe which would take into consideration the small numbers of animals in this region.

Setting up a new national alpaca enterprise: The UK alpaca national industry 1996-2004

J. Gaye

Alpacas of Wessex, Hingaston Gate, Marnhull, DT10 1NL Dorset, UK; johngaye@onetel.com

The aim of my paper is to evaluate the success to date of the British alpaca enterprise through comparison with two principle models of the USA and Australia.

Motivation of breeders

In the UK we have a population of about 12 –15,000 alpacas, a population which has grown from about 600 in 1996. There are probably about 300 – 400 actual owners of animals throughout Great Britain. Ownership varies between those who are really serious about breeding and improving the quality of their animals each year through to those who just have two gelded males as pets. This makes the creation of an industry working together with an aim very difficult to harmonise because there is not yet the financial incentive from the product unifying the goals and objectives of the breeders.

My observation of the industry in the States is that it is predominantly a life-style enterprise heavily dependent on extremely professional and slick marketing of that life style. In addition I am well aware how they have been lead into closing their registry in order to artificially hold up the values of their animals. In maintaining the value of their animals it works well but I do question its long-term viability particularly as I feel the gene pool is far too small for the health and well being of the overall national herd.

The Australians are rather different. There is a much more dynamic approach to the business in that they recognise that there must be a long term market for the fibre if alpacas are to be credible as a livestock industry. And the majority of them want to be recognised as such, not as hobby farmers or pet owners.

Marketing

In North America, Australia/NZ and Europe tens of thousands of dollars are spent each month promoting the animals and promoting individual herds. Both the Australians and the British have a long tradition of agricultural shows - local, regional and national. Marketing however has had some major spin-offs in the UK. Eight years ago alpaca garments were only known by a miniscule part of the shopping public. Now alpaca garments are increasingly available throughout the country. Still the most effective way of promoting our industry and our own breeding business is through the show classes being held at prestigious agricultural shows, being carried out in a thoroughly professional way.

The development of the end use of the fibre

To me this is the crux of the future of our individual industries. Both the British and the Australians recognised that there was no future in just marketing the raw fibre into the hand spinning market in their countries. Although there is a demand from hand spinners that demand is unlikely to grow at the same level as the supply that will be available. Both set up co-operatives with similar aims, which was to take the raw fibre from the breeders, process it, manufacture garments and market them as finished products. Both have been plagued with similar problems: too little capital, too little support from the industry, too little fibre to make the processing competitively costed, and a lack of product awareness in the market place. In addition, expertise in breeding animals did not necessarily provide expertise in dealing with textiles.

The infrastructure of support agencies

Extensive work has been done in the USA and Australia on research into camelid physiology and medicine, to an extent where many Americans veterinarians have been invited to present papers to breeders in Peru.

In the UK we have a growing network of vets, linked by the British Veterinary Camelid Society, taking an ever-closer interest in these animals. However apart from one very successful initiative there has been to date very little research into camelid medicine amongst the British veterinary community.

The breed societies vary a great deal and reflect something of the industry they represent. The Australians have an extremely good web site, they are actively promoting working into tracking the genetic histories of their members animals and have recently taken on support of the Fibre Coop.

The British have two societies which are still in settle down mode as they each come to recognise the strengths and roles of each other. Above all we need an infrastructure to educate and encourage breeders to work towards some common goals.

The breeding aims and the potential future of the industry

Selective breeding is vital as we cannot afford to stand still in our breeding programmes. It is important that there is a majority who are motivated to advance their herds and thus benefit the national enterprise. We must not let commercial competitiveness get in the way of working together, using the best males available wherever they may be placed.

Where is the British industry going? Are we going to pursue the Australian way, with the goal of a viable fibre market, or are we going down the American route where we promote the enjoyment and lifestyle to be gained in breeding these agreeable beasts.

To me the answer to this question lies in a number of questions:
- Can we establish a successful market for our product – fleece? Can we actively compete with the best fibre and alpaca garments coming from South America or possibly even from China.
- Can we continue to find people with vision who wish to breed up the quality of the animals through careful selective breeding techniques?
- Can we continue to identify and import animals from other countries for genetic improvement?
- Can we establish in the minds of the authorities, who might be able to provide some material support, that fibre production is a credible agricultural activity?
- Can we grow our national herd to a level where the processing of fleece becomes more economically viable?
- Can we persuade other European alpaca industries, or even the North American or Australian industries, to look to join with us in bulking up the fleece quantity?

Research on wild and domestic South American camelids: Comparison of two different approaches

Co-ordination: M. Gerken (Germany)
Chairpersons: W. Huanca (Peru), N. Renaudeau d'Arc (UK), C. Renieri (Italy) and C.R. Schmidt (Germany)
Reporter: M. Gerken (Germany) and G. Chiang (Peru)

Different scientific groups are working on either wild or domestic camelids. Accordingly, this roundtable offered a platform to combine their expertise. Particular attention was given firstly to the development of concepts of similarities and differences with regard to conservation and secondly to the scientific approaches to be adopted for research on both wild and domestic species.

Conservation programs for wild species

Schmidt (Germany): The Cites regulations help to stimulate, in an international context, public awareness concerning the world-wide problem of wildlife conservation. The legal restrictions on international marketing of vicuña fibre have resulted in increased animal numbers in their countries of origin. It is considered that conservation of the wild camelids should take place in South America. This is because the conservation programmes conducted by zoological organisations outside of South America cover populations which are too small to guarantee long-term conservation of the species.

Sustainable use of wild species

Renaudeau d'Arc (UK): The question of a sustainable use of the wild camelids in South America arises in the context of justification of the conservation of vicuña and guanaco. The term sustainability includes the three different aspects of economic, environmental and social development. Accordingly, sustainable use must consider the proper interests of the campesino population with priority given to the development of indigenous concepts. Wild and domestic species utilise the same natural resources (pasture, water), thus creating the potential for competition between the two groups of animals. Traffic in the wild animals through the borders of Chile and Peru continues to be a major problem.

Investigation in domestic camelids

Renieri (Italy): The difference in scientific approaches required to be taken for both wild and domestic animals lies in the interaction with humans. In domestic animals, artificial selection is the origin of the development of a large variation in phenotypes and genotypes. I see specific differences in breeding goals for the llama and alpaca domestic camelids. Alpaca should be selected for fibre with meat as valuable secondary product. Llama breeding should favour a dual purpose animal for both fibre and meat. The differentiation into fibre and meat types appears not very suitable for the long-term conservation of the llama as a domestic animal.

Huanca (Peru): For South American research in domestic camelids, the main attention should be given to fibre dynamics and reproduction. In South America, the quantity of the fibre is considered most important, while in Europe and Australia the main focus is on fibre quality. There are currently no national selection programmes for the improvement of fibre quality in Peru or Bolivia. In my opinion, it is necessary to start projects giving priority to programmes of genetic selection. It is apparent that highly advanced techniques of biotechnology such as embryo transfer are only of minor importance under the present conditions.

Renieri (Italy): Suitable methods require to be developed to provide effective selection in domestic camelids in South America. Nucleus breeding could usefully be practised even under the very harsh environments of the Altiplano. There is an example in Cailloma (Peru) where a Peruvian NGO supervises a herd of 800 alpacas. The results of the first and second shearings are used as parameters for the evaluation of the breeding animals.

Huanca (Peru): For many years, the best animals, which are winners in many fairs, are sold to other countries across the "Tripartito" zone between Chile, Peru and Bolivia. This practice has resulted in a considerable gene loss for South American countries.

Ayala (Bolivia): Experience in Bolivia has shown that breeders are interested in maintaining different coat colours in their herds. However, the white colour is favoured by the international fibre market resulting in the loss of genes controlling the fibre colour.

Transition from wild to domestic species

Schmidt (Germany): Keeping vicuñas or guanacos under captivity or semi-captivity in fenced areas provides the first step towards domestication. Fences prevent the free movement of animals thus interfering with the exchange of genes between populations and so counteracting natural selection. The alpaca has been already selected for fine fibre. We should not repeat such domestication by keeping vicuñas in captivity. In my opinion, wild animals should be considered as part of their natural environment. The wildlife management practice of "Chaccu" as practised in the Inca times is the best example of sustainable use of both wild and domestic species living on the same natural resources.

Renaudeau d'Arc (UK): The co-existence between wild and domestic animals depends on the carrying capacity of the respective ecosystems. The competition for the same scarce natural pasture could become a problem for the campesinos. In my consideration, sustainable management depends on the "benefits to the campesinos". The question is whether they could continue to survive under their conditions of limited resources.

Rey (Argentine): Sheep exert a strong negative influence on the vegetation on my farm in Patagonia. The burning of the vegetation in order to promote growth of grasses causes long-term damage to the ecosystem. Wild camelids such as guanacos could offer an interesting alternative to traditional sheep breeding in these regions.

In the following discussion, similar approaches for both wild and domestic species were outlined:
* Both types share the same environment.
* There are gaps with regard to the scientific knowledge on biology and physiology, and in particular on fibre biology. The relationship between photoperiod and the follicle growth should be considered to improve the quantity and quality of the fibre.
* Questions of conservation exist in both types with different priorities. Conservation of genes for fibre quality and colour are the main points to be considered in domestic species.
* Political support is needed at the rural level. The political instability of many South American countries hinders the development of both long-term conservation programmes for the wild species and large scale breeding programmes for domestic species.

The conservation of wild camelids requires support for the campesino population and the (economic) acknowledgement of their work. This complex task can only be solved by efforts from both national and international sources.

Breeders and keepers

Co-ordination: M. Trah (Germany) and M. Gauly (Germany)
Reporter: K. Dietze (Germany)

The main topic of this roundtable was "Health maintenance in new world camelids – diseases and prevention". An introductory survey on the recent situation in Germany was given by *Barbara Münchau (Germany)*. She focused on the increased appearance, and particular concentration in Germany, of Pseudotuberculosis which has been previously been described in new world camelids. Participants from other European countries as well as from Australia and South America could not confirm a similar observation.

The discussion of this bacterial borne disease lead to several recommendations:
*   mixed herds including sheep, should be avoided (sheep being the main vector of Pseudotuberculosis);
*   disease-positive animals have to be separated from the herd, since elimination (as practised with small ruminants) is often not an option;
*   newly purchased animals should be put in quarantine;
*   Erythromycin is an option for antibiotic treatment in order to decrease clinical signs;
*   Vaccination as a preventive measure should be discussed with the local veterinarian.

A special focus was then put on general questions of diagnostic tests to be applied when buying an animal and preventive treatments to be performed on a regular basis. Different routine diagnostic and treatment patterns were presented. These all included a regular vaccination for clostridial diseases (at least once a year) and anthelmintic treatments. The use of anthelmitics might be considered necessary up to four times a year. However, it should strongly depend on the parasite burden of the herd and its effectiveness should be checked in order to avoid the development of therapy-resistant parasites. In addition, vitamin E/selenium or vitamin ADE/selenium treatments were recommended, depending on pasture and soil quality, as well as optional IgG-tests for neonates to determine the sufficiency of colostral supply. Blood tests for brucellosis, leucosis and tuberculosis provide a further means of protecting the herd from the spreading of diseases by identifying disease-positive animals. When buying an animal, initial quarantining presents the best way to prevent new diseases from entering the herd. This practice has to be combined with the general vaccinations and anthelmintic treatments applied in that specific herd. Further diagnostic practises might be useful, especially to eliminate pseudotuberculosis and brucellosis.

In considering the spread of disease, one risk derives from the purchase of an animal carrying an infectious disease. A potential second risk arises from the often practised "mating-tourism", where mainly male animals are brought from one farm to another in order to provide mating services. Even though it is a routine farm procedure in Europe, North America and Australia, it has not currently been identified as a main source of disease transmission in new world camelids. In addition, it should always be considered as an option when choosing mating partners for animals of an individual herd.

A questionnaire was distributed to gather data on the actual health situation of new world camelids in European countries and participants were encouraged to send information to Dr. Münchau.

Latest information on diseases in camelids

B. Münchau
Rosengartenstr. 11, 72108 Rottenburg, Germany; Muenchaukameldoc@aol.com

Beginning with summer 2003, multiple abscesses of the lymph nodes and the skin, and fatalities caused by internal abscesses, have been observed and attested to with camelids in Germany (so called "Bubonic plague"). Clinical examinations predominantly diagnosed Pseudotuberculosis (caseous lymphadenitis and ulcerative lymphangitis), caused by *Corynebacterium pseudotuberculosis*, in some cases combined with *Staphylococcus dermatitis*.

Pseudotuberculosis is a known and world-wide-spread disease in small ruminants, with chronic progression. In camelids it was noticed in their countries of offspring. *C. pseudotuberculosis* (two biotypes) with toxic development causes cold, painless, caseous abscesses in lymph nodes, lymph vessels., reticular tissue, the lung and other organs (liver, milt, kidneys spinal marrow, brain, scrotum, testicles, epididymis). Observed were also septicemia, intoxications, arthritis, bursitis, mastitis, abortive birth, dermatitis. Other isolated bacteria obtained from abscesses: *Streptococcus spp., Staphylococcus spp., C. pyogenes, C. renale, C. equi, Shigella, E. coli, Pseudomonas,* furthermore *Actinomyces pyogenes* and *Histoplasma farciminosum* seem to be involved in the pathogenesis and out-break of pseudotuberculosis (Wernery and Kaaden, personal communication). Mortality rate among examined bactrian camels reached 15%.

The pathogens are excreted by faeces and above all, by the secretions of abscessing lymph nodes. They are eradicated by common disinfectants, when exposed to sunlight within 24 hrs, or when heated over 70°C (158°F), however they remain contagious for a very long time in soil, water, faeces, straw. The time span identified varies (up to 8 months). Infection are caused primarily through lesions of the skin and mucous membrane, but can also be oral and air born. Ticks can be excluded as transmitters. Transmissions on other animals (camelids, horses, mules, piglets) rodents, primates, birds and other wild animals as well as transmission from humans (anthropozoonosis) are described. Symptoms and progression vary and range from no symptoms to swellings of the lymph nodes and abscesses, discomfort when swallowing, broncheopneumonia and loss of weight, and death.

Diagnosis

Lymphadenopathy, toxicological and analytical tests on pathogens, serological (in sheep haemaglutination assay and ELISA).

Recommended therapy

Immediate isolation of infected animals, vigilant hygiene. Disinfection of stable, blankets, all utensils such as trimming machines. Douching of ruptured abscesses (*e.g.*, hydrogen peroxide), parenteral antibiotics after testing resistance. In Australia, tests with vaccines for sheep were experimentally conducted and have been successful. In Germany, this vaccine may only be administered with special permission.

Current status in new world camelids in Germany

We are familiar with multiple cases in different lama and alpaca stocks, in which pseudotuberculosis was diagnosed. The disease spread among the herds when co-mingled for breeding (up to 1½ years after a local case of pseudotuberculosis was cured) or when new animals were purchased, mostly acquired through animal dealers. Swellings and abscesses were seen from lips to toes and spread over the entire body. Cases of lameness and death were also observed.

The findings substantiate the observations pertaining to small ruminants: that pathogens remain contagious for a long time in soil or straw bedding and that clinically healthy animals, after having been in contact with the pathogens, may communicate the disease during their entire life span. They confirm the varying period of incubation over a couple of months (Steng & Sting, personal communication).

Bactrian camels and dromedaries

In multiple stocks of dromedaries and bactrian camels (imported, purchased from animal dealers, or acquired from breed) animals are affected with swellings of the lymph nodes and abscesses. In pus and biopsy material, *C. pseudotuberculosis, Staphylococcus aureus, enterococci, Enterobacter agglomerans*, aerobic spore-building bacteria and *Staphylococcus haemolyticus* were determined. As distinguished from literature, animals less than 3 years old are equally affected.

Further considerations

Increased co-mingling for breeding, and inter-mingling due to animal trading and show performances pose a risk for increased propagation. Even though the rate of mortality is low, the economic losses are high. We are therefore researching opportunities for early diagnosis, prophylaxis, therapy and medical examinations of herds. The serological assays in Germany (test for small ruminants) are currently not convincing, as acutely and chronically affected animals and pathogen excreting animals cannot be distinguished. An upward tendency of pseudotuberculosis in sheep has been noted as well. We are asking for your co-operation and observations on cases of these diseases. For better co-ordination, please confer directly with me.

Acknowledgement

Cordial thanks to Prof. Kaaden and his team (Inst. f. med. Mikrobiologie, Infektions- und Seuchenmedizin, Tierärztliche Fakultät, LMU München, Germany), and Dr. Wernery (Dubai), for their continuous advice and support, to the colleagues Dr Steng and Dr Sting (Schafsgesundheitsdienst Stuttgart, Germany) for their advice and their contributions granted in form of scientific studies, to all animal owners and veterinarians providing us with case studies, pictures and other results, helping us to clarify this problem.

Additional references

Behrens, H., 1987. Lehrbuch der Schafkrankheiten. Verlag Paul Parey, Berlin, pp. 77-78.

Dioli and Stimmelmayr, 1992. The one-humped Camel in Eastern Africa. In: H.J. Schwartz and Dioli (eds.), Verlag Josef Margraf, Weikersheim, pp. 212–216.

Nieberle and Cohrs, 1970. Lehrbuch der speziellen pathologischen Anatomie der Haustiere. Gustav Fischer Verlag, Jena.

Sheep Veterinary Society, Division of the British Veterinary Association: Caseous Lymphadenitis – Advice for SVS members.

Voigt, A. and F.-D. Kleine, 1973. Zoonosen. Gustav Fischer Verlag, Jena, pp. 164-165.

Weiss, E., 1999. Blutbildende Organe. In: Dahme, E. and E. Weiss (eds.), Grundriß der speziellen pathologischen Anatomie der Haustiere. Enke-Verlag, Stuttgart, pp.61-62.

Wernery, U. and O.R. Kaaden, 2002. Infectious Diseases of Camelids. Blackwell Wissenschafts-Verlag, Berlin, pp. 62 – 67.

List of participants

List of participants

Dipl. Ing. Rudi Afnan
Institut für Tierzucht und Haustiergenetik, Albrecht-Thaer-Weg 3, 37075 Göttingen, Germany
Tel.: +49-551-395360, Fax: +49-551-395587, E-mail: rudiafnan@yahoo.com

Erika Alandia Robles
Institut 480a, University of Hohenheim, Garbenstr. 17, 70593 Stuttgart, Germany
Tel.: +49-711-459 2477, Fax: +49-711-459 3290, E-mail: ealandia@uni-hohenheim.de

Prof. Dr. Francesco Ansaloni
Dipartimento Scienze Veterinarie, Via Fidanza, 15 - 62024 Matelica, MC, Italy
Tel.: +39-0737-4040408, Fax: +39-0737-404010, E-mail: francesco.ansaloni@unicam.it

Dr. Marco Antonini
Dipartimento di Scienze Veterinarie, Università di Camerino, Via Circonvallazione 93/95,
I-62024 Matelica (MC), Italy
Tel.: +39-0737-403451; Fax: +39-0737-403402, E-mail: antoninim@casaccia.enea.it

Dr. Celso Ayala Vargas
CECON-CIENCIA, P.O. Box 4303, Calle Policarpio Eyzaguirre 1417, Zona Callampaya, La
Paz, Bolivia
E-mail: celsoayalavargas@hotmail.com

Dr. Juan Ayala
Pasaje Salaverry 133 Surquillo, Lima 034, Peru
Tel.: +241-0480-9-9248568; E-mail: juedayala@hotmail.com

Alejandra von Baer
Casilla 31, Padre Las Casas, Chile
Tel.: +56-45-562912; Fax: +56-45-562912; E-mail: info@llamasdelsur.com

Dr. Matilde Basile
Dipartimento di Scienze Veterinarie, Via a Santa Croce, 25, I-80129 Napoli, Italy
Tel.: +39-0339-7402319, Fax: +39-0737-403402, E-mail: matilde.basile@unicam.it

Gianni Berna
ITALPACA, Associacione Italiana Allevatori Alpaca FAAZ, Niccone 173, I-06019 Umbertide,
Italy
Tel:: +39-075-9410931, Fax: +39-075-9410934; E-mail: info@italpaca.it

Dr. Sabine Bramsmann
Institut für Tierzucht und Haustiergenetik, Albrecht-Thaer-Weg 3, 37075 Göttingen, Germany
Tel.: +49-551-395610; Fax: +49-551-395587; E-mail: sbramsm@gwdg.de

Maria H. Bravo
222 County Route 164, Callicoon, NY 12723, USA
Tel. +1-845-482-3776; Fax: +1-845-482-4473; E-mail: mbravo@quintessence-inc.com

List of participants

Barbara Bruns
Große Theaterstr. 37, 20354 Hamburg, Germany
E-mail: svenwitt@compuserve.com

Prof. Dr. Christopher Cebra
Department of Clinical Sciences, Oregon State University College of Veterinary Medicine,
105 Magruder Hall, Corvallis, Oregon 97371, USA
E-mail: christopher.cebra@oregonstate.edu

Gloria Chiang
Av. Gral Canevarao 180, apto 306, Lince, Lima 14, Peru
E-mail: glomachiba@yahoo.com

Alvaro Claros
PRORECA, Av. Sánchez Lima 2340, La Paz, Bolivia
Tel.: +591-2-2126227, Fax: +591-2-425016, E-mail: proreca@entelnet.bo

Nina Collot D'Escury
La Foulerie, F-50160 Placy Montaigu, France
Tel.: +33-2-33 05 83 92, Fax: +33-2-33 05 83 92, E-mail: ninacollot@free.fr

Dr. Genaro Condori Choque
Av. Sánchez Lima 2340, La Paz, Bolivia
Tel.: +591 2-71955102, Fax: +591-2-2126227, E-mail: gencond@mailcity.com, gencond@
hotmail.com

Nadja Daghbouche
Europabüro, Universität Göttingen, Goßlerstr., 37075 Göttingen, Germany
Tel.: +49-551-39 9795, Fax: +49-551-39 12278, E-mail: nadja.daghbouche@goettingen.de

Dr. Gian Lorenzo D'Alterio
V.LE. Giustiniano Imp. 182, I-00145 Rom, Italy
Tel:. +39-0338-9140835, Fax:+39-06-5414510, E-mail: Gianlorenzod@yahoo.it

Klaas Dietze
Jahnstr. 26, 30982 Pattensen, Germany
Tel.: +49-5118-503786, E-mail: klaasdietze@aol.com

Dr. Ciara S. Dodd
BEPG School of Bioscience Cardiff University, Main Building, Park Place, Cardiff,
CF 10 3TL, UK
Tel.: +44-2920875073/5776, Fax: +44-02920874305, E-mail: doddcs@cf.ac.uk

Walter Egen
Alte Steige 34, 87600 Kaufbeuren, Germany
Tel.: +49-8341-73318, Fax: +49-8341-41194, E-mail: pichincha-llamas@online-service.de

Dr. Francesco Fantuz
Dipartimento di Scienze Veterinarie, Università di Camerino, Via Circonvallazione 93/95, I-
62024 Matelica, MC, Italy
Tel.: +39-0737-403427, Fax:+39-0737-403402, E-mail: francesco.fantuz@unicam.it

Karoline Favergeat
c/o Dr. Hubert Rey, 1, Place des Alliés, F-34500 Béziers, France
Tel.: +33-4-6749 2244, Fax: +33-4-6749 8575, E-mail: moulinmer@magic.fr

M.A. Lars Fehren-Schmitz
Historische Anthropologie und Humanökologie, Bürgerstr. 50, 37073 Göttingen, Germany
Tel. +49-551-393693; Fax.: +49-551-393645, E-mail: lfehren@gwdg.de

Dr. Angelika Freitag
Luetke Rott 25, 49549 Ladbergen, Germany
Tel:: +49-5485-1014, Fax: +49-5485-3509, E-mail: dr.angelika.freitag@web.de

Dr. Hugh Galbraith
School of Biological Sciences, University of Aberdeen, Hilton Campus, AB24 4FA Aberdeen, UK
Tel.: +44-1224-274232, Fax: +44-224-273731, E-mail: h.galbraith@abdn.ac.uk

Teresa J. Gatesman
Zimmermannstr. 60, 37075 Göttingen, Germany
Tel.: +49-551-2052500, Fax. +49-551-395587

Prof. Dr. Dr. Matthias Gauly
Institut für Tierzucht, und Haustiergenetik, Albrecht-Thaer-Weg 3, 37075 Göttingen, Germany
Tel.: +49-551-395602 Fax: +49-551-395587; E-mail: mgauly@gwdg.de

Dr. John Gaye
Hingaston Gate, Marnhull, DT10 1NL Dorset, UK
Tel.: +44-1258-821499, Fax: +44-1258-821300, E-mail: johngaye@onetel.com

Dr. Didier Genin
LPED, Université de Provence, Centre St. Charles, Case 10, 9 Place V. Hugo, 13331 Marseille Cedex 3, France
Tel.: +33-491-106357, E-mail: Didier.genin@up.univ-mrs.fr

Dr. Paul Gerber-Santesson
Den Gamla Möllegarden Gröstorp 9, 272 94 Simrishamn, Sweden
Tel.: +46-70-3050152, Fax: +46-414-10152, E-mail: paul.gerber@swipnet.se

Prof. Dr. Martina Gerken
Institut für Tierzucht und Haustiergenetik, Albrecht-Thaer-Weg 3, 37075 Göttingen, Germany
Tel.: +49-551-395603; Fax: +49-551-395587; E-mail: mgerken@gwdg.de

Prof. Dr. Luigi Grazia
Dipartimento di Protezione e Valorizzazione Agroalimentare, Università di Bologna, Via F.lli Rosselli, 107; 42100 Reggio Emilia, Italy
E-mail: luigi.grazia@unibo.it

Dr. Ilona Gunsser
Römerstr. 23, 80801 München, Germany
Tel.: +49- 89-347272, Fax: +49-89-391802, E-mail: Ilona.Gunsser@t-online.de

Dr. Mike Herrling
Upsteder Str. 31, 26409 Wittmund, Germany
Tel.: +49-4973-913511, Fax: +49-4973-913512, E-mail: info@avalon-alpacas.de

Dr. Wilfredo Huanca
Laboratory of Animal Reproduction, Faculty of Veterinary Medicine, University of San Marcos,
Circonvalacion Ave. Cdra. 29 - San Borja, 03-5034 Salamanca, Lima - 3, Peru
Tel.: +51-1-435 3348, E-mail: whuanca@appaperu.org

Dr. Ralph Kobera
Zeppelinstr. 1, 01454 Radeberg, Germany
Tel. +49-35206-21383, Fax: +49-35206-21383, E-mail: ralph.kobera@freenet.de

Kuno Kohn
Waliki GmbH, Leiferder Weg 8, 38122 Braunschweig, Germany
Tel.: +49-531-2872625, Fax: +49-531-2872636, E-mail:gschwark@waliki.de

Dr. Jerry Laker
Macaulay Institute, Craigiebuckler, Aberdeen AB15 8QH, UK
Tel.: +44-1224-38611; Fax: +44-1224-311556; E-mail: j.laker@macauly.ac.uk

Ing. Agr. Hugo Lamas
Instituto de Biologia de Altura, Avda. Bolivia 1661, 4600 San Salvador de Jujuy, Argentina
Tel.: +54-388-4223658; Fax: +54-388-4226110; E-mail: hlamas@inbial.unju.edu.ar, hlamas4@
yahoo.com.ar

Dr. Christine Lendl
Grasweg 2, 86459 Gessertshausen, Germany
Tel.: +49-8238-961814, Fax: +49-8238-961830

Dr. Ulrike Lippl
Seewenegg, 3703 Aeschi, Switzerland
Tel.: +41-33-654 09 13; E-mail: lippl.uli@bluewin.ch

Arnold Luginbühl
Seewenegg, 3703 Aeschi, Switzerland

Isabel Maria Madaleno
Tropical Institute, Andrade Street 8-2E, 1170-015 Lissabon, Portugal
Tel.: +351-21-8126329, Fax: +351-21-3150235, E-mail: Isabel-madaleno@clix.pt

Dr. André Markemann
Im Chausseefeld 5, 70599 Stuttgart, Germany
Tel.: +49-711-459 2256, Fax: +49-711-459 3290, E-mail: markeman@uni-hohenheim.de

Rodolfo Marquina Bernedo
Centro de Estudios y Promocion del Desarrollo (DESCO), Malaga Grenet 678, Umacollo,
Arequipa, Peru
Fax: +51-54-270144, Email: arequipa@descosur.org.pe

Dr. Eva Moors
Institut für Tierzucht und Haustiergenetik, Albrecht-Thaer-Weg 3, 37075 Göttingen, Germany
Tel.: +49-551-395613, Fax: +49-551-395587, E-mail: emoors@gwdg.de

Dr. Barbara Münchau
Rosengartenstr. 11, 72108 Rottenburg, Germany
Tel.:+49-7457-930067/949214, Fax:+49-7457-949215; E-mail:Muenchaukameldoc@aol.com

Dr. Carlos Pacheco Murillo
Urbanizacion las begonias E – 5, Distrito de Jose Luis Bustamante y Rivero, Arequipa, Peru
Tel.: +51-54-257043, Fax: +51-54-257043, E-mail: cpachecomg@hotmail.com

Dr. Nadine Renaudeau D'Arc
6 Trinity Street - NR 2 2BQ, GB-Norwich, UK
Tel.: +44-1603616337; E-mail: n.d-arc@uea.ac.uk

Prof. Dr. Carlo Renieri
Dipartimento di Scienze Veterinarie, Università di Camerino, Via Circonvallazione 93/95,
I-62024 Matelica (MC), Italy
Tel.: +39-0737-403436; Fax: +39-0737-403402; E-mail: carlo.renieri@unicam.it

Dr. Hubert Rey
1, Place des Alliés, F-34500 Béziers, France
Tel.: +33-4-6749 2244, Fax: +33-4-6749 8575, E-mail: moulinmer@magic.fr

Dipl. Ing. Alexander Riek
Institut für Tierzucht und Haustiergenetik, Albrecht-Thaer-Weg 3, 37075 Göttingen, Germany
Tel.: +49-551-395620; Fax: +49-551-395587; E-mail: ariek@gwdg.de

Dr. Jorge Rodriguez
Conopa, Los Cerezos 106, Lima 03, Peru
E-mail: webmaster@conopa.org

Dr. Tito Rodriguez Claros
Universidad Mayor de San Andrés, Agricultural Research Institute, Agronomy Faculty, P.O. 930,
Heroes del Acre 1850, La Paz, Bolivia
E-mail: rodriguezct01@hotmail.com

Dr. Christian R. Schmidt
Zoo Frankfurt, Alfred-Brehm-Platz 16, 60316 Frankfurt/Main, Germany
Tel.: +49-69-212 33727; Fax.: +49-69-212 37855; E-mail: christian.schmidt.zoo@stadt-frankfurt.de

Dr. Anja Schwalm
Stendaler Str. 11, 39590 Heeren, Germany
E-mail: anja.schwalm@agrar.uni-giessen.de

Gerhard Schwark
Waliki GmbH, Leiferder Weg 8, 38122 Braunschweig, Germany
Tel.: +49-531-2872625, Fax: +49-531-2872636, E-mail:gschwark@waliki.de

List of participants

Helga Schwark
Waliki GmbH, Leiferder Weg 8, 38122 Braunschweig, Germany
Tel.: +49-531-2872625, Fax: +49-531-2872636, E-mail:gschwark@waliki.de

Susan Tellez
6648 Marshall Place, Beaumont, TX, USA
Tel.:+1-409-656-2140, E-mail: sztellez@aol.com

Oscar Toro Quinto
Centro de Estudios y Promocion del Desarrollo (DESCO), Malaga Grenet 678, Umacollo,
Arequipa, Peru
Fax: +51-54-270144; E-mail: arequipa@descosur.org.pe

Dr. Michael Trah
Drosselweg 2, 71522 Backnang, Germany
Tel.: +49-7191-85600, Fax: +49-7191-85291; E-mail: trah@gmx.de

Susanne E. Ulbrich
Institute of Physiology, Technical University of Munich, 85354 Freising, Germany
Tel.: +49-8161-71 5550, Fax: +49-8161-71 4204; E-mail:ulbrich@wzw.tum.de.

Dr. Jane Vaughan
PO Box 406, 3226 Ocean Grove, Australia
Tel.: +61-3-5254 3365, Fax: +61-3-5254 3365; E-mail: vaughan@ava.com.au

Sven Witt
Große Theaterstr. 37, 20354 Hamburg, Germany
E-mail: svenwitt@compuserve.com

Denis Wolf
Institute of Parasitology, Justus Liebig University, Rudolf-Buchheim-Strasse 2, 35392 Giessen,
Germany
e-mail: christian.bauer@vetmed.uni-giessen.de

Prof. Dr. Clemens Wollny
Institut für Tierzucht und Haustiergenetik, Kellnerweg 6, 37079 Göttingen, Germany
Email: cwollny@gwdg.de

Maria Wurzinger
University of Natural Recources and Applied Life Sciences, Vienna, Department of Sustainable
Agricultural Systems, Gregor-Mendel-Straße 33, A-1180 Wien, Austria
Tel.: +43-1-47654 3273, Fax: +43-1-47654 3254, E-mail: maria.wurzinger@boku.ac.at

Printed in the United States
⸙ Baker & Taylor Publisher Services